Societal Responses
to Regional Climatic Change

Published in cooperation with
the Environmental and Societal Impacts Group,
National Center for Atmospheric Research

Societal Responses to Regional Climatic Change

Forecasting by Analogy

EDITED BY

Michael H. Glantz

Westview Press
BOULDER & LONDON

A Westview Special Study

Published in 1988 in the United States of America by Westview Press, Inc., 5500 Central Avenue, Boulder, Colorado 80301, and in the United Kingdom by Westview Press, Inc., 13 Brunswick Centre, London WC1N 1AF, England

Library of Congress Cataloging-in-Publication Data
Societal responses to regional climatic change.
  (A Westview special study)
  1. Climatic changes—Social aspects—North America.
2. Environmental impact analysis—North America.
I. Glantz, Michael H. II. Series.
QC981.8.C5S634 1988   304.2'5'097   88-10732
ISBN 0-8133-7639-4

Printed and bound in the United States of America

∞ The paper used in this publication meets the requirements of the American National Standard for Permanence of Paper for Printed Library Materials Z39.48-1984.

10   9   8   7   6   5   4   3   2

This book is dedicated to two scientists whose careers have been devoted to an improved understanding of atmospheric processes for the benefit of humanity.

William W. Kellogg
National Center for
Atmospheric Research,
Boulder, Colorado

Gordon McKay
Atmospheric
Environment Service,
Downsview, Ontario

Contents

Acknowledgments

I would like to acknowledge the strong and constant support of the U.S. Environmental Protection Agency, specifically the support of Dennis Tirpak and Joel Smith in the Office of Policy Analysis. I would also like to thank Barbara Brown, Jan Stewart, and Regina Gregory for the excellent work they have done in the coordination and preparation of this manuscript. Special thanks go to Maria Krenz for her tireless efforts of support for this entire research activity, including this publication. I also wish to acknowledge the interest and cooperation of the participants of the workshop (held in Boulder, Colorado, in late June 1987) and the contributors to this publication. The constant support of the National Center for Atmospheric Research is gratefully acknowledged. NCAR is operated by the University Corporation for Atmospheric Research under sponsorship of the National Science Foundation.

The views expressed in these chapters are those of the individual contributors and do not necessarily reflect the view of the supporting organizations.

Michael H. Glantz

1

Introduction

Michael H. Glantz

The summer of 1988 will be a memorable one in the annals of American agricultural history. Persistent severe drought conditions plagued many parts of the North American continent. In addition to the lack of moisture for agricultural, municipal, and industrial purposes, temperatures were excessively high for protracted periods of time, with many new records being set in various parts of the continent. Corn production was drastically reduced in the eastern parts of the Great Plains. Farmers in the Great Lakes region who supplied hay to cattle owners in the southeastern United States a few years ago were now recipients of hay from those they had once helped. Water rationing was put into effect in parts of California. Climate has surely been on the minds of policymakers as special meetings were convened to discuss the economic and environmental impacts of droughts and high temperatures and as President Reagan visited some of the drought-affected areas declaring his support for a bipartisan multibillion-dollar drought relief package.

Perhaps *the* catalyst to political action was not so much the drought's impacts on agriculture but its impacts on navigation. For the first time in several decades extremely low flow conditions in the Mississippi River system brought barge traffic to a halt. Dredging operations were expanded and accelerated in order to try to keep river channels open. Requests were made for water releases from Lake Michigan in order to keep river traffic moving on the Mississippi. This request resurrected domestic as well as international political concerns about the allocation of the water resources of the Great Lakes between nations, states, and between competing uses.

Adding to the political concerns about the impacts of climate on environment and society were statements by prominent scientists about the implications of these droughts for the future. James Hansen, scientist at the Goddard Institute for Space Studies (GISS), noted that based on his assessment of global temperature records, 1988 would prove to have been the warmest year on record and that the three other record-setting warmest years also occurred in the 1980s. He suggested that this was proof positive that the carbon dioxide/trace-gases-induced global warming, referred to as the greenhouse effect, was already in progress and that societies everywhere were seeing the first signs (i.e., impacts) of such a warming.

Whether these droughts and high temperatures are the first signs of the manifestations of a global warming is of less concern at this moment than the need to improve our understanding of the interrelationships between climate and society. Climate is, in fact, always changing. Societies are also undergoing constant change. It is important to know how well societies cope with climate variability today so that we can be better prepared to cope with climate variability and climate change in the future.

This volume is an outgrowth of a project undertaken by the Environmental and Societal Impacts Group (ESIG) at the National Center for Atmospheric Research (NCAR) for the U.S. Environmental Protection Agency's Office of Policy Analysis in order to identify societal responses to extreme climate-related events in North America. The project is based on the premise that in order to know how well society might prepare itself for a future change in climate (the characteristics of which we do not yet know), we must identify how well society today can cope with climate variability and its societal and environmental impacts. It is also based on the premise that while the climate of the future may not be like the climate of the present, societal responses to climate change in the near future will most likely be like those of the recent past and the present.

There has been considerable speculation about what the warming of the global atmosphere by several degrees Celsius will do to regional climate and to human activities presently attuned to that climate. The basis for this speculation (and that is all it is at present) comes from general circulation model (GCM) output as a result of sensitivity studies associated with a CO_2 doubling. It has also come from historical analogues such as the Climate Optimum that occurred earlier in this millenium. Other climate analogues include the Altithermal as well as epochs tens of thousands of years ago when the earth's atmosphere was much warmer than it is at present.

The GCMs are better at dealing with temperature than they are with precipitation. Although the outputs from the various GCMs do not necessarily agree with each other on temperature, they do fall within a well defined range. There is considerable disagreement, however, as to how a global warming might translate into precipitation changes at the regional and local levels. This disagreement has not hindered speculation about regional and local climate changes and their socioeconomic impacts.

GCM-generated scenarios have a considerable amount of scientific credibility in the non-scientific community, including policymakers. However, potential changes in climate several decades into the future that are being suggested by the scientific community are beyond the scope of experience (and possibly education) of policymakers. Thus, the suggested changes lack political and social credibility and this leads to a real problem for policymakers. Either policymakers accept them unquestioningly as the product of an objective scientific community that lays all its information on the table, including questions of uncertainty and perhaps even dissenting views, or they disregard them as speculative ventures of scientists.

Scenarios about future worlds based on human experience— analogies or case scenarios—have the political and social credibility that computer-generated scenarios lack. Policymakers or, more broadly, decisionmakers who have been directly involved in problems generated by climatic anomalies in the recent past have already been using that experience as a guide to dealing with current issues. Such experience is being passed on to prospective decisionmakers through education, just as the experiences of the Great Plains drought in the 1930s have been carried from one generation to the next. Yet, when compared to scientific models, the experience-based scenarios seem to lack scientific credibility. They are often discounted with such statements as "the past is no guide to the future." This belief has been reinforced by the view that, with a changing climate, the past will not be representative of the future.

The purpose of looking back is to determine how flexible (or rigid) societies are or have been in dealing with climate-related environmental changes. Societies everywhere have already shown the propensity to prepare for the last climate anomaly by which they were affected. However, such anomalies seldom seem to recur in the same place, with the same intensity, or with the same societal impacts. We must be aware of past events but we must not get drawn into preparing for them. Our decisions today must take into consideration the need to maintain as much flexibility as practicable in the face of future unknowns.

Forecasting the future by analogy can be a fruitful approach to improve our understanding of how well society is prepared to cope with the presently unknown regional characteristics of a potential climate change some decades in the future. However, we must not expect analogues to tell us what that future will be. No forecasting system has been successful in that endeavor. Analogues can, however, help us to identify societal strengths and weaknesses in coping with extreme meteorological events so that we can reinforce those strengths and reduce the weaknesses.

Both approaches (modeling and identifying contemporary analogues) have their good points and their shortcomings in producing scenarios. It is important that they both be taken into consideration at the same time in order to maintain an air of reality about discussions of future climate change. Either set of scenarios considered by itself can be misleading; focusing on one particular scenario of the future would most likely lead to very different policy responses from those involving a focus on a different but equally plausible scenario. We must find a way to combine the strengths of the two approaches.

A set of case studies has been selected for evaluation with the expectation that such studies could identify strengths and weaknesses of existing social systems with respect to their ability to respond to extreme changes in the environment. The case studies were selected for a variety of reasons: they represent geographic and sectoral diversity, diversity with respect to level of government responsibility. Each of the contributors has a special expertise in the topic that he or she addressed. They were asked to respond to questions about how well society dealt with a particular climate-related extreme event; who in society had primary responsibility for taking action; what general suggestions might improve the situation if a similar event were to occur in the future, and so on.

The case studies are preceded by a section with chapters presenting the historical context of the CO_2 problem, the political context of the problem, the notion of the case scenario approach and the constraints of relying exclusively on scenarios generated by general circulation models for societal aspects of climate impacts studies.

Kellogg's chapter provides a brief history of the global warming issue, presenting a clear picture of the rise and fall and rise again in the awareness of the impacts of the burning of fossil fuels on atmospheric temperatures. It is important to note that in the early decades of this century, until about the mid-1950s, the heating up of the atmosphere was not looked on as an adverse change for society. Some scientists thought, for example, that a warming would thwart the oncoming of the next ice age.

Glantz's chapter addresses the question of why, despite the large body of scientific literature on the global warming issue, there has to date been relatively little political action. It discusses a range of factors that support taking action as well as a range of factors that constrain action on this long-term, low-grade but cumulative environmental change.

The chapter by Jamieson assesses the concept of scenario and the case scenario approach to "forecasting by analogy," discussing the strengths and weaknesses of this method. He identifies some of the misperceptions about the use of the term scenario in the scientific community in attempts to gain a glimpse of the future.

Katz's chapter discusses the use of scenarios generated by general circulation models (GCMs) and looks at the limitations on the value of such scenarios for assessment of the societal implications of a global warming. Katz also discusses advantages and disadvantages of using recent contemporary extreme meteorological events as analogues to the regional aspects of a carbon dioxide/trace-gases-induced global warming.

The chapter by Glantz and Ausubel, written in 1984, represents a first attempt to forecast by analogy the societal responses to the impacts of climate change. A shorter version of this paper appeared in *Environmental Conservation* in 1984. The authors note that model projections, historical reconstructions, and paleoecological assessments have suggested that the U.S. Great Plains would most likely become drier in the event of a global warming.

That region is already plagued by a change in its water balance because of the relatively rapid drawdown of the water in the Ogallala Aquifer on which farmers have come to depend.

The next section of the book consists of 10 case studies of the impacts on environment and society of extreme climate-related events. The cases provide examples of such impacts on different human activities in various parts of North America.

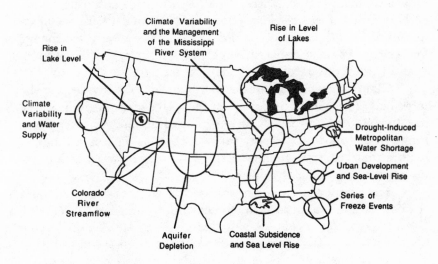

They are representative of studies that might be undertaken for the purpose of forecasting the societal impacts of and responses at some time in the future to a changing climate. The cases include rising levels of the Great Lakes (S. Cohen), rising levels of the Great Salt Lake (P. Morrisette), the effects of sea-level rise on Charleston, SC (M. Davidson), sea level rise and coastal subsidence in Louisiana (M. Meo), Mississippi River navigation system (W. Koellner), the Colorado River Compact (B. Brown), changes in water resources in the Sacramento Basin (P. Gleick), water management in Northern Virginia (D. Sheer), the depletion of the Ogallala Aquifer (D. Wilhite), and recurrent freezes and Florida citrus production (K. Miller).

The concluding section summarizes the highlights of the preceding chapters and serves as an "executive summary." The information in this section is drawn from the chapters, as prepared by the contributing authors with the general findings prepared by the editor.

2

Human Impact on Climate: The Evolution of an Awareness

William W. Kellogg

INTRODUCTION

It is common knowledge that the climate inside our large cities is several degrees warmer than that of the surrounding countryside. This comes as no great surprise, since we can feel the sun-warmed pavements in summer, and we are aware—perhaps with a twinge of guilt—of all the heat that leaks into the outdoors in winter. Thus, on a local scale our activities have clearly modified the climate in which many of us live.

But is it possible to change the climate of our whole enormous planet? Everyone knows that there have been major climate changes in the distant past with the waxing of ice ages. These major changes must have been due to combinations of natural influences that affected the heat balance and temperature of the

This chapter is an abridged version of W.W. Kellogg's article of the same title which appeared in Climatic Change, 10 *(1987) 113– 36. It is reprinted in its abridged form with permission of the editor of* Climatic Change *and the author.*

William W. Kellogg is a Senior Scientist (retired) at the National Center for Atmospheric Research and a member of the Environmental and Societal Impacts Group. Before coming to NCAR, he was Head of the Planetary Sciences Department of the Rand Corporation. He is a past president of the American Meteorological Society and of the Meteorological Section of the American Geophysical Union. His research has been involved in the physics of the upper atmosphere, meteorological satellites, the atmospheres of Mars and Venus, and most recently the theory of climate change.

entire earth. How can little creatures like us compete with those titanic forces that drive the winds of the atmosphere and the ocean currents?

Only 25 years ago, few scientists would have expressed much concern over this question. If prodded a bit, one could easily show that the total power output of all human activities (about 8×10^{12} W) was utterly insignificant (about 1/10,000) compared to the radiant heat absorbed from the sun. This simple argument seemed to prove that we could not influence the heat balance of the earth or its climate.

But there are always a few independent spirits who are not willing to accept obvious answers without searching for other possibilities. Can there be *leverage points* in the complex system that determines climate—leverage points that we can reach? (In the modern technological era a better analogy might be a *servo*, of the sort that allows a pilot to steer a supertanker with the turn of a knob.)

As it turns out, there are indeed some crucial leverage points in the climate system. For one thing, humanity has been changing the face of the earth for many thousands of years. It is common knowledge that most of the dense forests that blanketed northern Europe at the time of the Roman conquests have been cleared for agriculture, and also gone are the forests that used to extend from the Black Sea to the Persian Gulf. Many decades of severe soil erosion in Africa (and elsewhere), coupled with poor agricultural practices and deforestation, have permitted the spread of deserts, and there is good reason to believe that deserts tend to perpetuate their own dry climate (e.g., Bryson and Baerreis, 1967; Kellogg and Schneider, 1977).

It is often hard to distinguish between ecological changes that can be directly attributed to human activities and those that were caused by natural processes, but there is no doubt about the fact that the surfaces of large areas of the earth have a different heat and water balance than they used to have, before humanity had a major influence on the vegetation cover. This must surely have had an influence on regional climates, though the effects are hard to quantify (SMIC, 1971; Flohn, 1975). It probably even had a small effect on the global climate (Sagan et al., 1979).

Another way by which humanity has influenced the climate is by changing the composition of the atmosphere. The layer of smog

and smoke that hangs over virtually every big city is clear evidence
of this fact. It is not so obvious, but nevertheless an established
fact, that such air pollution changes the heat balance and rainfall of
a region and consequently the climate. This is due to the fact that
the particles, or "aerosols," generated by industry, automobiles,
and burning debris absorb solar radiation, and this warms the smog
layer while preventing some sunlight from reaching the surface.
The result is a more stably stratified lower atmosphere, one that
inhibits convection and rainfall over a city (SMIC, 1971; Kellogg,
1980). Downwind, on the other hand, there is evidence that rainfall
may actually increase due to the role that some industrial particles
play as cloud condensation nuclei or freezing nuclei (Dettwiller and
Changnon, 1976). However, this is mainly a *regional* effect.

The human influence on *global* climate that is receiving by
far the most attention these days is the increase of atmospheric
carbon dioxide from the burning of fossil fuels, and also of other
trace gases contributed by our intensifying industrial and agricultural activities. The fact that the carbon dioxide concentration in
the atmosphere is increasing can be observed, as will be detailed
shortly. Since carbon dioxide is just one of several trace gases that
absorb infrared radiation emitted by the earth's surface (and all
of them are increasing), adding more of the absorber prevents a
fraction of the outgoing radiation from escaping to space. This
warms the lower atmosphere, and consequently the surface as well.
The process is often referred to as the "greenhouse effect," though
the analogy to a greenhouse is far from perfect.

HISTORY OF UNDERSTANDING OF CARBON DIOXIDE AND THE GREENHOUSE EFFECT

Since this account concerns the evolution of our awareness
of the potential for anthropogenically-induced climate change, it
is pertinent to point to the earliest inklings in the scientific literature. Revelle (1985) and Ausubel (1983) have recently published
excellent reviews of the early efforts to define the role of carbon
dioxide in the atmosphere, from the points of view of both our depletion of the fossil fuel reserves of the earth, and also the role of
carbon dioxide in sustaining plant life and in regulating the surface
temperature. We have drawn on those reviews in this brief summary, but restrict the discussion to the climatic aspects of carbon
dioxide.

According to Revelle, the first scientist to propose that the gases in the atmosphere could retard "heat" radiation (infrared) from escaping to space and thereby warm the surface was the prominent Frenchman, Jean Baptiste-Joseph Fourier (1768–1830). He pointed out that this effect was similar to the glass in a "hothouse," though it seems that he misunderstood the mode of action of a greenhouse.

Subsequently, John Tyndall in England actually measured the absorption of infrared radiation by carbon dioxide and water vapor and showed that these atmospheric constituents could significantly raise the earth's surface temperature (Tyndall, 1863). He was apparently the first to make an important additional deduction, namely that glacial periods may have been caused by a decrease in atmospheric carbon dioxide.

Just before the turn of the century, two remarkable studies appeared that greatly advanced our appreciation of the effects of carbon dioxide changes on climate. It was realized that the concentration of carbon dioxide was probably increasing, as humanity took carbon out of the earth in the form of coal, petroleum, or natural gas and burned it. Svante Arrhenius, a Swedish scientist who received one of the first Nobel prizes in chemistry, attempted to calculate the effect on the average surface temperature of a doubling of carbon dioxide. He used S. Langley's measurements of infrared radiation from the moon as it passed through the atmosphere at different angles above the horizon and at different humidities for his estimates of the atmospheric absorption due to both carbon dioxide and water vapor, and he combined these with independent measures of carbon dioxide to estimate its current optical depth. He then calculated that doubling the carbon dioxide would raise the average surface temperature by 5–6 K, the larger warming occurring at high latitudes. A decrease of carbon dioxide to two-thirds of the present would cause a cooling of 3 to 3.4 K, he wrote. In these calculations he took into account the likely increase of total water vapor content with increasing temperature, an important "feedback mechanism" (though that term would not be invented until some 50 years later). On the basis of his calculations he deduced correctly that "if the quantity of carbonic acid [carbon dioxide] increases in geometric progression, the augmentation of the temperature will increase nearly in arithmetic progression" (Arrhenius, 1896, 1908).

The American geologist Thomas C. Chamberlin (who became president of the University of Wisconsin) built on the insights provided by Arrhenius to "frame a working hypothesis of the cause of glacial epochs on an atmospheric basis" (Chamberlin, 1899). Much of his argument deals with the changes in sea level relative to the continents and the effects these would have on the weathering of silicate rocks and atmospheric carbon dioxide depletion, a matter also treated by Arrhenius. Chamberlin suggested a mechanism involving the effects on sea level of continental ice sheets and, among other things, showed that this could result in the cyclic behavior of glaciations and interglacials in the last million years or so as the balance between carbon dioxide removal by weathering of rocks and replenishment by volcanic emissions was altered. (An alternative theory based on cyclic perturbations of climate due to changes in the earth's orbit and axis of rotation was to be suggested by Milankovitch (1930) some 30 years later, but nevertheless Chamberlin's hypothesis involving carbon dioxide is a reasonable explanation of cyclic climatic variation throughout later geologic time.)

It is interesting to note that an important aspect of this hypothesis, namely the role that the oceans play as a major reservoir of carbon dioxide, was recognized by Chamberlin and studied by one of his students, C.F. Tolman (1899). Although relatively little was known then about the exchange rates of water masses within the oceans and their influence on storage times of carbon dioxide, the early insight provided by Tolman on this aspect of the question was notable.

Thus it seems that many of the important pieces of the carbon dioxide/climate puzzle were in place by 1900, though of course there were serious gaps in the knowledge of rates of removal and replenishment of carbon dioxide, and also of many other factors that influence the climate system. Furthermore, it would have been virtually impossible at that time to foresee that the worldwide use of fossil fuels would grow exponentially at a rate of about 4 percent per year for the next 73 years, and even after that continue to increase at a substantial rate. When Arrhenius made his calculations in the 1890s of the effects of doubling or tripling the atmospheric concentration of carbon dioxide, the time when that would occur must have seemed very far in the future indeed.

Nevertheless, the intriguing idea that humanity could raise the earth's temperature seems at first to have attracted surprisingly little attention in the scientific community and even less in the public media. It was treated in several papers and books in the 1920s and 1930s, notably by biologists or ecologists who were more interested in the global carbon cycle and the squandering of a virtually irreplaceable natural resource, namely, our store of fossil fuels. The implication that climate could be impacted seems to have been of less interest or else utterly ignored.

The importance of the carbon dioxide–climate issue was beginning to be recognized and addressed by a larger community by the 1950s and, for example, John von Neumann (1955) wrote about the possibility of "climate control." In 1957 Roger Revelle and Hans Suess, two scientists at the Scripps Institution of Oceanography, made a statement in an article in *Tellus* that has since been repeated many times: "Human beings are now carrying out a large-scale geophysical experiment," namely testing the greenhouse effect of carbon dioxide by actually changing its atmospheric concentration. They also pointed out that the newly added carbon dioxide would probably remain in the atmosphere for many centuries because of the slowness with which the oceans could absorb it (Revelle and Suess, 1957).

In the post-International Geophysical Year (IGY) period (1957), there was a growing general awareness of what might be taking place as a result of our "geophysical experiment." In 1963 the Conservation Foundation sponsored a meeting on this topic, and its report stated the situation more clearly than any before it: "It is estimated that a doubling of the carbon dioxide content of the atmosphere would produce a temperature rise of 3.8 degrees (Celsius)"—though the time scale involved is left unspecified (Conservation Foundation, 1963).

Just two years later, in 1965, the august President's Science Advisory Committee (PSAC) published under the White House seal a report of its Environmental Pollution Panel entitled, *Restoring the Quality of Our Environment* (PSAC, 1965). (The Panel's Chairman was John Tukey, and Roger Revelle was Chairman of the Sub-Panel on Atmospheric Carbon Dioxide. Other members of that Sub-Panel were Wallace Broecker, Harmon Craig, Charles David Keeling, and Joseph Smagorinsky.) The PSAC report deals with many aspects of air and water pollution, but it will also be remembered as a first public recognition in a U.S. government document that climate change could be caused by human activities and that this would have important consequences for the world.

THE 20TH CENTURY RECORD OF CARBON DIOXIDE AND OTHER GREENHOUSE GASES AND CONJECTURES ABOUT THE FUTURE

It was during the IGY that Revelle's colleague at the Scripps Institution of Oceanography, Charles David Keeling, started his now-famous nearly continuous monitoring of carbon dioxide at the Mauna Loa Observatory in Hawaii and at the South Pole. While there have been short interruptions in the record (for example, at the South Pole in 1963 for about one year), these two stations have given us the best picture available of the rise of carbon dioxide from 1958 onward (Keeling et al., 1984). This is shown in Figure 1. During the 1960s and 1970s there were other monitoring stations set up by the National Oceanic and Atmospheric Administration's (NOAA's) Global Monitoring for Climate Change program, and now the extensive global effort to monitor atmospheric concentrations of trace gases is being encouraged and coordinated internationally by the World Meteorological Organization (WMO), the International Council of Scientific Unions (ICSU), and the United Nations Environment Programme (UNEP) (Wallen, 1980; Gammon et al., 1985).

It should again be emphasized that carbon dioxide is not the only trace gas whose atmospheric concentration is increasing as a result of human activities, and that most of those other gases are long-lasting in the atmosphere and are also good absorbers of infrared radiation. They are often referred to together as "greenhouse gases," since the presence of all of them tends to warm the lower atmosphere. Some of the other gases in question are the chlorofluorocarbons (CFCs) used as propellants in spray cans and also in refrigerators and air conditioners, methane, nitrous oxide, and ozone. Since their concentrations in the atmosphere are increasing even more rapidly than carbon dioxide (Rasmussen and Khalil, 1986), it is expected that *early in the next century the contributions of all those other gases to a global warming could nearly match that of carbon dioxide alone* (Ramanathan et al., 1985). For this reason

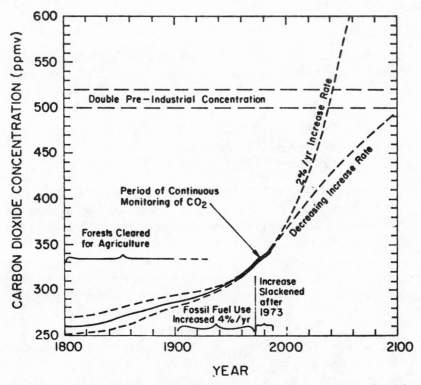

Figure 1. Past and future changes in atmospheric carbon dioxide
concentration. Some increase probably occurred in the
19th century due to extensive clearing of forests and
the conversion of biomass to carbon dioxide (a process
still going on), though there were few direct measurements in that period. During most of this century fossil
fuel burning and the consequent production of carbon
dioxide increased at a rate of 4 percent/yr, but in 1973
the OPEC oil embargo and a worldwide recession resulted in a definite slackening of the rate of increase (see
Clark, 1982). A good guess is that future consumption
of fossil fuels will lie somewhere between a continuing
2 percent/yr increase ("high") and a linearly decreasing rate of change such that 50 years from now we will
return to the present level of consumption ("low"). For
the high case the concentration of carbon dioxide will
be double that of the pre-Industrial Revolution level
before the middle of the next century; for the low case
it may be about 100 years later.

they are all being monitored at stations scattered throughout the world, from Point Barrow (Alaska) to the South Pole Station.

What have we learned about the trends of carbon dioxide concentration (and those of the other greenhouse gases) from all these measurements? When Keeling started his measurements during the IGY, the concentration at the South Pole and Mauna Loa was between 312 and 313 parts per million by volume (ppmv), and now it has climbed to about 345 ppmv, as shown in Figure 1. Before the Industrial Revolution and the widespread burning of fossil fuels, and before we had embarked on the large-scale clearing of forests for agriculture in the 19th century, it is estimated from air samples trapped in ice cores that the concentration of carbon dioxide was 250 to 270 ppmv—that is an increase of 20 to 30 percent in less than 200 years (Oeschger et al., 1985; Gammon et al., 1985). So there is no doubt at all that we have been raising the level of carbon dioxide in the atmosphere by our activities, both industrial and agricultural.

Though it may be getting ahead of our story, it is natural that we should try to estimate what will take place in the future, assuming that humanity will continue to burn fossil fuels and that more carbon dioxide will be added to the atmosphere each year. This will obviously depend on what humanity will require in the way of energy and the methods used to generate it, whether fossil fuel, nuclear, solar, wind, or other "renewable energy sources." We have no very rational way to predict this, so one procedure is to invoke two possible scenarios of fossil fuel use, a very "high" and a very "low" scenario, with the real course presumably lying somewhere in between. I have chosen the "high" scenario to be a continuation of the 2 percent per year rate of increase that has prevailed since 1973 (Rotty and Masters, 1985; Rotty and Marland, 1986); for the "low" scenario I have assumed a linear decrease in the rate of increase of fossil fuel use such that in 50 years it has returned to the present annual consumption (Kellogg, 1979). It is further assumed that the fraction of new carbon dioxide remaining in the atmosphere (the "airborne fraction") remains at its average historical level of about 55 percent. The results are also shown in Figure 1.

Note that the time when the atmospheric concentration reaches twice its preindustrial concentration (a kind of milestone invoked by Arrhenius in 1896 and still used by climate modelers) is in the first half of the 21st century for the "high" scenario, but sometime after 2100 AD for the "low" scenario. This provides a rough time scale for the growth of carbon dioxide. However, also note that the temperature change in that future period will presumably be *hastened* by the addition of other greenhouse gases as already emphasized (Ramanathan et al., 1985); and *slowed* for a few decades by the thermal inertia of the oceans (Schneider and Thompson, 1981; Hoffert and Flannery, 1985).

RECENT STUDIES OF HUMAN EFFECTS ON CLIMATE

The general picture of what was understood in the post-IGY period and the outlines of the theory of a greenhouse gas-induced climate change have been described above, not necessarily in a historical or chronological order. Consider now the situation as it existed in the late 1960s. The PSAC Report of 1965 had called the attention of the world to the distinct possibility that the earth could become warmer as a result of human activities, and a handful of scientists on both sides of the Atlantic were beginning to develop a physical theory to explain the behavior of the complex system that determines climate (e.g., Manabe and Wetherald, 1967; Sellers, 1969; Budyko, 1969).

The time was ripe then to address the question of global environmental problems and the scope of human activities in a more systematic and quantitative way. The most monumental effort up to that time to actually mount a study of this kind was that led by Carroll L. Wilson, M.I.T. Professor of Management. He organized a distinguished Steering Committee to plan what came to be called the *Study of Critical Environmental Problems*, or "SCEP" for short. The SCEP took place for the entire month of July 1970 at Williams College in Williamstown, Massachusetts, and involved approximately 40 scientists and professionals drawn from over a dozen different disciplines. The associate director of the study was William H. Matthews of the M.I.T. Political Science Department.

The major objective of SCEP, as stated in the preface of the report by Wilson and Matthews (SCEP, 1970), was "to raise the level of informed public and scientific discussion and action on global environmental problems," and this it did. There were seven work groups in all, each one dealing with an environmental discipline, or else gathering global statistics—mostly economic and industrial—that could be used by the disciplinary working groups.

As I was Chairman of the Work Group on Climatic Effects, I will confine my remarks to that part of the study. It is fair to say that virtually every member of that work group came away with a heightened awareness of the subject with which we were dealing, and a sense of the totality of human impacts on the environment. On the subject of carbon dioxide and climate, we were emphatic about its importance but noncommittal about the magnitude or time scale of the change in store for the world. I believe that what we concluded was about as strong a statement as could have been made at that time, to wit:

> Although we conclude that the probability of direct climate change in this century resulting from CO_2 is small, we stress that long-term potential consequences of CO_2 effects on the climate or of social reaction to such threats are so serious that much more must be learned about future trends of climate change. Only through these measures can societies hope to have time to adjust to changes that may ultimately be necessary (SCEP, 1970, 12).

We also spoke out on the subject of air pollution by particles, on the effects of cirrus clouds from jet aircraft, on surface changes such as deforestation, and on the effects of supersonic transports (SSTs) on climate and the ozone layer. The media representatives that came to Williamstown to be briefed on the SCEP results were most intrigued by the last item, since there was already a debate going on in Congress about the federal appropriation for building two prototype SSTs, and they were relatively less interested in climate change.

It was clear, at any rate, that SCEP had indeed "raised the level of ... scientific discussion ... on global environmental problems," and Carroll Wilson decided to seize the opportunity to organize a followup that would involve the *international* scientific community. As usual, he was successful. Again he obtained financial support from a variety of governmental organizations, private foundations, and corporations; and the Royal Swedish Academy of Sciences and the Royal Swedish Academy of Engineering Sciences offered to be our hosts. The meeting took place at a conference center in Wijk, near Stockholm, for three weeks in July 1971. William Matthews was again the associate director, and I participated as one of the "secretaries"—the other was G.D. Robinson, a former President of the British Royal Meteorological Society.

This time the discussion focused more sharply on the question of climate change, and the report was entitled *The Study of Man's Impact on Climate*, or "SMIC" for short. It was to provide "an authoritative assessment of the present state of scientific understanding of the possible impacts of man's [*sic*] activities on the regional and global climate" (SMIC, 1971). More than a decade and a half later one can look back at this book with some perspective, and it is fair to say that it did indeed serve as an authoritative source of information on virtually all aspects of the question of climate change and many related subjects—though bit by bit some things in it were overtaken by new advances of science and have become outdated, as will be pointed out.

It is particularly notable that the following year (1972) witnessed the first United Nations Conference on the Human Environment (in Stockholm) and the SMIC Report was said to be "required reading" for all participants. One useful outcome of the conference was the decision to found a new organization devoted to the preservation of the global environment, the United Nations Environment Programme.

Toward the end of the study, after most of our conclusions were out on the table, I was intrigued by the apparent impasse in the assembled group between two opposing schools of thought: the climate "coolers" and the climate "warmers," if you will. It depended on whether you came from the atmospheric particle or aerosol camp, or the carbon dioxide and infrared-absorbing gases camp.

It was generally thought in 1971 that industrial and agricultural aerosols served to both absorb and scatter sunlight back to space, wherever they were, and that this would mean less sunlight reaching the surface and a net *cooling*. (Now that we know a lot more about the radiative influence of aerosols, we recognize that this is not necessarily true and a greatly oversimplified view, e.g.,

Kellogg et al., 1975; Kellogg, 1980; Coakley et al., 1983). As already explained, the carbon dioxide-and-other-infrared-absorbing-gases people recognized the greenhouse effect and its *warming* influence. There were some other human activities that could tip the balance as well, such as clearing forest land for agriculture, irrigating deserts, creating large artificial bodies of water, and so forth, but these did not seem to us to be quite as important in the short term as the changes in the atmosphere.

We decided to call an evening meeting to thrash out a consensus, and to decide (if we could) whether we would predict a net cooling or a warming in the decades ahead due to human activities. It would clearly be useful if we could make such a prediction with some degree of conviction. However, the impasse prevailed, much to my disappointment. There were just too many honest differences of opinion and not enough facts at hand to resolve them.

Additionally, there was a clear reluctance on the part of many of those colleagues (and of others not at Wijk) to make any predictions at all about the future—to "stick out one's neck." Scientists are trained to be cautious about jumping to conclusions too fast, and furthermore we will always be awed by the complexity of the planetary climate system and aware of our inability to understand all its interactions. In the end the SMIC Report states that it is definitely within humanity's power to change the global climate, but there is no further indication of what will probably happen— just what *could* happen, and possibly with serious consequences.

That was the fairest statement that could be made by the climatological community in 1971. And it helped to fuel the fires of climate research around the world, especially in the United States. A quantitative theory of climate was emerging, primarily as a result of early research in the United States and the Soviet Union, and our computers were increasingly able to handle complex numerical models of the climate system. Bit by bit the pieces of the climate picture were fitted into place.

Such a process is necessarily a gradual one, with many individual contributions and much discussion and some confusion at each stage, but as usual there were certain milestones that served to mark the progress. One was the International Symposium on Long-Term Climate Fluctuations, sponsored by the WMO and held in Norwich, England, in the summer of 1975 (WMO, 1975). Among other revelations at the symposium was the finding that low-lying industrial aerosols, and also smoke particles from slash-and-burn agricultural practices, absorb sunlight quite strongly and do *not* cause a cooling of the lower atmosphere when they are over land—which is, of course, where most of this form of air pollution is found (Kellogg et al., 1975). This seemed to remove the impasse encountered a few years earlier at Wijk, and left the greenhouse warming to dominate the stage.

In 1977 I was asked by the WMO to prepare a Technical Note entitled *Effects of Human Activities on Global Climate* (Kellogg, 1977). I relied heavily on the SMIC Report, of course, but was able to bring together a great deal of further information that was simply unavailable in 1971. Notable were some better estimates of the greenhouse warming for a given increase in carbon dioxide and other related factors. These were worked out by such people as Suki Manabe of the Geophysical Fluid Dynamics Laboratory (GFDL) in Princeton (a SMIC participant), William Sellers of the University of Arizona, Stephen Schneider of NCAR (also a SMIC participant), and Michael Budyko of the Main Geophysical Observatory in Leningrad (another SMIC participant). In that report I combined the effect of increasing carbon dioxide on global temperatures with a range of estimates of future fossil fuel consumption (see Figure 1) and drew a "temperature-versus-time scenario" curve out to the year 2050. Even though this exercise was accompanied by many caveats and warnings not to take this prediction too seriously, it seems to have captured the imagination of a great many people both inside and outside the climatological community. A more recent version of this temperature scenario is reproduced here as Figure 2.

A major milestone was a report of a 1979 summer study sponsored by the U.S. National Academy of Sciences and held at Woods Hole, Massachusetts, at which a blue-ribbon group of meteorologists (and an oceanographer) sought to determine whether the various climate models that were being depended on to relate the greenhouse warming to increased carbon dioxide were scientifically based, and whether they were reliable enough to be believed as simulators of the behavior of the real climate system. This ad hoc committee was chaired by Jule G. Charney, then head of the M.I.T. Meteorology Department, and a scientist who had not up to then become directly embroiled in the carbon dioxide-climate change debate. Its conclusion was that there was no good reason to

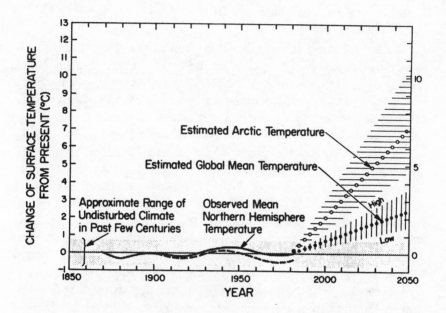

Figure 2. Estimates of past and future temperature variations.
The future changes of global mean surface temperature
are based on a 3°C warming for a doubling of carbon
dioxide concentration, and the possible range between
a high and a low scenario is the same as in Figure 1.
Changes in the Arctic are expected to be about three
times larger than the average. The shaded area shows
the range within which the earth's temperature has remained for the past 1,000 years or more. The dashed
line is the hypothetical global temperature record that
might have occurred if there had been no increase in
carbon dioxide or other infrared absorbing gases. The
rate of global temperature change shown here will be
slowed by the thermal inertia of the oceans and hastened by the contributions of other trace gases to the
greenhouse effect. (Source: Kellogg and Schware, 1981,
1982.)

reject the finding that a doubling of carbon dioxide concentration from the pre-Industrial Revolution value of about 260 ppmv would cause an average global warming of 3 ± 1.5 K, and that such a warming could occur sometime in the 21st century (NAS, 1979). (See Figures 1 and 2.)

The "Charney Report," as it is often called, did not add much to our fund of knowledge about the theory of climate change. However, it did provide some much-needed assurance that our pictures of how the planet could respond to a carbon dioxide change were scientifically sound and believable—at least, to the extent that human ingenuity and computer capacity would allow.

That same year a much larger international gathering was convened to deal with all aspects of climate variability and change. The occasion was the historic World Climate Conference, organized and hosted by the WMO in Geneva, in February 1979. Several hundred scientists and dignitaries attended this week-long conference, and the last two days were devoted to meetings of smaller working groups to review the terms of reference of the proposed international World Climate Programme. Throughout this week a conference statement was discussed and drafted and redrafted, and at the final plenary session the chairman, Robert M. White, was able to obtain agreement on this statement.

The statement is too long to repeat here, but one of its most significant points was as follows:

> It is possible that some effects on a regional and global scale may be detectable before the end of this century and become significant before the middle of the next century. This time scale is similar to that required to redirect, if necessary, the operation of many aspects of the world economy, including agriculture and the production of energy. Since changes in climate may prove to be beneficial in some parts of the world and adverse in others, significant social and technological readjustments may be required (WMO, 1979, 714).

The statement went on to call on all nations to unite in efforts to understand this climate change and to plan for it—and perhaps to lessen its effects. It detailed many of the climate-related problems that had to be solved by scientific research. It concluded with the thought that such international cooperation could only prosper in a world at peace, a rather obvious and most appropriate observation.

THE ACTIVISTS AND THE MUDDLERS-THROUGH

Notice that the statement of the World Climate Conference does not call for international action to *prevent* the future climate change, though this point was discussed at the time. In every group there are activists whose instinctive response to an apparent threat is to counterattack, and by opposing end the threat. This is a particularly enticing notion when one is faced with such a convincing scenario of a future earth warmed by carbon dioxide and other greenhouse gases—an insight that humanity has never experienced before (neglecting the story of Joseph in Egypt as recounted in *Genesis*).

It is highly pertinent to note that those measures taken to prepare for a gradual climate change, assumed to be inevitable, are usually measures that make good sense anyway. They are things that will help to cope with short-term climatic events as well, such as floods, droughts, hot and cold spells, and they will generally increase our ability to adapt to change.

The debate between the activists or intervenors and the adaptors or muddling-through camps can be illuminated by analyzing the conclusions of two recent prestigious studies. The first is the U.S. National Research Council (NRC) report entitled *Changing Climate* (NRC, 1983a), and the second is the report of the WMO, UNEP, and ICSU International Conference held at Villach, Austria, in October 1985, bearing the imposing title *International Assessment of the Role of Carbon Dioxide and of Other Greenhouse Gases in Climate Variations and Associated Impacts* (WMO, 1985).

The NRC report is an excellent review of where we stand in our understanding of climate change and its many implications, and, of course, it calls for more research on the subject. It discusses the alternative policies suggested to deal with climate change. The chapter by Thomas C. Schelling, a Harvard University professor, is a particularly incisive statement on the options available to individuals and nations to prevent or adapt to the climate change in store for the world. But when it comes to a recommendation about what should be done about the current situation there is no clarion call for immediate action. Rather: "In our judgment, the knowledge we can gain in coming years should be more beneficial than a lack of action will be damaging .... Our stance is conservative: we believe there is reason for caution, not panic" (NRC, 1983a, 61 and xiii).

The statement of the Villach conference, while still calling for more research (of course), has a somewhat different ring. It notes that many important decisions are being made "based on the assumption that past climatic data, without modification, are a reliable guide to the future," which is no longer valid in view of the "significant warming of the global climate in the next century." The final conference recommendation does not have so much to do with science as with future policy:

> Support for the analysis of policy and economic options should be increased by governments and funding agencies. In these assessments the widest possible range of social responses aimed at *preventing* or adapting to climate change should be identified, analyzed and evaluated. These assessments should be initiated immediately and should employ a variety of available methods. Some of these analyses should be undertaken in a regional context to link available knowledge with economic decisionmaking and to characterize regional vulnerability and adaptability to climatic change. Candidate regions may include the Amazon Basin, the Indian subcontinent, Europe, the Arctic, the Zambezi Basin, and the North American Great Lakes (WMO, 1985, 3, italics added).

The statement went even further in suggesting that action might indeed be taken to avert the climate change:

> While some warming of climate now appears inevitable due to past actions, the rate and degree of future warming could be profoundly affected by governmental policies on energy conservation, use of fossil fuels, and the emission of some greenhouse gases (WMO, 1985, 1).

We should add parenthetically that, as in virtually all large scientific debates, there is a small but vociferous handful of dissenters from the consensus of the scientific community. They are apparently skeptical of the idea that a global warming is indeed taking place. Whatever motivates them, they are applauded by the nonscientific public sector that finds the thought that we are tampering with the earth's climate distasteful—perhaps a kind of tribal guilt feeling that we are fooling with Mother Nature. One of the more outspoken of these dissenters argues that all the climate modelers are exaggerating the effect of the carbon dioxide greenhouse effect by about an order of magnitude, and adds the happy thought that the added carbon dioxide will encourage plant growth (the last is probably true, but irrelevant to the climate issue) (Idso, 1980, 1982). Unfortunately, considerable effort has had to be expended to demonstrate that such arguments are fallacious (e.g., Schneider et al., 1980; Ramanathan, 1981; Smagorinsky, 1983).

If any further justification were needed to support the contention that the climate change predicted by the theoretical models is actually taking place, we can turn to the temperature record itself. The most recent analysis of the global mean surface temperature from 1861 through 1984, taking proper account of both land observations and those at sea, shows that there has been a slow warming trend of 0.6 to 0.7 K in that period (Jones et al., 1986). This is entirely consistent with the models' results based on the increase of carbon dioxide concentration in the same period, and there is no other convincing theory to explain this warming trend (WMO, 1982).

FUTURE CHANGES, SCENARIOS, AND IMPACTS

At this point it is important to say a word about the crucial homework left for climatologists and those trying to estimate the impacts of future climate change. While a good deal can be said about the global aspects of the increase of carbon dioxide and the other infrared-absorbing gases resulting in the greenhouse effect, this kind of information is simply not very useful to political or managerial decisionmakers. It is not enough to tell them that the world will be warmer, when what they need to know is the kind of changes to be expected here—wherever "*here*" turns out to be. And it is the changes in rainfall, temperature, and soil moisture on a regional scale that determine where certain things can grow and where agriculture and forestry and tourism will flourish or fail.

Note that the regional scale approach was urged in the recommendation of the Villach Conference, quoted above.

Rainfall, as all meteorologists know very well, is an elusive phenomenon, hard to predict and hard even to describe after the fact. It has great variability even under nonchanging conditions, since it is determined by a complex set of processes that include the ebb and flow of cold and warm air masses, the presence of mountain barriers, the amount of moisture acquired while the air was flowing over water, conditions favorable for cloud formation and convective lifting, and so forth. Thus, changes in patterns of rainfall are a *secondary* result of changes in atmospheric heat balance. It is therefore not surprising that the climate models do not agree on the rainfall changes that may occur in a region, even though this may be required by planners.

This requirement poses a very great challenge to the climatological community. How can we get better information about the regional changes to be expected? Our mathematical models of the global climate system are getting better and better, as both human ingenuity and increasingly fast computers allow us to take more physical factors and interactions into account. However, the spatial resolution of the best current climate models is still about 400×400 kilometers (roughly the size of the grid used to carry out the computations), and this barely begins to give us some insight into what might happen regionally (Hansen et al., 1984; Manabe et al., 1981; Manabe and Wetherald, 1986; Washington and Meehl, 1984; Mitchell, 1983). Moreover, it is most frustrating to note that, when one compares the results of several experiments with different climate models to determine the effects of increased carbon dioxide, the regional changes of temperature and rainfall are often inconsistent (Schlesinger and Mitchell, 1985; Kellogg and Zhao, 1988). Thus, climate models are apparently not yet at the stage where they can be relied upon to give us all the answers that we want, but it is certainly not too early to glean what we can from them.

There is another avenue that has been followed to get some further inklings of future climate changes, and these involve looking at past warm periods to see what happened to rainfall and the geographical patterns of vegetation (Aspen Institute, 1978). We can look at anomalously warm years in this century, when the meteorological observation networks of the world were fairly well established, and we find that indeed warm years or seasons did have characteristic anomalies in rainfall distributions (Wigley et al., 1979; Williams, 1980; Jäger and Kellogg, 1983). We can also look at the period some 4,500 to 8,000 years ago, the so-called Altithermal (or Hypsithermal) Period, when the world as a whole was warmer (especially in summer), and try to reconstruct the patterns of natural vegetation, lake levels, streamflows, and so forth, and from all that information get an idea of the regional differences between then and now (Nicholson and Flohn, 1980; Flohn, 1979, 1982; Kellogg, 1982; Butzer, 1980; Kutzbach and Street-Perrott, 1985; Street-Perrott and Harrison, 1984). When these three approaches are used together, one can begin to see where there is some degree of agreement between them (for a regional application of this approach, see Glantz and Ausubel, this volume). Figure 3 is an attempt to map the degree of agreement on future changes of soil moisture.

Some of my climatologist colleagues will not approve of these rather speculative ideas, feeling that our understanding of the climate system and the changes that it may undergo is not yet complete enough. I would have to agree in part. The important point is that we now foresee a global warming, though the time scale is not firm, and we are convinced that such a drastic climate change will bring with it major readjustments at all levels of society. Some educated speculation is in order so that we can begin to take the measure of these readjustments, and so that we can do some intelligent long-range planning.

This planning process has already begun, and there are some steps that could be taken now to mitigate the damage—steps that make sense whether or not one believes in the inevitability of climate change (e.g., Kellogg and Schware, 1982). However, no one seems to be very good at long-range planning, and relatively few activities demand a planning horizon of several decades. Exceptions where future climate impacts *can* be anticipated are, for example, large-scale water management (building of dams and irrigation projects) and the forest products industry (multimillion-dollar tree plantations that will be harvested in 50 years or more) (Kellogg, 1983). In spite of the lip service given to studies of climate change and its societal impacts, *we have yet to see an important governmental or industrial decision that actually acknowledged the climate change factor*. The educational process is under way.

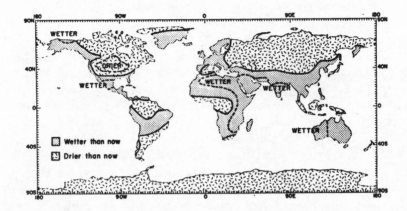

Figure 3.   An attempt to present a "best guess" or scenario of
possible soil moisture patterns on a warmer earth in
terms of regions where it may be wetter or drier than
now in summer; in the tropics this probably applies to
the rainy season. It is based on climate model experiments and reconstructions of past conditions (see text).
Where there seems to be some agreement between the
various approaches the area has been designated by a
dashed line and a label. (Source: Kellogg and Schware,
1981; Kellogg, 1982.)

Indeed, this educational process has been hastened and encouraged by several intergovernmental and scientific bodies, notably WMO, UNEP, the European Community, and ICSU. In the
United States the Department of Energy has been sponsoring a
multiyear research program that has recently climaxed in a series of
excellent scientific "state-of-the-art reviews" and a final Statement
of Findings (SOF) report addressed to Congress and the public.
According to the director of this program:

The SOF will summarize what we know and do not
know and the degree of certainty of our knowledge. It
will also present the rationale for further studies. These
studies will be needed to provide an accurate scientific

basis for assessments of the potential impacts of energy-related activities (Koomanoff, 1985, viii).

The perception of the need to study the climate system of the earth has re-awakened the scientific community to the verity of the thought that our planet must and can be studied as one interacting whole. We now have earth satellites and global telecommunications networks to observe it, and (equally important) supercomputers capable of digesting the vast flow of new data and organizing it. There is also the inevitable realization that the biosphere must be studied along with the atmosphere, the oceans, the land surfaces, the polar regions, and the sun if we are to fathom the behavior of the system that governs our planetary environment. All this has led to a new initiative that is rapidly gaining momentum both inside and outside of scientific circles, the International Geosphere-Biosphere Program (IGBP) (NRC, 1983b; Malone and Roederer, 1985; UCAR, 1985; NASA, 1986; NRC, 1986). In the United States it has received considerable governmental attention under the banner of Earth Systems Science, "a program for global change." The question of climate change is, of course, only one aspect of the alterations that are taking place on every continent, such as the depletion of rainforests in the tropics, the spread of deserts, the loss of topsoil in cultivated regions, the extermination of species of plants and animals every year, and so forth (Kellogg, 1986).

CONCLUSION

The world has progressed a long way in the century since Tyndall, Arrhenius, and Chamberlin made those first tentative suggestions about humanity's ability to change the global climate. By far the biggest advances in the evolution of an awareness that this could really be true were in the 1970s. That was when the "level of informed public and scientific discussion" seems to have experienced a dramatic upswing, and climate research made rapid progress on all fronts.

Carroll Wilson apparently prophesied that this was going to happen, and he was able to make it a self-fulfilling prophecy. The SCEP and SMIC studies, which he conceived and skillfully organized with the help of William Matthews and a host of friends and colleagues, were both events which were well timed, as it turned out. They achieved their goals magnificently, and inspired a generation of young scientists who would make the study of the theory of climate their life's work—and also some older ones (like myself) who would then seriously turn to research on the impacts of climate change.

REFERENCES

Arrhenius, S., 1896: On the influence of carbonic acid in the air upon the temperature of the ground. *Philosophical Magazine, 41*, 237–71.

Arrhenius, S., 1908: *Worlds in the Making.* New York, NY: Harper & Brothers.

Aspen Institute, 1978: *The Consequences of a Hypothetical World Climate Scenario Based on an Assumed Global Warming Due to Increased Carbon Dioxide.* Report of Symposium and Workshop. Boulder, CO: Aspen Institute for Humanistic Studies.

Ausubel, J.H., 1983: Historical note. In *Changing Climate*, Report of the Carbon Dioxide Assessment Committee, Board on Atmospheric Sciences and Climate, National Research Council. Washington, DC: National Academy Press, 488-91.

Bryson, R.A., and D.A. Baerreis, 1967: Possibilities of major climate modification and their implications: Northwest India, a case for study. *Bulletin of the American Meteorological Society, 48*, 136–42.

Budyko, M.I., 1969: The effect of solar radiation variations on the climate of the earth. *Tellus, 21*, 611-9.

Butzer, K.W., 1980: Adaptation to global environmental change. *Professional Geographer, 32*, 269–78.

Chamberlin, T.C., 1899: An attempt to frame a working hypothesis of the cause of glacial periods on an atmospheric basis. *Journal of Geology, 7*, 545–61.

Clark, W.C. (Ed.), 1982: *Carbon Dioxide Review: 1982.* New York, NY: Oxford University Press.

Coakley, J.A., Jr., R.D. Cess, and F.B. Yurevitch, 1983: The effect of tropospheric aerosols on the earth's radiation budget: A parameterization for climate models. *Journal of Atmospheric Science, 40*, 116–38.

Conservation Foundation, 1963: *Implications of Rising Carbon Dioxide Concentration of the Atmosphere.* New York, NY: The Conservation Foundation.

Dettwiller, J., and S.A. Changnon, Jr., 1976: Possible urban effects on maximum daily rainfall rates at Paris, St. Louis, and Chicago. *Journal of Applied Meteorology, 15*, 517–9.

Flohn, H., 1975: History and intransitivity of climate. In *The Physical Basis of Climate and Climate Modeling*, GARP Publ. Ser. No. 16. Geneva, Switzerland: World Meteorological Organization, 106–18.

Flohn, H., 1979: A scenario of possible future climates—natural and man-made. In *Proceedings of the World Climate Conference*, WMO No. 537. Geneva, Switzerland: World Meteorological Organization, 243–68.

Flohn, H., 1982: Climate change and an ice-free Arctic Ocean. In W.C. Clark (Ed.), *Carbon Dioxide Review: 1982.* New York, NY: Oxford University Press, 143–79.

Gammon, R.H., E.T. Sundquist, and P.J. Fraser, 1985: History of carbon dioxide in the atmosphere. In J.R. Trabalka (Ed.), *Atmospheric Carbon Dioxide and the Global Carbon Cycle.* Report DOE/ER-0239. Washington, DC: U.S. Department of Energy, 25–62.

Hansen, J., A. Lacis, D. Rind, G. Russell, P. Stone, I. Fung, R. Ruedy, and J. Lerner, 1984: Climate sensitivity: Analysis of feedback mechanisms. In J.E. Hansen and T. Takahashi (Eds.), *Climate Processes and Climate Sensitivity.* Maurice Ewing Series No. 5. Washington, DC: American Geophysical Union, 130–63.

Hoffert, M.I., and B.P. Flannery, 1985: Model projections of the time-dependent response to increasing carbon dioxide. In M.C. MacCracken and F.M. Luther (Eds.), *Projecting the Climatic Effects of Increasing Carbon Dioxide.* Report DOE/ER-0237. Washington, DC: U.S. Department of Energy, 149–90.

Idso, S.B., 1980: The climatological significance of a doubling of the earth's atmospheric carbon dioxide concentration. *Science, 207*, 1462–3.

Idso, S.B., 1982: *Carbon Dioxide: Friend or Foe?* Tempe, AZ: IBR Press.

Jäger, J., and W.W. Kellogg, 1983: Anomalies in temperature and rainfall during warm Arctic seasons. *Climatic Change, 5,* 39–60.

Jones, P.D., T.M.L. Wigley, and P.B. Wright, 1986: Global temperature variation between 1861 and 1984. *Nature, 322,* 430–4.

Keeling, C.D., A.F. Carter, and W.G. Mook, 1984: Seasonal, latitudinal, and secular variations in the abundance and isotope ratios of atmospheric CO_2. *Journal of Geophysical Research, 89,* 4615–28.

Kellogg, W.W., 1977: *Effects of Human Activities on Global Climate.* Tech Note No. 156 (WMO No. 486). Geneva, Switzerland: World Meteorological Organization.

Kellogg, W.W., 1979: Influence of mankind on climate. In F.A. Donath, F.G. Stehli, and G.W. Wetherill (Eds.), *Annual Reviews of Earth and Planetary Science.* Palo Alto, CA: Annual Reviews, Inc., 63–92.

Kellogg, W.W., 1980: Aerosols and climate. In W. Bach, J. Pankrath, and J. Williams (Eds.), *Interactions of Energy and Climate.* Dordrecht, The Netherlands: D. Reidel Publishing Co., 281–96.

Kellogg, W.W., 1982: Precipitation trends on a warmer earth. In R.A. Reck and J.R. Hummell (Eds.), *Interpretation of Climate and Photochemical Models, Ozone and Temperature Measurements.* New York, NY: American Institute of Physics, 35–46.

Kellogg, W.W., 1983: Future climate on a warmer earth. In W.F. Miller (Ed.), *Water: A Resource in Demand.* Proceedings of Symposium on Future Climate and Potential Impacts on Natural Resources Management, Texas A&M University, August 1982, Southern Cooperative Series Bulletin 288. Mississippi State, MS: Mississippi Agricultural Experiment Station, Mississippi State University, 2–8.

Kellogg, W.W., 1986: The changing health of Planet Earth: Observing the earth from space. *EOS, 67,* 816–9.

Kellogg, W.W., J.A. Coakley, and G.W. Grams, 1975: Effect of anthropogenic aerosols on the global climate. In *Proceedings of the WMO/IAMAP Symposium on Long-Term Climatic Fluctuations.* WMO Doc. 421. Geneva, Switzerland: World Meteorological Organization, 323–30.

Kellogg, W.W., and S.H. Schneider, 1977: Climate, desertification, and human activities. In M.H. Glantz (Ed.), *Desertification*. Boulder, CO: Westview Press, 141–64.

Kellogg, W.W., and R. Schware, 1981: *Climate Change and Society: Consequences of Increasing Atmospheric Carbon Dioxide*. Boulder, CO: Westview Press.

Kellogg, W.W., and R. Schware, 1982: Society, science, and climate change. *Foreign Affairs, 60*, 1076–109.

Kellogg, W.W., and Z.-C. Zhao, 1988: Sensitivity of soil moisture to doubling of carbon dioxide in climate model experiments. Part I: North America. *Journal of Climate, 1*, 348–66.

Koomanoff, F.A., 1985: Foreword. In M.C. MacCracken and F.M. Luther (Eds.), *Projecting the Climatic Effects of Increasing Carbon Dioxide*, Report DOE/ER-0237. Washington, DC: U.S. Department of Energy, v–vii.

Kutzbach, J.E., and F.A. Street-Perrott, 1985: Milankovitch forcing of fluctuations in the level of tropical lakes from 18 to 0 kyr BP. *Nature, 317*, 130–4.

Malone, T.F., and J.G. Roederer, 1985: *Global Change*. Cambridge, England: Cambridge University Press.

Manabe, S., and R.T. Wetherald, 1967: Thermal equilibrium of the atmosphere with a given distribution of relative humidity. *Journal of the Atmospheric Sciences, 24*, 241–59.

Manabe, S., R.T. Wetherald, and R.J. Stouffer, 1981: Summer dryness due to an increase of atmospheric CO_2 concentration. *Climatic Change, 3*, 347–86.

Manabe, S., and Wetherald, R.T., 1986: Reduction in summer soil wetness induced by an increase in atmospheric carbon dioxide. *Science, 232*, 624–8.

Milankovitch, N., 1930: Mathematische klimalehre und astronomische theorie der klimaschwankungen. In W. Köppen and R. Geiger (Eds.), *Handbuch der Klimatologie I(A)*, Berlin: Gebrüder Borntraeger, 1.

Mitchell, J.F.B., 1983: The seasonal response of a general circulation model to changes in CO_2 and sea temperature. *Quarterly Journal of the Royal Meteorological Society, 109*, 113–52.

NAS (National Academy of Sciences), 1979: *Carbon Dioxide and Climate: A Scientific Assessment*. Washington, DC: Climate Research Board, National Academy of Sciences.

NASA (National Aeronautics and Space Administration), 1986: *Earth System Science: A Program for Global Change (Overview).* Washington, DC: NASA Advisory Council, NASA.

Nicholson, S.E., and H. Flohn, 1980: African environmental and climatic changes and the general atmospheric circulation in late Pleistocene and Holocene. *Climatic Change, 2,* 313–48.

NRC (National Research Council), 1983a: *Changing Climate.* Report of the Carbon Dioxide Assessment Committee, Board on Atmospheric Sciences and Climate. Washington, DC: National Academy Press.

NRC (National Research Council), 1983b: *Toward an International Geosphere-Biosphere Program.* Report of a Workshop. Washington, DC: National Research Council.

NRC (National Research Council), 1986: *Global Change in the Geosphere-Biosphere: Initial Priorities for an IGBP.* Washington, DC: U.S. Committee for an International Geosphere-Biosphere Program, National Research Council.

Oeschger, H., B. Stauffer, R. Finkel, and C.C. Langway, 1985: Variations in the CO_2 concentration of occluded air and of anions and dust in polar ice cores. In E.T. Sundquist and W.S. Broecker (Eds.), *The Carbon Cycle and Atmospheric CO_2: Natural Variations Archean to Present.* Geophysical Monograph 32. Washington, DC: American Geophysical Union, 132–42.

PSAC (President's Science Advisory Committee), 1965: *Restoring the Quality of Our Environment, Report of the Environmental Pollution Panel.* Washington, DC: President's Science Advisory Committee, The White House.

Ramanathan, V., 1981: The role of ocean atmosphere interactions in the CO_2 problem. *Journal of the Atmospheric Sciences, 38,* 918–30.

Ramanathan, V., H.B. Singh, R.J. Cicerone, and J.T. Kiehl, 1985: Trace gas trends and their potential role in climate change. *Journal of Geophysical Research, 90,* 5547–66.

Rasmussen, R.A., and M.A.K. Khalil, 1986: Atmospheric trace gases: Trends and distributions over the last decade. *Science, 232,* 1623–4.

Revelle, R., 1985: Introduction: The scientific history of carbon dioxide. In E.T. Sundquist and W.S. Broecker (Eds.),

The Carbon Cycle and Atmospheric CO_2: Natural Variations Archean to Present. Geophysical Monograph 32. Washington, DC: American Geophysical Union, 1–4.

Revelle, R., and H.E. Suess, 1957: Carbon dioxide exchange between atmosphere and ocean and the question of an increase of atmospheric CO_2 during the past decades. *Tellus, 9,* 18–27.

Rotty, R.M., and G. Marland, 1986: Fossil fuel combustion: Recent amounts, patterns, and trends of CO_2. In J.R. Trabalka and D.E. Reichle (Eds.), *The Changing Carbon Cycle: A Global Analysis.* New York, NY: Springer-Verlag, 474–90.

Rotty, R.M., and C.D. Masters, 1985: Carbon dioxide from fossil fuel combustion: Trends, resources, and technological implications. In J.R. Trabalka (Ed.), *Atmospheric Carbon Dioxide and the Global Carbon Cycle.* Report DOE/ER-0239. Washington, DC: U.S. Department of Energy, 63–80.

Sagan, C., O.B. Toon, and J.B. Pollack, 1979: Anthropogenic albedo changes and the earth's climate. *Science, 206,* 1363–8.

SCEP (Study of Critical Environmental Problems), 1970: *Man's Impact on the Global Climate: Report of the Study of Critical Environmental Problems.* Cambridge, MA: The MIT Press.

Schlesinger, M.E., and J.F.B. Mitchell, 1985: Model projections of equilibrium response to increased CO_2 concentration. In M.C. MacCracken and F.M. Luther (Eds.), *Projecting the Climatic Effects of Increasing Carbon Dioxide.* Report DOE/ER-0237. Washington, DC: U.S. Department of Energy, 81–148.

Schneider, S.H., W.W. Kellogg, and V. Ramanathan, 1980: Letters to the editor. *Science, 210,* 6–7.

Schneider, S.H., and S.L. Thompson, 1981: Atmospheric CO_2 and climate: Importance of the transient response. *Journal of Geophysical Research, 86,* 3135–47.

Sellers, W.D., 1969: A global climate model based on the energy balance of the earth-atmosphere system. *Journal of Applied Meteorology, 8,* 329–400.

Smagorinsky, J., 1983: Effects of carbon dioxide. In *Changing Climate,* Report of the Carbon Dioxide Assessment Committee, Board on Atmospheric Sciences and Climate, National Research Council. Washington, DC: National Academy Press, 266–84.

SMIC (Study of Man's Impact on Climate), 1971: *Inadvertent Climate Modification: Report of the Study of Man's Impact on Climate*. Cambridge, MA: The MIT Press.

Street-Perrott, F.A., and S.P. Harrison, 1984: Temporal variations in lake levels since 30,000 Yr BP: An index of the global hydrological cycle. In J.E. Hansen and T. Takahashi (Eds.), *Climate Processes and Climate Sensitivity*. Geophysical Monograph 29. Washington, DC: American Geophysical Union, 118–29.

Tolman, C.F., Jr., 1899: The carbon dioxide of the ocean and its relations to the carbon dioxide of the atmosphere. *Journal of Geology, 7*, 585–601.

Tyndall, J., 1863: On radiation through the earth's atmosphere. *Philosophical Magazine, 4*, 200–7.

UCAR (University Corporation for Atmospheric Research), 1985: *Opportunities for Research at the Atmosphere/Biosphere Interface*. Boulder, CO: UCAR.

von Neumann, J., 1955: Can we survive technology? *Fortune*, June issue, 106–8 and 151–2.

Wallen, C.C., 1980: Monitoring potential agents of climate change. *Ambio, 9*, 222–8.

Washington, W.M., and G.A. Meehl, 1984: Seasonal cycle experiment on the climate sensitivity due to a doubling of CO_2 with an atmospheric general circulation model coupled to a simple mixed-layer ocean model. *Journal of Geophysical Research, 89*, 9475–503.

Wigley, T.M.L., P.D. Jones, and P.M. Kelly, 1979: Scenario for a warm, high-CO_2 world. *Nature, 283*, 17–20.

Williams, J., 1980: Anomalies in temperature and rainfall during warm Arctic seasons as a guide to the formulation of climate scenarios. *Climatic Change, 2*, 249–66.

WMO (World Meteorological Organization), 1975: *Proceedings of the WMO/IAMAP Symposium on Long-Term Climatic Fluctuations* (Norwich, U.K., 18–23 August 1975). Report WMO No. 421. Geneva, Switzerland: World Meteorological Organization.

WMO (World Meteorological Organization), 1979: *Proceedings of the World Climate Conference* (Geneva, 23–13 February 1979). Report No. 537. Geneva, Switzerland: World Meteorological Organization.

WMO (World Meteorological Organization), 1982: *Detection of Possible Climate Change.* Report of the JSC/CAS Meeting of Experts (Moscow, October 1982). W.W. Kellogg and R.D. Bojkov (Eds.), World Climate Programme Report No. 29. Geneva, Switzerland: World Meteorological Organization.

WMO (World Meteorological Organization), 1985: *International Assessment of the Role of Carbon Dioxide and of Other Greenhouse Gases in Climate Variations and Associated Impacts* (Villach, Austria, 9–15 October 1985). WCP Newsletter No. 8. Geneva, Switzerland: World Meteorological Organization.

3

Politics and the Air Around Us: International Policy Action on Atmospheric Pollution by Trace Gases

Michael H. Glantz

INTRODUCTION

After more than a decade of rather intense discussion and new scientific information on the warming of the global atmosphere as a result of increased emissions of greenhouse gases and of deforestation, one can still find hawks, doves and owls on this issue within the scientific community. "Hawks" believe that the evidence of a CO_2/trace-gases warming is very convincing and that the warming is already under way. "Hawks" represent the true believers. "Doves" (a dwindling number) feel that the greenhouse warming scenario is yet another doomsday scenario that will most likely fail to materialize. They often point to the failure of earlier doomsday scenarios for the environment to make their point. They tend to highlight the existing scientific uncertainties with regard to the global warming issue. Doves also include those who believe

This chapter is a revised version of a background paper prepared at the request of the Deputy Director of the U.S. National Science Foundation to be used for discussion at a meeting in Venice, Italy, in May 1988 of directors of science foundations or academies of the United States, Canada, France, Germany, Japan, Sweden, Italy and the United Kingdom.

Michael H. Glantz is head of the Environmental and Societal Impacts Group at the National Center for Atmospheric Research. He received his Ph.D. in political science from the University of Pennsylvania. His main areas of research relate to the interactions between climate and society with a special focus on drought.

in societal ingenuity (the ability of society to respond or adapt to crises). There are activists among both the hawks and the doves. The "owls" have yet to make up their minds on the issue. They believe that, while the existing scientific evidence appears to be very convincing, there are some important pieces to the scientific puzzle that need to be put into place, such as the regional impacts of a global warming or the role of cloud feedback mechanisms.

The weight of evidence, despite some crucial, persistent scientific uncertainties, has shifted among scientists in favor of the position of the hawks. Carbon dioxide and other trace gases are observed to be increasing in the atmosphere and are apparently heating up the lower atmosphere to unprecedented levels (for recent decades) and at seemingly unprecedented rates. There is an increasing number of scientists in many countries (the Netherlands, the United States, Canada, Germany, Sweden, and the USSR, among others) who have become hawkish on the global warming issue and as a result have become more vocal about their views.

It appears that societies have embarked for some time on the course of unwittingly altering the chemistry of the atmosphere in such a way as to change our global climate. Revelle and Suess (1957) noted thirty years ago that humankind had embarked on a "large-scale geophysical experiment" by pursuing human activities that will ultimately alter climate in unknown ways. Today, however, we are on an advertent course. We have identified the problem and have speculated considerably about its societal, environmental, and political impacts. There is a growing awareness that, even if we stopped burning fossil fuels and manufacturing certain trace gases today, the problem would still be with us for decades into the future, as the greenhouse gases continue to remain in the atmosphere and as the oceans slowly respond to the warming. In other words, societies are already committed to at least some measure of human-induced global warming. Some groups have become deeply concerned about how future generations will be affected by the global warming.

All of this problem recognition notwithstanding, solutions have not been pursued, either by individuals, nations, or by international organizations, for a variety of reasons (discussed below). It could be that the debate on CO_2 is only now heating up to the level where policies might be developed and actions taken to cope with the specter of an unprecedented global warming (e.g., International Conference on the Changing Atmosphere, held in Toronto, 27–28 June 1988). It is important, however, to put this recent resurgence of interest in the effects of increasing atmospheric carbon dioxide into perspective. Warnings by scientists regarding a global warming as a result of fossil fuel burning have been around since at least the end of the 1800s and early 1900s. The global warming "crisis" has apparently crept up on the international community over a period of a century (e.g., Ausubel, 1983; Kellogg, this volume). The question arises: when will decisionmakers ever have enough information to take deliberate action on this issue (inaction is also a form of action)?

Hawks, doves and owls can also be found in the policymaking community. For example, several U.S. Senators (e.g., Chafee, Gore, Wirth) as well as U.S. Representatives (e.g., Brown, Schneider, Sheuer, Skaggs) are seriously concerned about global warming and its potential impacts on the United States and the world, and so are now carrying the banner for action on this issue. Several hearings have been held by various subcommittees in the U.S. Congress on this topic (e.g., U.S. Senate, 1987).

In the international arena, the Executive Director of the United Nations Environment Programme, M. Tolba, has been concerned about the lack of international action on the global warming issue. Dutch policymakers, for example, worry about the sea-level rise aspect of the global warming as a large part of their country is below sea level. But there are also owls (perhaps the largest part of the international political community) and doves on the global warming issue. Because different communities are involved and within each community there are different perceptions of the seriousness of this issue, relative to other, perhaps more immediately pressing problems, it is no wonder that no general agreement about the issue has as yet formed either within nations or between them.

THE PROBLEMS

When considering a response to the greenhouse warming, societies everywhere and the people who govern them are faced with several problems. One problem can be represented by metaphors that we have been culturally taught to live by. These are norms that we take into account, at least subconsciously, before we make decisions. These norms, metaphors, or adages are frequently used by societies or by individuals as a general guide to their actions (Lakoff and Johnson, 1980). For example, "look before you leap" is sound advice that is often provided in response to the need to make a certain decision that requires careful consideration. But there is also the adage that "one who hesitates is lost," suggesting that a risk taker must make bold decisions. These two metaphors are in apparent conflict. Applying either of these adages in the absence of the other to the global warming issue would lead to totally different responses.

Another problem confronting decisionmakers relates to when to act on the CO_2 issue. Three recent reports of possible policy responses to the global warming issue take very different positions. Mintzer (1987) has compared these studies. A 1983 study by Seidel and Keyes said it was too late to respond to a global warming because we have already committed ourselves to a warming. A National Research Council study in the same year suggested that nothing needed to be done in the near-term and that a "wait and see" approach seemed to be called for (Mintzer, 1987, 43). A more recent study by Mintzer suggests that it is not too late to restrict CO_2 emissions and that "the longer the delay before preventive policies are identified, agreed upon, and implemented, the more extreme the policies imposed to stay within prudent bounds will have to be" (1987, 43). Clearly, there is no agreement on when it would be necessary to make public policies as a direct response to a global warming.

The timing of response to the global warming is not the only problem at hand. Another problem is having to decide when there is enough information to act. The scientific uncertainties surrounding the CO_2/trace-gases issue are not insignificant; but there will always be some unresolved scientific issues on the causes and impacts of the global warming. How then should scientists as well as decisionmakers treat those uncertainties? In addition, it is not necessarily true that more scientific information will lead to more scientific consensus. The issue of a global warming is much clearer today than at any time in the past few decades, but those remaining scientific uncertainties must be put into proper perspective (are they major or minor?), lest they be used to paralyze any action, even educational, on the global warming issue.

Yet another problem with taking action relates to the fact that the CO_2/trace-gases warming is, as noted earlier, part of a process; from the mining of coal or the extraction of oil and natural gas, to their combustion, to their transport in the atmosphere, to their impacts on climate, to the impacts of the changes in climate on environment and society. The question frequently arises about where in the process societies should intervene in a purposeful way. This is an important question, because the answer to it will determine the kinds of policies that should be pursued, and the kinds of evidence needed to convince decisionmakers to pursue those policies. It will also determine the relative importance as well as direction of social science research needs. Related to the concern about where to intervene in the process is all the talk about strategies, that is, whether to focus attention primarily on prevention, mitigation, or adaptation to the yet-to-be-identified regional and local changes that would most probably accompany a global warming.

Some observers (Clark, 1985; Mintzer, 1987; Speth, 1987) have suggested that we move directly from scientific findings on the global warming issue to the making of policy. That may be an appropriate response at the emissions end of the process, if the reason for reducing the use of fossil fuels relates to resolving other problems as well (not just the global warming one), such as the push toward energy conservation or toward alternative sources of energy. If, however, the global warming issue is the only major driving force for an attempt to reduce fossil fuel combustion (or to reduce the other activities that abet the warming such as deforestation), then moving from scientific findings directly to policy formation will not necessarily work. What would be needed for most decisionmakers and for the general public would be reduced scientific uncertainty about the global warming issue in general, and, specifically, more "proof" about the regional impacts of the global warming as well as better information about society's ability to cope with climate variability and climate change.

ACTORS

The actors in the CO_2/trace-gases issue are many. They include producers, consumers and policymakers: the individuals who clear forested areas for agricultural purposes, automobile drivers, public service companies, coal miners, corporations that manufacture trace gases, government agencies that seek to develop their national economies, national decisionmakers, non-governmental organizations, international agencies, and so forth. These actors are directly or indirectly involved in the production of CO_2 and other important radiatively active trace gases.

Yet another set of actors includes the physical, social, and policy scientists who have been meeting to discuss the global warming issue. They are actively trying to come to grips with this issue and their ranks are divided in terms of speculation about the regional implications of the warming and what might be done about those implications.

There have been several meetings of these actors in recent years to present their perspectives on the global warming issue, to question the science of the issue, to discuss potential impacts as well as possible policy options to cope with (prevent, mitigate, adapt to) the global warming; some of these meetings were held in Villach, Austria, in 1980, 1983, 1985 and again in September 1987. The most recent Villach meeting focused on developing policy options for responding to climatic change (WMO, 1988). In June 1988 the Canadian government convened a major international political conference entitled "The Changing Atmosphere: Implications for Global Security."

Interest in the policymaking aspects of the global warming issue is presently at an all-time high. The issue has achieved a new level of popular concern in the United States, partly because of Congressional hearings which have been held on this topic throughout 1987. More dramatically, the topic was graphically portrayed on the covers of the 19 October 1987 issue of *Time* magazine and the 11 July 1988 issue of *Newsweek*. Devastating drought occurred across North America in the summer of 1988. Dr. James Hansen, scientist at the Goddard Institute for Space Studies, believes that these droughts were the first conclusive manifestations of a global warming (Begley et al., 1988). International attention for this issue has been heightened by the news releases about the drawing up of the ozone protocol and by the mounting news releases about the greenhouse effect in market- and planned-economy countries, as well as in developed and developing countries.

WINNERS AND LOSERS

Regardless of what opposing view one might take with respect to the distant future, a global warming of several degrees Celsius will prove to be a boon to some and a bane to others, at least in the short term. Some areas will become drier while others will become wetter. Some areas will seemingly benefit from, and therefore desire, those changes while others will not. Clearly, there will be winners and losers. This perception about the potential benefits as well as adversities associated with the yet-to-be-identified regional impacts of a global warming will clearly serve to constrain international cooperation on the CO_2 issue.

CO_2 and Economic Development

If, for example, China proceeds with its economic development plans, coal will be its main source of energy. There may prove to be little that can be said to convince the Chinese to delay their plans for development in order to attempt to reduce CO_2 emissions to the atmosphere. If China does cut back on its dependence on fossil fuels, in the absence of a viable replacement, it will surely be a "loser" with respect to development. If, however, it continues to use, or to increase its use of, fossil fuels it may or may not be a loser from the effects of the climate change.

To get the global community to consider alternatives will take creative thinking such as a suggestion that one of the industrialized countries, such as the United States, would cut back on its production of CO_2 (this could be partially achieved by more efficient fossil fuel use) so that China could continue on its development course unfettered; or that one of the industrialized countries would compensate China for not increasing its use of fossil fuels.

A look at the USSR points out a similar dilemma. On the one hand Soviet scientists are among the international "whistle-blowers" on the deleterious effects of a CO_2/trace-gases-induced global warming. On the other hand, there are ministries within the USSR that are in the midst of carrying out major plans to develop Siberia and the fossil fuel resources in that region. There seems to be no tie-in within the USSR between these groups and no resolution of their opposing objectives and concerns. This is an example of the sensitivities of bureaucratic units within the same government, each with its own jurisdiction over a geographic or functional area and each responding to different sets of interests. A similar situation can be observed in the United States with respect to the development of oil reserves in Alaska.

The example of acid rain in the United States is somewhat analogous to CO_2 and underscores the existence of bureaucratic (as well as political) jurisdictional disputes about whether a particular environmental problem is real and whether any consensus might be reached among those affected by the problem. Coal miners in West Virginia have a different view about the seriousness of acid rain in the northeastern part of the United States than do those who live in the impacted region. Likewise, policymakers in the industrial Ohio Valley do not readily admit to their constituents' involvement in the acid rain problem nor are they willing to take steps that might prove financially harmful to the region's inhabitants.

Thus, even within countries and in regions within these countries there could be winners and losers if the use of fossil fuels and hence the emissions of CO_2 continue even at present rates. How to restrain the activities of the potential winners and to compensate the potential losers are important issues that need to be addressed in both national and international policymaking circles. While more efficient energy use can result in some reduction of CO_2 emissions, the alternatives such as shifting to a dependence on solar, wind, or nuclear energy may be costly—too costly for the developing countries. Are the industrialized countries willing (or able) to financially support such a shift within the developing world? Are the industrial countries willing to cut back their own fossil fuel use (by energy conservation and the use of alternative energy sources) in favor of the developing countries? After all, wasn't it the industrialized countries that loaded the atmosphere with CO_2 and the other radiatively active trace gases in order to develop their economies? Shouldn't the developed countries reduce some of their fossil fuel dependence so that the developing countries can use fossil fuels to develop their economies? These difficult questions must be addressed.

TO ACT OR NOT TO ACT

The list of reasons for not doing anything about global warming is long: there are often other more pressing issues on any given day, the scientific uncertainties are important (e.g., the lack of knowledge at this time about regional specificity concerning the impacts of a climate change and about the role of cloud feedback mechanisms), conflicting interests within and among societies, the absence of an imminent "dread factor" (Slovic et al., 1980), problems associated with a common property resource, diffuse impacts on a global basis, perceptions of the problem as a long-term, low-grade one, the legacy of the supersonic transport (SST) issue of the early 1970s (British and French scientists versus American scientists), short time horizons of decisionmakers (versus time horizon of the impacts of a greenhouse warming), population increase, discounting the past as well as the future, viewing the warming as a process as opposed to an event, the problems associated with alternative energy sources, the potential for present "climate-related" losers to become "climate-change-related" winners and vice versa, the use of an inappropriate metaphor, and so forth.

The list of reasons that should push toward some degree of international action or at the least cooperation on the global warming issue is ostensibly much shorter: concern about the fate of the global environment, the view that with a climate change "nobody really wins," stress placed not on the short-term impacts but the longer-term benefits of action for the global community, a call for energy conservation (a good idea for some actors in its own right), the view that international agreements such as the Montreal "Resolution on a Protocol Concerning Chlorofluorocarbons" for the protection of the ozone layer (henceforth referred to as the ozone protocol) have been drawn up on extremely controversial issues that could serve as analogues to those interested in developing such a protocol for the global warming issue, individual and societal fear of or opposition to change, and so forth.

Each of these reasons—pro and con—will be discussed briefly.

CONSTRAINTS ON ACTION TO COMBAT
A GLOBAL WARMING

Issue Competition

At any given point in time there will be issues competing for the resources (i.e., time, attention, and funds) of decisionmakers. Usually those issues that require most immediate action are considered first and those issues that are not necessarily associated with the normally defined criteria of a crisis (short time to act, high costs for not acting, etc.) are put aside. Each day there are new traditionally defined crises emerging and as a result those "less burning" issues become continually delayed, until at some time in the future they too are perceived to require immediate attention. In the mid-1970s, U.S. economist Robert Heilbroner expressed this view, noting that

> The problem of global thermal pollution, for all its awesome finality ... stands as a warning rather than as an immediate challenge. Difficulties of a much more matter-of-fact kind—resource availability, energy shortages, the pollution resulting from noxious by-products of industrial production—are likely to exert their throttling effect long before the fatal, impassable barrier of irreversible climate damage is reached (1975, 54–5).

There are many reasons why the global warming issue may not be perceived as a crisis in the traditional sense of the term. Although societal actions continue to aggravate the CO_2/trace-gases-induced global warming today, there is a time lag (on the order of decades if not centuries) before we can clearly identify the impacts of our actions. In addition, there is enough scientific uncertainty surrounding this issue so that action can effectively be blocked, even by a minority opinion.

The CO_2 aspect of today's global warming issue has been with us since the turn of the century (Arrhenius, 1896, 1908). It was debated then and faded away only to reappear as a scientific issue in the late 1930s (Callendar, 1938). It faded away once again, probably overridden by more pressing economic and political issues, and re-emerged as a scientific issue in the mid-1950s (Plass, 1956; Revelle and Suess, 1957), only to fade away once again. Its re-emergence in the 1970s, however, is unlike earlier ones. Now there is concern about the environmental and societal implications of a global warming and about the fact that the rate of change in the content of CO_2 and other trace gases in the atmosphere is much larger now than it was at the turn of the century (WMO, 1979, 1983; NRC, 1983).

In the past, concern about such issues was minimal and it was felt (based on speculation) that the impacts would only be beneficial. For example, Callendar wrote that

> In conclusion it may be said that the combustion of fossil fuel, whether it be peat from the surface or oil from 10,000 feet below, is likely to prove beneficial to mankind in several ways, besides the provision of heat and power. For instance the above mentioned small increases of mean temperature would be important at the northern margin of cultivation, and the growth of favorably situated plants is directly proportional to the carbon dioxide pressure.... In any case the return of the deadly glaciers should be delayed indefinitely (1936, 236).

Even the articles written in the 1950s were neutral with respect to the implications for society of a global warming.

As far as the scientific aspects of the global warming issue are concerned (with regard only to the burning of fossil fuels), it appears that most of the aspects discussed today (except for the role of the clouds, and perhaps the confounding effects associated with volcanic activity) were being discussed at least fifty years ago; the role of the oceans, the role of land clearing, the differential impacts according to latitude, even the projected temperature increases, and so forth. While we have acquired considerably more knowledge through modeling and paleoclimatic investigations, the scientific premise of CO_2-induced global warming remains essentially the same as it was in the 1930s, as expressed by Callendar.

Conflicting Time Horizons

Social, economic, and political time horizons do not mesh well with the time scales on which certain environmental problems take place. While political leaders in the United States are elected either every two, four, or six years, environmental problems of the long-term, low-grade but cumulative kind evolve over much longer time scales. The mismatch between these time horizons often enables a given set of decisionmakers to delay responding to a call for action. The mismatch of time horizons also enables other "more pressing" issues, the outcomes of which can more immediately affect a policymaker's tenure in office, to up-stage the environmental problems considered to have dire consequences some time in the distant future. For any given issue, the general public and the media have a relatively short attention span.

Discounting the Future

While each generation tends to value its own consumption more highly than that of succeeding generations, some are willing to make great sacrifices to enhance the well-being of future generations and others are not. The conflicting values placed on decisions that might affect future generations can act as a constraint for action regarding the global warming issue. Contrary to popular belief, not everyone harbors good will toward societies that will exist decades or centuries in the future. Heilbroner addressed this issue, writing that

> On what private, 'rational' considerations, after all, should we make sacrifices now to ease the lot of generations whom we will never live to see? There is only one possible answer to this question. It lies in our capacity to form a collective bond of identity with those future generations. ... There are many who would sacrifice much for their children; fewer who would do so for their grandchildren. *Indeed, it is the absence of just such a bond with the future that casts doubt on the ability of nation-states or socio-economic orders to take now the measures needed to mitigate the problems of the future* [italics added] (1975, 115).

Discounting the Past

There is also a tendency to discount the importance or value of past experience, both in the science and the impacts related to the global warming issue. For example, some argue that the findings of Arrhenius in 1896 or Callendar in 1938 were not as certain as the results of current research, because today we have the best scientific and most current information available. Weighing only the results of present research makes the problem of global warming seem like a much more recent issue than it really is.

With respect to social impacts and discounting the past some observers assume that past experiences in dealing with climate change and climate variability can provide little insight into the societal impacts of a global warming. They say this because changes associated with a warming are suggested to be beyond anything encountered in human experience. Thus they are skeptical of mitigation or adaptation measures that may be proposed based on past experience. Not everyone agrees, however, that these past experiences should not serve as a basis for present and future action, as witnessed by the contributions to this volume.

Scientific Uncertainties

Several important scientific uncertainties remain. As far as policymakers are concerned, it was only ten years ago that U.S. Congressional Hearings were held on climate change. At that time Congress, at the behest of the scientific community, was mainly concerned about the societal and environmental implications of a global cooling. Such publications as *Ice, The Cooling, Fire or Ice*, and *Weather Conspiracy: The New Coming Ice Age* appeared in bookstores. Studies of regional scenarios related to a global cooling were also undertaken at that time (e.g., Goldsmith, 1977; CIA, 1974). A turning point in the concern from global cooling to global warming occurred with the International Symposium on Long-Term Climate Fluctuations, held in the summer of 1975 (Kellogg, in this volume, Kellogg et al., 1975).

Global warming is expressed in terms of global mean temperatures. National, state, and local decisionmakers need to know, however, what that global mean will translate into at the regional and local levels. As of today, the general circulation models (GCMs) of the atmosphere have a poor record at producing scenarios (especially for precipitation patterns) for regions of the globe. Despite the fact that these models represent the best science and the best hardware available they do not yet produce (for the policymaking community) credible or reliable regional scenarios. Other approaches to the development of reliable and credible regional scenarios are presently being developed to complement those being produced by the GCMs (e.g., Parry et al., 1987).

Problems with Alternative Energy Sources

We can find supporters of solar energy joining hands with supporters of nuclear energy to identify the continued burning of fossil fuels as a major contributor to atmospheric pollution by trace gases. Their cooperation ends there. Clearly, the replacement of fossil fuels with nuclear power could substantially reduce CO_2 emissions to the atmosphere. In practice, however, reliance on nuclear energy remains highly controversial and its development highly political. The technical problem of long-term storage of waste products has not as yet been solved. The cost of nuclear-generated electricity has been much higher than originally projected and plants have not worked nearly as well as expected. Thus it is much more expensive than energy generated from fossil fuels. Every time there has been a nuclear accident—such as at Three Mile Island (USA) or Chernobyl (USSR)—the plans for reliance on this form of energy become less popular as well as less certain. The technology of nuclear power generation is also very sophisticated and capital intensive, and would not be useful for transfer to developing countries. Solar energy, while used for hot water heating and electricity generation in some remote areas, is still in the experimental stage for large-scale electricity generation and not available as yet on a commercial basis. It appears that it will not be able to compete economically with fossil fuels, at least in the near future. Thus, more efficient use of fossil fuel energy seems the only present option other than the status quo. Lack of alternative energy sources that are both politically and technically acceptable is a constraint to the reduction of CO_2 emissions.

Societal Cleavages

Society is divided in many ways, along political, cultural, economic, and social lines. A change in the status quo (e.g., the dependence of some regions on the mining of coal or the extraction of oil or natural gas and of other regions on the burning of it) will be opposed by some groups in society while favored by others. Even the seemingly harmless suggestion that with energy conservation everyone wins is rent with conflict. For example, corporations that "sell" energy and workers whose livelihoods depend on the production of energy would most assuredly disagree. Even shifts within a country among fossil fuels, while reducing CO_2 emissions (coal produces the most CO_2 and natural gas the least, per unit of energy produced), would lead to political conflict between affected groups. In a democratic society it is not easy for long existing societal cleavages to be healed, allowing consensus for action on the global warming to be formed.

Obviously there are conflicting interests between nations as well. For example, while it is in the interest of the Chinese to develop their coal reserves to finance their economic development plans, or while it is in the interest of the oil producing countries to encourage the use of their oil, it will not be in the interest of low-lying coastal countries to suffer from the projected sea-level rise associated with a global warming. How, then, might a consensus for action be formed when these international actors have such diametrically opposed views and, in a global sense, conflicting priorities?

Population Increase

Populations are increasing around the globe, with rates in some developing countries approaching 4 percent per year. Growing populations require additional resources of food, water, land and energy. Furthermore, it is anticipated that as poor countries improve their economies, much more of their energy will be supplied from fossil fuels. Adding their rapidly growing populations to the expected increase in affluence (with improved economies) means that their share of global CO_2 production will increase.

Populations in the industrialized countries increase slowly or not at all; but because per capita consumption of energy and other resources is extremely high, these countries contribute the major share of global CO_2 emissions today. Stabilizing or reducing global CO_2 emissions, then, raises a number of questions. What are the issues that surround the lowering of per capita energy consumption in the industrialized countries while permitting it to increase in the developing world? What is the best strategy for lowering energy consumption in developed countries? Is it possible to raise incomes in developing countries without following the same paths of pollution that the now-industrialized countries had followed?

The population debate is a highly charged political issue about which much has already been written. With regard to the global warming issue, it acts, in the short term, as a constraint on taking universal action to reduce CO_2 emissions for two reasons. First, the scale of fertility declines and consequent population stabilization required to reduce total global demand for CO_2/trace-gases-producing commodities is likely to take many decades to accomplish even under the most favorable conditions. Second, it is necessary to reduce total demand for CO_2/trace-gases-producing commodities in both industrialized and developing countries—first by improving efficiency and second by designing an energy economy that does not rely on fossil fuels.

Warming-as-Process Versus Warming-as-Event

Some people tend to look at the global warming primarily as a process; a long-term, low-grade, cumulative process in which today's amounts of CO_2 and other trace gases in the atmosphere are much like yesterday's and will be similar to tomorrow's, but that in a year or so, the gradual change will be measurable; and in several years that change will become even more noticeable. They believe that the time to act is early in the process because of the nature of the greenhouse gases; they can remain in the atmosphere for many decades or longer.

Others, however, focus their attention on the warming primarily as an event. A doubling of CO_2, for example, will at some time in the future mean higher temperatures, higher or lower rainfall amounts depending on the region of the globe, higher evaporation rates, higher sea level, and so forth. Those who perceive the warming as an event are looking for signs of change before they will join others in a call for action. Those signs of change may be subtle ones such as a separation of the climate change signal from the variability of normal climate conditions; or they may be major ones such as rapid rise in sea level, a disintegration of the West Antarctic ice sheet, or agricultural activities in such relatively inhospitable regions as Siberia or northern Canada.

These conflicting perceptions of the global warming issue inhibit action on dealing with the issue.

Diffuse Impacts

The regional and local impacts of a global warming have not yet been identified, but even when the findings of the scientific assessments are presented (more rain, less rain, higher temperatures, higher sea levels, etc.), many countries will probably not believe that such seemingly small increases in temperature or unknown but seemingly not-so-severe changes in rainfall could be "all that harmful" and may in fact believe that they could prove to be generally beneficial. Sea level rise is one impact that seems to possess some credibility among coastal countries and low lying areas but the rates of increase and the ultimate level of increase remain controversial. At this time the scenarios that have been suggested for the future with a warmer earth's atmosphere are numerous, often conflicting, and quite possibly unreliable as forecasts of a future world. The use of global projections, therefore, with little reliable information on regional impacts, minimizes the "dread factor" that some scientists have attempted to associate with the global warming issue.

Lack of Dread Factor

There have been several attempts to identify a "dread factor" for the CO_2/trace-gases issue, but none of these attempts has met with success. For example, in the late 1970s there was considerable attention by a few scientists (and the media) given to the prospect of the disintegration of the West Antarctic ice sheet and the drastic (5–8-meter) rise in sea level that would accompany the disintegration. Although debate on the role of the West Antarctic ice sheet in the sea-level issue remains, such a rapid rise in sea level has been discounted. Another attempt at finding a "dread factor" emerged during Congressional hearings when it was suggested that a CO_2 doubling in the atmosphere (over pre-Industrial Revolution levels) would most likely occur in the first few decades of the 21st century. But soon less attention and importance was placed on a CO_2 doubling, because it was assumed that no rapid changes could be associated with such a doubling. A recent attempt at identifying an "ozone hole" equivalent (i.e., "dread factor") appeared in an article in *Nature* (Broecker, 1987), in which the author suggested that "there is now clear evidence that changes in the Earth's climate may be sudden rather than gradual." In other words, policymakers should be prepared to cope with the impacts of step-like changes in regional climates around the globe. Yet, this too is speculation about how the course of global warming might proceed. The most recent attempt to identify a CO_2/trace-gases-related dread factor was made by a prominent scientist, who has suggested that four of the hottest years on record have occurred in the 1980s (Begley et al., 1988). This claim has yet to be validated by the scientific community.

Technological Fix

Many people have a blind faith in technology, holding the belief that human ingenuity in times of crisis will rise to the occasion and develop a way to make the global warming problem "go away." Such an attitude encourages procrastination on taking immediate or near-term action on a long term, low grade but cumulative environmental change.

The Supersonic Transport (SST) Legacy

In the early 1970s the U.S. scientific community was concerned that the exhaust from a large fleet of high-flying SSTs would ultimately deplete stratospheric ozone, thereby allowing more ultraviolet radiation to reach the earth's surface, leading to an increase in the incidence of skin cancer. This scientific issue became controversial within the United States as the SST issue became politicized, with political leaders (including the U.S. Department of Transportation and President Nixon) and some scientists on one side of the issue and environmentalists and other scientists on the other side (see Glantz et al., 1982, for a discussion of this issue).

This scientific issue was one of international concern as well, as scientists in the United States, Great Britain, France, and the USSR took opposing scientific points of view. These governments (except for the United States) had already decided to support the development of a high-flying fleet of SSTs (the Franco-British Concorde and the Soviet Tupolev), and their scientists, using essentially the same pool of technical information as used by the Americans, came up with an opposing position on the impacts of such an SST fleet. An article written by an American scientist appeared in *Science* in 1971 (Johnston, 1971) supporting the view of ozone destruction by a large, high-flying SST fleet. In 1973 an article appeared in *Nature* (Goldsmith et al., 1973) representing the British scientific view, noting that the impacts on stratospheric ozone of a fleet of Concordes would be relatively minimal. An assessment of this controversy in international science was written some years later and was appropriately entitled *The Bias of Science* (Martin, 1979). There is a legacy in this experience for the CO_2 issue. Those countries that wish to "do something" about CO_2 emissions, as well as those countries that do not wish to take action (for whatever reason), could search for, and most likely find, reasons to support their actions in the existing scientific literature.

The "Nothing to Lose" Syndrome

In many regions of the world the inhabitants might welcome a change in their climate. One scientist has even suggested that the increase of carbon dioxide in the atmosphere would help to close the gap between rich and poor nations. Indian scientist, J. Bandyopadhyaya (1983), in his book *Climate and World Order* has noted that, in contrast with the tropical regions, the temperate climate has supplied the industrialized countries with many permanent natural advantages and that climate has been and continues to be one of the major factors behind their accelerated development. While the industrialized countries are trying to "preserve the global climate status quo by fighting against 'climate variability' " by trying to inhibit the upper level of the global warming, it should be in the interest of the developing nations to change their tropical climate and to modify "the global climate dichotomy." Bandyopadhyaya based his argument on the view that a more uniform global climate would help accelerate the economic growth of the developing countries without adversely affecting the industrialized world. The preceding thoughts represent a perception that today's "climate-related" losers could become tomorrow's "climate-change-related" winners.

60

Inappropriate Metaphors

Societies are faced with many environment-related problems; ozone depletion, acid rain, air pollution, nuclear winter, and so forth. The images that the names of these environmental problems conjure up are negative ones. When it comes to the greenhouse warming metaphor, however, the image is a positive one. People have been taught for generations that greenhouses are good. They protect vulnerable plant life from harsh climatic conditions. They provide an environment in which plants can thrive through inhospitable seasons under the most favorable conditions. But today the metaphor of a greenhouse has been applied to the global warming situation. Now we are supposed to relate a "greenhouse gas" or a greenhouse effect to an adverse, harmful situation. Perhaps the use of the greenhouse metaphor has become part of the problem in getting people to take the global warming issue seriously. What may be needed is a metaphor that portrays an adverse condition, a phrase such as the "trace-gases pollution" problem. If the issue is referred to in this way, it might be taken more seriously by the general public and by the media as a threatening environmental change.

FACTORS FOSTERING COOPERATION TO COMBAT A GLOBAL WARMING

Concern about the Fate of the Global Environment

Environmental movements blossomed around the globe in the 1960s. Concern focused on the fate of the global environment including the overuse and misuse of natural resources. Many of the trends adversely affecting environmental quality identified in this era have continued into the 1980s more or less unabated: deforestation, desertification, soil erosion, population growth, atmospheric and oceanic pollution, among others. Carbon dioxide and stratospheric ozone depletion are just the latest issues to be taken up by environmental interest groups. There is a precedent of international cooperation on addressing such global environmental issues.

An environmental problem can be viewed as global in several respects: global in cause, global in effect, global in interest. This global interest can be witnessed by the series of international conferences convened by the United Nations in the 1970s to address problems of common concern: Conference on the Environment (Stockholm, 1972), World Food Conference (Rome, 1974), World Population Conference (Bucharest, 1974), Habitat (Vancouver, 1976) Water (Buenos Aires, 1977), Desertification (Nairobi, 1977), Science and Technology (Vienna, 1979). The concepts of "One Earth" and "Spaceship Earth" have certainly caught on as symbols of the human predicament among many segments of society.

Nobody Wins with Climate Change

The argument that nobody really wins with a global warming is based on the view that the sum of impacts of a climate change will be detrimental to humanity. If, for example, more rain appears in arid and semiarid regions, will the corresponding increase in food production compensate for a possible loss in formerly productive regions such as the U.S. Great Plains (a region from which food aid comes in times of recurrent, intense drought in arid and semiarid regions around the globe)? The same kind of argument can be made within countries as well. Agricultural productivity may benefit from a warming but what will happen to coastal areas as a result of the attendant sea-level rise? Many important coastal cities of the world would be at risk to sea-level rise of several meters (e.g., Barth and Titus, 1984; NRC, 1985; Bardach, 1987; Davidson, in this volume). In addition, perceived short term benefits must be weighed against the longer-term costs and benefits.

Energy Conservation

As a result of the increased cost of fossil fuels during the 1970s, energy is being used considerably more efficiently today, with the developed nations using less energy than was predicted in the early 1970s. There remains a large potential for using energy even more efficiently. The investments necessary to achieve this are cost effective at present energy prices (Krenz, 1980). Therefore, the developed nations can further reduce their fossil fuel use and thereby reduce CO_2 emissions. This could allow developing nations to use more fossil fuels without increasing the present level of global CO_2 emissions. The relatively simple technologies required for more efficient energy use can, of course, also be shared with the developing nations, thus further reducing the fossil fuel requirements for their development. The primary motivation for efficient use of fossil fuels would not need to be the reduction of CO_2 emissions, since it would also benefit the balance of trade of energy-importing nations as well as conserve a finite, relatively scarce, resource.

Precedents for International Agreements

A paper trail of international agreements exists on issues that at one time were viewed as extremely controversial and seemingly intractable. Perhaps the most recent political one is the agreement reached in December 1987 and signed in June 1988 between President Reagan and Secretary General Gorbachev on the reduction of intermediate nuclear forces. With respect to the environment, the most recent agreement is the "Resolution on a Protocol Concerning Chlorofluorocarbons" signed in Montreal in September 1987. Therefore, one should not be deterred in one's efforts to reach some kind of international understanding on the global warming issue. In fact, at the International Conference on the Changing Atmosphere (held in Toronto, 27–28 June 1988), Canadian Prime Minister Mulroney initiated a process to develop an international law of the atmosphere in response to the CO_2/trace-gases-induced global warming.

The Ordeal of Change

People often fear change. This could be a powerful factor in bringing nations together to discuss, if not actively combat, global warming. True, with a global warming there will be winners and losers, at least in the short term, but who will they be? True, different regions will be affected in varying ways by a warming, but will they be affected in positive or negative ways with respect to the way these societies operate today? In many instances, if given the choice, people would favor "the devil they know" over "the devil they don't know." A supporting argument could be made that it might be much easier to reach agreement on action to combat a CO_2/trace-gases-induced global warming before the winners and losers have been clearly identified. Once winners and losers have been identified as such, the losers will seek change but the winners will oppose it. (See Brown, this volume, for an example of this situation with regard to the Colorado River Compact).

SOME INSTITUTIONAL ASPECTS OF
THE GLOBAL WARMING ISSUE

Among those who consider the global warming issue an ominous environmental change that needs to be dealt with on more than an ad hoc basis, there is debate over whether existing institutions are appropriate for dealing with the issue. What institutional mechanisms are needed for dealing with the global warming issue, an issue that most observers consider a long-term, low-grade cumulative environmental problem? Some observers suggest that new institutional arrangements are required, such as the recently developed ozone protocol.

It is useful to examine this analogy further. In 1987, representatives of 24 nations and the European Community negotiated an international protocol on the control of the manufacture and use of CFCs. The proposed treaty, which must be ratified by at least 11 countries before it takes effect, calls for cutting world consumption of CFCs by 50 percent (of the 1986 level) by 1999 (C&EN, 1987, 1557). This has been suggested as a prototype of a process that might be pursued for controlling CO_2 emissions to the atmosphere.

CFCs are used as spray can propellants, refrigerants, cleaning solvents, foam-blowing agents, and for other industrial purposes. The CFCs are inert chemical compounds that eventually diffuse to all parts of the atmosphere. In the stratosphere the chemical molecules are broken down by ultraviolet rays, freeing the chlorine atom to enter into catalytic reactions with ozone molecules. Chlorine in the stratosphere is an effective "ozone eater." With the destruction of stratospheric ozone more ultraviolet reaches the earth's surface, thereby increasing the incidence of skin cancer (and other health problems as well) among humans in addition to bringing about other deleterious effects, many of which are yet unknown, to living organisms. Moreover, it now appears that, incrementally, CFCs along with other trace gases (excluding for the moment CO_2) are contributing to the global warming. Some observers suggest that these gases taken collectively might equal and perhaps surpass the contribution of CO_2 to the global warming in the next few decades (e.g., NRC, 1983; Ramanathan et al., 1985).

The value of the actions suggested in the protocol to reduce CFC emissions notwithstanding, there may be little similarity between attempts to control CFC emissions and attempts to control

CO_2 emissions to the atmosphere. In other words, the recent ozone protocol may be a "misplaced analogy" to the CO_2 issue. Despite apparent similarities in their impacts on the global atmosphere (i.e., their respective contributions to the global warming), there are important differences between these issues. For example, there is a "dread factor" that the public can relate to that is associated with the CFCs; increased CFCs decrease stratospheric ozone and therefore increase the incidence of skin cancer. With a reduction in stratospheric ozone there may be no winners. The risk of skin cancer will increase for everyone although certain populations are more at risk than others. Opponents of any restrictive action on CFC production point out that every day people are increasing their risk to skin cancer by moving from northern locations to southern ones (e.g., from Minnesota to Texas) or from low altitudes to higher ones; this is tantamount to an increase in exposure to harmful ultraviolet radiation. However, moving from one location to another involves individual choice; the consequences of ozone depletion allow no such choice.

For the CFCs there is an additional "dread factor"—the ozone hole. This factor, perhaps more than the others, has catalyzed policymakers (if not the protocol negotiators) to take action. To some the ozone hole over the Antarctic is a "warning shot across the bow"; if no appropriate action is taken now, a similar hole could develop in other parts of the atmosphere, over inhabited regions. Any near- or mid-term "dread factors" associated with the CO_2 issue are less clear. Furthermore, there is no CO_2 equivalent now to an "ozone hole" catalyst for action. The most recent catalyst for action on the global warming issue may prove to be a situation such as the increase in the frequency of extreme climatic conditions, as exemplified by James Hansen's belief that the four warmest years during the period of instrumental records (for global air surface temperatures) may have all occurred in the 1980s (Begley et al., 1988).

What will a warming do to climate and human activities at the local and regional level—a little wetter, a little drier, a little warmer, a little colder, higher agricultural yields, lower agricultural yields? The "dread factor" associated with inaction on the issue of reducing the use of fossil fuels remains unclear. In other words, the costs (or benefits) of inaction are not clearly identifiable. It is important to note that nonintervention (inaction) is also a form of intervention (action). Inaction may imply that the "dread factor" is not perceived to be high and therefore there is inadequate backing for action; or that a conclusion has been reached that there is time to act in the future, or that the costs of inaction at this time will not be high, or that the costs involved in viable action would be too high (e.g., shifting dependence from fossil fuel to solar or nuclear). Also, while individuals are quite aware (in a qualitative sense) of how they benefit from the use of fossil fuels, they are less aware of the need for CFCs. (Many people relate CFCs to propellants in hairsprays and underarm deodorants or containers for fast-food store hamburgers; alternatives have been found or are at least available in these instances in the United States.)

CFCs are the source of two environmental problems; a global warming and a depletion of stratospheric ozone. In light of this, another interesting question must be raised, one that might shed light on the use of the recent protocol to control CFCs as a model for future attempts to achieve a CO_2 protocol: Would the 24 national representatives have agreed to the protocol negotiated in Montreal (or even met to discuss it) if the CFCs had been associated only with a global warming problem? Most probably not. I would venture to say that the skin cancer relationship to the depletion of ozone (a factor known and discussed since the mid-1970s) and the rapid sharp decrease in ozone over the Antarctic region served as the underlying cause and the catalyst, respectively, for public support for action to combat ozone depletion.

Both CO_2 and CFC emissions are part of broader processes. One can find analogies and disanalogies between these two issues depending on the stage in the processes that one wishes to compare. For example, they could be analogous with respect to the impacts of both of these trace gases on climate. They are not analogous, however, in comparing their sources; only a small number of companies produce CFCs whereas innumerable individuals as well as factories are burning fossil fuels in one form or another, producing CO_2 emissions.

Before accepting the analogy with the ozone protocol as a guide for trying to achieve an international agreement on CO_2 emissions, one must make sure that a more appropriate analogy does not exist. If we can identify such an analogy, we might be able to gain some insight on whether and how to proceed at the national and international levels to secure an international accord of some sort on limiting future levels of atmospheric carbon dioxide and other radiatively active trace gases. What similarities, for example, might the CO_2 issue (and process) share with such international processes as the development of the atmospheric test ban treaty, the nuclear non-proliferation treaty, or attempts to develop a New International Economic Order?

Several organizations are involved in the policy aspects of global warming, some more directly than others: in the United States these include but are not limited to the Environmental Protection Agency, the Department of Energy, the National Climate Program Office, the National Science Foundation, the World Resources Institute, Woods Hole Oceanographic Institute, the American Association for the Advancement of Science, the National Center for Atmospheric Research, the Geophysical Fluid Dynamics Laboratory, Goddard Space Flight Center, Goddard Institute for Space Studies, NASA, the National Academy of Sciences, and so forth. Internationally, the organizations that are active in the CO_2/trace-gases warming issue include the United Nations Environment Programme, the World Meteorological Organization, the International Institute for Applied Systems Analysis, the Beijer Institute, the European Community, the International Council of Scientific Unions, and so forth. Clearly, there are numerous groups, centers, and institutes involved in this issue, in addition to scores of researchers at various universities around the world.

The International Council of Scientific Unions (ICSU), as well as the U.S. National Academy of Sciences and the U.S. Department of Energy, have raised concern about the scientific aspects of the problem. The Beijer Institute, the International Institute for Applied Systems Analysis, the World Resources Institute, the Hubert Humphrey Public Policy Institute, the U.S. Congress, the Environmental Protection Agency, and the U.S. National Academy of Sciences, have voiced concern about the need for some sort of policy action on this issue. What is it that is missing? What is not being done with the current arrangements that should be done? These are questions that need to be answered before we get involved in creating new national or international institutions to cope with the global warming issue.

Before we continue to search for new policy options and institutional arrangements in order to "deal" with the global warming issue, it would be wise to ask what it is that we would like such an international institution to do. Is it to be used as a forum where different views on the issue can be expressed and evaluated? Is it to be an advocacy institution (i.e., an interest group), carrying the message about the adverse impacts of a global warming to policy- and decisionmakers around the globe? Should such an institution be used to educate societies and policymakers about the potential consequences of a global warming? What is it that the new institution might do that could not effectively be done through existing institutional arrangements? We must avoid a "rush to judgment" on creating new international institutional arrangements to deal with the CO_2/trace-gases-warming issue until we have investigated the potential contributions and roles of existing institutions.

CONCLUSIONS

Why do we have a scientific community warning us of an impending climate crisis on the one hand and a lack of concerted response within countries and among them on the other? It appears (from the list of pro and con reasons for taking action on the global warming issue) that the reasons that constrain action are seemingly more specific and numerous than those reasons that promote action. Even the call to various policymakers to unite to combat an excessive global warming seems to have been falling (generally speaking) on deaf ears.

The past few decades have been filled with doomsday scenarios; limits to growth, stratospheric ozone depletion as a result of supersonic transport, the China syndrome, the food, water and energy crises, nuclear winter scenarios, and the like. Many of those doomsday projections have not yet materialized. Most likely they were made to draw attention to impending crises rather than to suggest that we were already committed to such scenarios. The global warming issue is the latest of this type of projection.

Even if this one is *the* real doomsday scenario, will anyone believe it? How does one build a sustainable, broad-based coalition to create a better earth or a more wholesome environment? How might we (humankind) circumvent these constraints and create an atmosphere whereby this problem can merit more sustained (as opposed to sporadic) and serious (as opposed to nominal) attention by policymakers, the public, and the media? How can we convince individuals, organizations, governments to take actions that would at least slow down, if not prevent, unprecedented levels of global warming from occurring—actions that would benefit humankind even if the global warming were not to occur for some as yet unforeseen reason? Other concerns that need to be addressed are as follows:

- What will be considered enough information to prompt action?

- When will we have that information?

- What are the identifiable catalysts to social action that might be analogous to the "ozone hole"?

- Where in the process of the global warming should societies act?

- What kinds of tradeoffs are societies willing to make to cut down global CO_2 emissions?

- What is the appropriate process to pursue to reach an international agreement on the global warming issue?

- How can the sustained interest in this issue be maintained among the public, the policymakers, and the media?

- How important should the remaining "known" scientific uncertainties be as obstacles to actions on the global warming issue?

- What kind of credible and reliable social science research is needed to better understand this issue and societal responses to it?
- How can we develop reliable and credible regional scenarios that might occur under conditions of a warmer atmosphere?

These are but a few of the questions that need to be addressed if humankind is to understand and respond to scientific information on the global warming issue. These questions must be addressed in many different forums and by many different groups. The call to action by any one group will most likely require corroboration by similar pronouncements from different groups representing different interests.

ACKNOWLEDGMENTS

I would like to thank the following people for their critical assessments of early drafts of this discussion paper: John Moore (National Science Foundation), John Steele (Woods Hole Oceanographic Institution) John Firor (National Center for Atmospheric Research (NCAR)), Robert Chervin (NCAR), Judith Jacobsen (NCAR), Dale Jamieson (University of Colorado), D.J. Fisk (Department of Environment, UK), Mario Pedini (Italy), Ralph Cicerone (NCAR), Pierre Lafitte (France), Nobuo Egami (National Institute for Environmental Studies, Japan), Paul Crutzen (Max Planck Institute, Germany), and Maria Krenz (NCAR). I would also like to thank Jan Stewart and Regina Gregory for their continued support during the preparation of several drafts of this manuscript.

REFERENCES

Arrhenius, S., 1896: On the influence of carbonic acid in the air upon the temperature of the ground. *Philosophical Magazine, 41*, 237.

Arrhenius, S., 1908: *Worlds in the Making*. New York, NY: Harper & Brothers.

Ausubel, J.H., 1983: Historical note. In *Changing Climate*, Report of the Carbon Dioxide Assessment Committee, Board on Atmospheric Sciences and Climate, National Research Council. Washington, DC: National Academy Press, 488–91.

Bandyopadhyaya, J., 1983: *Climate and World Order*. New Delhi, India: South Asian Publishers.

Bardach, J.E., 1987: Developing policies for responding to future climate change. Keynote paper at the workshop on "Developing Policies for Responding to Climate Change" (Villach, Austria, 28 September–2 October 1987).

Barth, M.C., and J.G. Titus (Eds.), 1984: *Greenhouse Effect and Sea Level Rise: A Challenge for this Generation*. New York, NY: Van Nostrand Reinhold.

Begley, S., with M. Miller and M. Hager, 1988: The endless summer? *Newsweek*, 11 July, 18–20.

Bolin, B., B.R. Döös, J. Jäger, and R.A. Warrick (Eds.), 1986: *The Greenhouse Effect, Climatic Change, and Ecosystems*. SCOPE 29. Chichester, UK: John Wiley and Sons.

Broecker, W.S., 1987: Unpleasant surprises in the greenhouse? *Nature, 328*, 123–6.

C&EN (Chemical and Engineering News), 1987: Landmark ozone treaty negotiated. 24 September, 1557.

Callendar, G.S., 1938: The artificial production of carbon dioxide and its influence on temperature. *Quarterly Journal of the Royal Meteorological Society, 64*, 223–40.

Glantz, M.H., J. Robinson, and M.E. Krenz, 1982: Improving the science of climate-related impact studies: A review of past experience. In W. Clark (Ed.), *Carbon Dioxide Review: 1982*. New York, NY: Oxford University Press, 55–93.

Goldsmith, E., 1977: The future of an affluent society: The case of Canada. *The Ecologist, 7*, 160–94.

Goldsmith, P., A.F. Tuck, J.S. Foot, E.L. Simmons, and R.L. Newson, 1973: Nitrogen oxides, nuclear weapon testing, Concorde and stratospheric ozone. *Nature, 244*, 545–51.

Heilbroner, R., 1975: *An Inquiry into the Human Prospect*. New York, NY: W.W. Norton & Co.

Johnston, S.T., 1971: Reduction of stratospheric ozone by nitrogen oxide catalysts from supersonic transport exhaust. *Science, 173*, 517–22.

Kellogg, W.W., J.A. Coakley and G.W. Grams, 1975: Effects of anthropogenic aerosols on the global climate. In *Proceedings of the WMO/IAMAP Symposium on Long-Term Climatic Fluctuations*. WMO 421. Geneva, Switzerland: World Meteorological Organization, 323–30.

Krenz, J.H., 1980: *Energy: From Opulence to Sufficiency*. New York, NY: Praeger Publishers.

Lakoff, G., and M. Johnson, 1980: *Metaphors We Live By*. Chicago, IL: University of Chicago Press, 181–214.

Martin, B., 1979: *The Bias of Science*. O'Connor, Australia: The Society for Social Responsibility in Science.

Mintzer, I.M., 1987: *A Matter of Degrees: The Potential for Controlling the Greenhouse Effect*. Washington, DC: World Resources Institute.

NRC (National Research Council), 1983: *Changing Climate*. Report of the Carbon Dioxide Assessment Committee, Board on Atmospheric Sciences and Climate. Washington, DC: National Academy Press.

NRC (National Research Council), 1985: *Glaciers, Ice Sheets, and the Sea Level: Effects of a CO_2-Induced Climatic Change*. Washington, DC: National Research Council.

Parry, M.L., T.R. Carter, and N.T. Konijn (Eds.), 1987: *The Impact of Climatic Variations on Agriculture*. Volume 2. *Assessments in Semi-Arid Regions*. Dordrecht, The Netherlands: Reidel.

Plass, G.N., 1956: The carbon dioxide theory of climatic change. *Tellus, 8*, 140–54.

Plass, G.N., 1959: Carbon dioxide and climate. *Scientific American, 201*, 3–9.

Ramanathan, V., R.J. Cicerone, H.B. Singh, and J.T. Kiehl, 1985: Trace gas trends and their potential role in climate change. *Journal of Geophysical Research, 90*, 5547–66.

Revelle, R., and H.E. Suess, 1957: Carbon dioxide exchange between atmosphere and ocean and the question of an increase of atmospheric CO_2 during the past decades. *Tellus, 9*, 18–27.

Seidel, S., and D. Keyes, 1983: *Can We Delay a Greenhouse Warming?* Washington, DC: U.S. Environmental Protection Agency.

Slovic, P., B. Fischhoff, and S. Lichtenstein, 1980. Facts and fears: Understanding perceived risk. In R.C. Schwing and W.A. Albers, Jr. (Eds.), *Societal Risk Assessment: How Safe is Safe Enough?* New York, NY: Plenum Press, 181–214.

U.S. Senate, 1987: Hearings before the Subcommittee on Science, Technology, and Space and the National Ocean Policy Study of the Committee on Commerce, Science, and Transportation on Global Climate Change due to Manmade Changes in the

Earth's Atmosphere. 16 July 1987. Washington, DC: U.S. Government Printing Office.

WMO (World Meteorological Organization), 1979: *World Climate Conference*. Extended summaries of papers presented at the conference (Geneva, February 1979). Geneva, Switzerland: WMO.

WMO (World Meteorological Organization), 1986: *Report of the International Conference on the Assessment of the Role of Carbon Dioxide and of Other Trace Gases in Climate Variation and Associated Impacts* (Villach, Austria, 9–15 October 1985). WMO Report 661. Geneva, Switzerland: WMO.

WMO (World Meteorological Organization), 1988: *Developing Policies for Responding to Climate Change*. A summary of the discussions and recommendations of the workshops held in Villach (26 September–2 October 1987) and Bellagio (9–13 November 1987). WMO/TD–No. 225. Geneva, Switzerland: WMO.

4

Grappling for a Glimpse of the Future

Dale Jamieson

INTRODUCTION

Most of us are interested in knowing the future. We want to know the future in order to control it. We want to control the future in order to benefit from it, or to mitigate or avoid harms that we would otherwise suffer. As Francis Bacon wrote in the sixteenth century, "knowledge itself is power" (Passmore, 1974, 18).

Throughout history, many methods have been employed for trying to glimpse the future. These methods have ranged from reading tea leaves to running computer models. Not all of these methods have been equally honored, however. The ancient Hebrews regarded fortune-telling as a sin against God, while holding that prophecy is one of God's greatest gifts. Our current cultural demarcation between honored and dishonored ways of trying to glimpse the future can be seen in the following news report from St. Augustine, Florida.

> An ordinance that outlaws predicting the future for pay should be rewritten because it might apply to doctors and stock brokers, as well as targeted palmists....

Dale Jamieson is Director of the Center for Values and Social Policy, and Associate Professor of Philosophy at the University of Colorado, Boulder. He is the author of nearly forty articles, reviews, and book chapters on various issues in moral, social, and political philosophy, and the philosophy of science and technology. One of his areas of continuing research is environmental ethics and the foundations of environmental policy.

"We have made it illegal in the city to predict the future for compensation," said Dobson [the city attorney] ... the ordinance could apply to "the account executive ... who thinks tomorrow Ford (stock) will go up." (Associated Press, 1987)

The methods of doctors and stock brokers are honored ways of trying to glimpse the future, while those of palmists are not.

Among the most honored ways of trying to glimpse the future is scientific forecasting (or "futurology"). Futurologists give us long-term predictions about economic cycles, food production, energy supplies, the likelihood of war, and more. Some even speculate on the very fate of the earth.

Scientific forecasting is supposed to be different from fortune-telling, and from just plain guessing. In some ways it is. Scientific forecasting employs computers, manipulates large data sets, and enjoys the respectability conferred by science. Yet in other ways scientific forecasting, at least in areas which involve human behavior, does not seem much different from fortune-telling. Consider, for example, the fate of macroeconomic forecasts in the first half of the 1980s.

The final issue of *Business Week* provides some casual, but revealing, information on the predictive performance of the major modelling services over the previous year.... The December 28, 1981, issue offers a brief column entitled "How the Forecasters Went Wrong in 1981," and this should have been taken as a disclaimer for the feature story, "Scanning a Brighter but Hazy Future." The "brighter future" for 1982 turned out to be the most severe recession of the post-World War II period.... At the end of 1982 we had "Why the Forecasters Really Blew It in 1982" and the main fare offered "Slow Motion Recovery..." as a prediction for 1983. Growth in real GNP for 1983 turned out to be a robust 6.5 versus an average prediction from the models of only 3.7 percent (McNown, 1986, 363).

It is not surprising that some see the distinction between futurologists and palm readers as a distinction without a difference. According to Lester Thurow, "futurology is the intellectual's version of going to a palm reader" (Traub, 1979, 24).

This volume is directed toward glimpsing the future. In particular we are concerned to give plausible answers to the following complex question: How are humans likely to respond to environmental changes at the regional level brought about by a carbon dioxide/trace-gases-induced global warming?

In this chapter I will sketch some aspects of an approach to developing plausible answers to this question. This approach, employed by most of the contributions to this volume in varying degrees, is referred to as the *Case Scenario Approach* (CSA). Before sketching the CSA, I will try to explain why I think scientific forecasting may not be a promising approach to developing answers to our question.

SCIENTIFIC FORECASTING

Scientific forecasting in climatology depends on large-scale computer models called General Circulation Models (GCMs). Existing GCMs are inadequate in many ways. One important problem concerns lack of resolution: they cannot support reliable predictions of precipitation and temperature at the regional or seasonal level (Wigley et al., 1986, 287). In addition, existing GCMs focus on mean values. They can tell us how different inputs into the atmospheric system may affect mean temperature and precipitation. Yet for many purposes, information about climate variability may be more important than information about mean values.

There are further problems that must be solved if GCMs are to be significantly more helpful in forecasting climate change (see also Katz, this volume). It is often thought that climatic events that are unprecedented in the historical record are evidence of climate change. Yet such events may instead be testimony to the inadequacy of the historical record. Extreme events can be due to variability that is part of a long-term stable climate regime. Because of the shortness of our time horizon and our failure to pay enough attention to variability, we may sometimes mistake variability for climate change. Moreover, it is conceptually unclear how climate change can be distinguished from variability. Claims about climate change are made relative to some time frame, and there is no widely accepted theory about what time frames should be preferred. Not enough attention has been paid to this problem, yet predictions of climate change and societal and environmental responses to them rest on our ability to distinguish real change from variability.

Models have their strengths as well as their weaknesses. If they are not used in mindless or derivative ways, they force us to be clear about our assumptions. Rather than remaining tacit and unacknowledged, assumptions must be made explicit if they are to figure in a computer model. In addition, because the computers employed in these modeling efforts are so powerful, they can show us consequences of our assumptions that otherwise might remain hidden. Finally, although GCMs are not advanced enough to support detailed predictions, they do help illuminate interactions between different elements in the climate system. For this reason GCMs can be used to perform a function that is analogous to sensitivity analysis. Models can help us see the range of possible climatic changes. They can also show how sensitive these changes are to shifts in various inputs and relationships.

However helpful computer models may be, it is unlikely that scientific forecasting will be able to produce good answers to our question. Our question—How are humans likely to respond to the regional impacts of a global warming?—is difficult to answer for many reasons. One reason is that good answers will involve data from both the physical sciences and the human sciences. Scientific forecasting is most at home in the physical sciences where causal laws can be framed that support deterministic predictions. There is increasing agreement among philosophers and reflective methodologists that attempts to model the human sciences on this conception of the physical sciences have largely failed (Fiske and Shweder, 1986). Indeed some have argued that such attempts are doomed from the outset (Popper, 1957; Winch, 1958; Roth, 1987).

The question that we are dealing with is even more complex than many other questions that arise in the human sciences. If global warming is now occurring, it is not just something that is happening *to* people: People are implicated in bringing it about. In response to global warming, we can expect various modulations of human behavior. These modulations will in turn affect atmospheric conditions, and this in turn will affect human behavior, and so on. One consequence of this feedback between human behavior and atmospheric conditions is that in order to answer our question, we must gain some insight into the interactions between climate and behavior. Rather than being an "impacts" study, such projects might better be viewed as a study of the interactions between climate and human behavior.

Although scientific forecasting does not provide a promising approach to answering our question, research with climate models is important. The present-day scientific consensus presupposes that global warming is in fact occurring, and much of the evidence for this involves research with climate models. If we are to understand human responses to global warming, however, we must go beyond what we can learn from the models.

THE CONCEPT OF A SCENARIO

In the previous section I tried to explain why scientific forecasting is not likely to give us plausible answers to the question we are addressing. In this section I will develop one of the fundamental concepts involved in the CSA: the concept of a scenario. The CSA is devoted to constructing scenarios about human responses to global warming, based on analogies with other cases of human responses to recent extreme or prolonged environmental changes. It is important to be clear about what scenarios are.

The notion of a scenario is widely used in the climatological literature. Unfortunately, it is often used in a vague or misleading way. The concept of a scenario is a rich one, and of great utility in a number of different areas of investigation.

The *Oxford English Dictionary* defines "scenario" as "a sketch or outline of the plot of a play, giving particulars of the scenes, situations, etc." The *Random House College Dictionary* supplies a second sense: a scenario is "an outline of a natural or expected course of events." These definitions suggest two characteristics of scenarios that are worth considering.

First, scenarios are sketches or outlines of stories rather than abstract sets of statements or propositions. Scenarios are narratives that typically have beginnings, middles, and ends. They are constructed in order to serve some purpose, and they are told from a point of view. They bring together diverse information, and engage our imagination.

Second, scenarios about the future are stories about a natural or expected course of events. They are not predictions, nor are they fantasies; they are plausible stories. Forward-looking scenarios are intended to be useful, and to fulfill this function they must focus on past or present facts, activities, trends, tendencies, or dispositions, and extrapolate them into the future. A scenario need not be true in order to be useful. There may be a number of different ways that the future could go, each of which can reasonably be seen as "natural" or "expected." A single person may imagine several plausible futures, and different people may have different conceptions, all reasonable, of what constitutes a "natural" or "expected" course of events. A multiplicity of plausible yet mutually inconsistent scenarios reflecting these facts may be useful in guiding our thinking about the future.

Like other stories, scenarios can be constructed from a variety of materials. Myths, legends, and anecdotes can all provide the makings of scenarios. So can historical facts, the results of sociological research, and the outputs of computer models.

Climatologists often distinguish two kinds of scenarios: those based on GCMs and those based on historical analogues. Although both GCMs and historical analogues can provide the makings of scenarios, there is nothing about these sources themselves that ensure that what is based upon them are scenarios. GCMs typically are used to produce abstract characterizations of future states of the atmosphere. These are not, strictly speaking, scenarios; and we have already discussed their shortcomings with respect to answering the question that concerns us. The search for historical analogues is in the spirit of the CSA, but it may not yield scenarios, and this approach has some other serious shortcomings that shall be discussed later in the section "Traps and Pitfalls."

The CSA, like the search for historical analogues, relies on analogy. Before considering the advantages and disadvantages of the CSA it is important to get clear about the nature of analogical reasoning, the engine that drives the CSA.

ANALOGY AND SCENARIO CONSTRUCTION

The CSA relies on analogical reasoning in the following way. A speaker claims that some actual case X is analogous to a hypothetical case Y with respect to some set of characteristics. On the basis of these characteristics, a story is told about Y that is projected from the story of X.

This volume reports a number of actual cases. For example, stories are told about human responses to decreasing ground water levels in the Ogallala Aquifer region, and increasing water levels in the Great Lakes, the Great Salt Lake, and in the coastal regions of Louisiana (Wilhite, Cohen, Morrisette, and Meo, respectively, this volume). In each case what is attempted is to identify some important respects in which the cases are analogous to ways that societies might respond to environmental changes in a warmer world.

These analogies work at different levels. In some cases analogies are drawn between physical changes that have occurred, and those that might result from global warming, for example sea-level rise and coastal subsidence in Louisiana (Meo, this volume). In other cases the analogies are on the social side, for example drought in northern Virginia (Sheer, this volume). In these cases people and institutions respond analogously in both the actual and hypothetical cases, although the character of the physical changes to which they are responding is quite different.

Viewed abstractly, this sort of scenario construction may seem very complex. In a way it is. Yet it is easier to construct plausible scenarios based on analogy in particular cases than it is to generalize methods for doing so.

Recently there have been some attempts to formalize analogical reasoning. Skorstad et al. (1987) have developed a computer program that is supposed to sort sound analogies from unsound ones. They assume that sound analogies are those based on common causal or structural elements, and unsound ones are those based on appearance or other superficial properties. A proposed analogy between the solar system and an atom would be sound, while a proposed analogy between a blade of grass and a house would be unsound.

This will not do, however. The proposed analogy between the solar system and an atom would fail if what we were focusing on was an electron's ability to jump orbits, or the fact that planets often have satellites. On the other hand the proposed analogy between a blade of grass and a house might succeed in the context of a discussion about color. A house may have more in common with a tent or a Winnebago with respect to its causal and structural properties, but this is not sufficient for supposing that an analogy between houses and tents is always sound while an analogy between houses and blades of grass is always unsound. Pragmatic considerations, such as context, are very important.

The proposal of Skorstad et al. (1987) does not take pragmatic considerations into account. Its failures are inherent in any purely structural approach to analogical reasoning. We can begin to see this by comparing analogical reasoning with deductive reasoning.

Consider the following deductive argument: If P then Q, P therefore Q. This argument is valid in a straightforward sense: Any case in which the premises are true is one in which the conclusion is true as well. Instances of this argument are sound when the premises are true.

Consider the following analogical argument: X is analogous to Y, Y has property a, therefore X has property a. This argument is not valid. In some cases true premises lead to a true conclusion, but in some cases they do not. Consider the following example.

Teaching is analogous to picking fruit.
Picking fruit involves long hours and low pay.
Therefore, teaching involves long hours and low pay.

In this case the premises are true, and so is the conclusion. But consider another argument of the same form.

Teaching is analogous to picking fruit.
Most of those who pick fruit are undocumented workers.
Therefore, most of those who teach are undocumented workers.

In this case the premises are true but the conclusion is false. These examples show that analogical arguments are not formally valid. Unlike deductive reasoning, for analogical reasoning to be acceptable other considerations besides formal validity must be involved.

One reason why it is difficult to assess analogical reasoning is that everything is analogous to everything else in some respect or other. Computers and coffee grinders are analogous in that both are machines. Long's Peak and Picasso's *Guernica* are analogous in that both are physical objects.

It might be thought that similarity-counting is a way out of this difficulty. Sound analogies, it might be suggested, are those drawn between things which share many properties. Unsound analogies are those drawn between things which share few properties. Unfortunately this suggestion will not work. Goodman (1972) has shown that by employing some logical "tricks" it can be demonstrated that any two objects have an indefinite number of similarities.

Good analogical reasoning does not concern the number of similarities two objects share, but rather the significance of the similarities. Good analogical reasoning fixes on important similarities, the contemplation of which supports correct inferences and interesting insights. Identifying important similarities involves pragmatic considerations regarding contexts, interests, and purposes. These considerations cannot be taken up in any purely structural account.

The assessment of analogical reasoning is difficult at best. Perhaps that is why the SCOPE report on the "greenhouse effect" (Bolin et al., 1986) discusses climate analogues but does not directly address analogical reasoning. Perhaps the thought is that identifying analogues is more scientifically respectable than reasoning by analogy.

If we are told that X is an analogue of Y, this suggests that being an analogue of Y is a property that X has. By investigation we can discover that X has this property. Analogical reasoning, on the other hand, is something that people do. Analogues are discovered while analogies are proposed. The surface grammar suggests that analogues are somehow "in the nature of things" while analogies are conjectures put forward by speakers (for the distinction between surface grammar and deep structure see Chomsky, 1965). This is deceptive, however. Analogues are nothing more than reified analogies. Analogues are what speakers claim are analogous to something else. Since the language of analogues suppresses reference to the reasoner, it may appear to be more "scientific" and less "subjective." Grammatical suppression does not alter the substance of the matter, however. Analogues are established by people engaged in analogical reasoning. "Subjectivity" is as present in the discussion of analogues as it is in analogical reasoning.

It is difficult to frame general principles for assessing sound analogical reasoning. For this reason no formal procedure can be given for constructing plausible scenarios based on analogy. Still in this area, as in so many others, our inability to generalize a method does not impair our ability to find our way with particular cases.

In the next two sections some of the problems and prospects of the case scenario approach will be discussed. Some of these primarily concern analogical reasoning, while others mainly concern scenario construction. Since both are involved in the CSA, I will not try to separate out various concerns. Moreover, since I cannot claim to offer an algorithm for constructing scenarios based on analogy, what I have to say should be construed as (hopefully helpful) advice.

ADVANTAGES OF THE CASE SCENARIO APPROACH

The advantages of the CSA over other approaches have been hinted at, but now it is time to make them explicit. There are four advantages that I will discuss. These concern wealth of detail, integration of a broad range of knowledge, multiplicity of perspectives, and communicability and usability.

Wealth of Detail

Whatever the future is like, it will be just as specific and detailed as the present. When we wake up in the morning, and get ready for work, our thoughts and actions will be affected in various ways, both gross and subtle, by climatic conditions. Our attempts to anticipate the ways in which this will be so will be more successful if we can in some way capture the texture and detail of the life we may lead. The best way of doing this is through stories rather than predictions or speculations that are in the form of abstract sets of sentences. Indeed, it has been claimed by some that "we have no other way of describing 'lived time' save in the form of narrative" (Bruner, 1987, 12; see also Ricoeur, 1984).

A story conjures up a complete world. It presents us with a slice of the life that is led in that world. The teller of the tale may not say everything we want to know, but we can be sure that at least some of our questions have answers. In some cases we may have no idea what they are, while in other cases plausible answers are suggested by what is said. Stories are about worlds, but they do not wholly constitute them. The contours of the story should not be mistaken for the contours of the world.

Consider an example. There are details about James Bond that Ian Fleming does not relate. Yet Fleming's portrait of Bond is sufficiently rich for us to identify plausible answers to some unanswered questions, while rejecting other answers as implausible. Bond is not a devout Catholic nor is he an active member of the communist party. He does not have a disfiguring scar on his face nor is he sickly and sallow in appearance.

The information contained in the stories we construct about human responses to global warming will outrun what can be obtained by mapping analogies between the cases and the stories. The case studies, which are stories of real events, are rich enough to permit us to fill in the missing pieces in our scenarios about the future. Scenario construction involves projection. It is this projection in narrative form which gives scenarios their wealth of detail.

Integration of a Broad Range of Knowledge

The question we are trying to answer is a very broad one. Human responses to global warming may range from modifications of individual behavior, to new developments in international law. Virtually everything which can affect, or be affected by, human behavior bears on this question.

Scenarios are tightly woven and finely textured, at least compared to abstract sets of sentences and arrays of quantitative data. Although a story can be constructed by extrapolation from some single (or few) isolated trend(s), the resources of narrative permit the telling of more complex tales. A scenario can integrate different trends and tendencies and illuminate their interactions (Martino, 1983, 148). For this reason a good scenario can give us a feel for what life in the future might be like, and make vivid for us what effect on everyday life various possible solutions may have. Good scenarios engage our imagination, and help us break through the constraints enmeshed in our methodologies and enshrined in our systems of representation and expression.

One special advantage of stories is that they can incorporate knowledge and belief that we would have a very difficult time making explicit. It is a challenge for any sort of cognitive activity to figure out how to employ knowledge that cannot be made explicit. Since computer models are constituted by and operate on explicit statements, they have no way of employing inexplicit knowledge. This has been one of the great stumbling blocks in the development of artificial intelligence (Dreyfus, 1985).

Scenarios bring inexplicit background knowledge into play in a number of ways. Our judgments that particular scenarios are implausible, far-fetched, or totally wild, often rest on inexplicit background beliefs. Once these judgments have been made, sometimes the inexplicit beliefs can be made explicit and evaluated accordingly. Aspects of plausible scenarios may also draw on inexplicit beliefs about the future. Thinking about a particular case may move us to a future scenario in a way that might be described as "intuitive." Still there may be nothing mystical or unscientific about this. We may be exploiting relationships between the terms of the analogy that we cannot make explicit.

Multiplicity of Perspectives

Scenarios are stories. They are told by different tellers from different perspectives for different purposes. Burke (1945) provides a structure which may help us see how a multiplicity of stories can arise. (For a somewhat different analysis, focusing on "doomsday" stories, see Krieger, 1987).

According to Burke, stories involve at least the following elements: an agent, an action, a goal, a setting, and an instrument. The telling of a story is motivated by a trouble. A trouble arises when there is a lack of fit between two or more elements of the story; for example, an agent is not at home in a setting, or the instrument at hand is not suitable for achieving some stated goal (see also Bruner, 1987).

Stories about human responses to global warming can be told from the point of view of a number of different agents: individuals as producers, consumers, or householders; collectives such as neighborhoods, communities, towns, businesses, unions, or clubs; governments at the local, state, or federal levels. The actions are whatever is bringing about global warming and what is being done in response to it. The setting is the planet Earth and its resources. Reasonable people have very different views regarding the action and the setting. The goal could be prevention, mitigation, or adaptation. The instruments are the policies or behaviors that are available for trying to realize the goal. These range from doing nothing to imposing various informal or formal sanctions or incentives.

When the structure of a story is laid out in this way, it is easy to see how a multiplicity of stories can arise, each with its own claim to plausibility. We are therefore able to tell a number of different stories, each individually plausible, yet collectively inconsistent.

This can be valuable for a number of reasons. By inviting a multiplicity of stories we can gain a number of different glimpses of the future. Taken together, a multiplicity of scenarios may provide a much richer picture of the future than any single scenario. Different scenarios may illuminate different features of the future. Or we might see inconsistent scenarios converging on a similar future, or similar scenarios suggesting very different futures. Since scenarios are not predictions, we are not compelled to search for *the* single "right" scenario among the many that might be produced. *When it comes to scenario construction, a multiplicity of collectively inconsistent scenarios is a resource rather than a weakness.*

Communicability and Usability

The scenarios that are the output of the CSA are usable for policy purposes in ways that other kinds of technical reports often are not. Predictions, charts, graphs, and computer projections do not in themselves lead to understanding or provide guides for action, however precise and accurate they may be. Appropriately, there is massive anecdotal evidence for the conclusion that decisionmakers often act on the basis of an especially moving story, even if the story is uncharacteristic and the decision appears to be insensitive to well-documented facts (Bruner, 1987; Martino, 1983, 148; Neustadt and May, 1986; Tversky and Kahneman, 1986).

We can see in our own lives how much we have been influenced by stories. For example, we learn not to lie from the story of George Washington and the cherry tree, and we learn to be persistent from the story of the tortoise and the hare. Stories can convey values as well as facts, and integrate them in ways that invite reflective judgment.

Stories are also conducive to communication. Because quantitative data carry the aura of authority, they often seem to quash discussion rather than stimulate it. This is especially unfortunate when we are concerned with complex problems the parameters of which are very uncertain. Stories, on the other hand, invite question and discussion. Different stories about how the future might go represent different points of view. When these stories are brought into conversation they can shape and modify each other. We often have less resistance to appreciating a story than to appreciating an argument or assertion. Yet appreciating a story that is different from the one we would tell involves understanding another point of view. This characteristic of scenarios has made them popular in business for purposes of mediation and long-range planning (Huss and Horton, 1987).

TRAPS AND PITFALLS

There are dangers to be avoided when using the case scenario approach. I will discuss three: lack of definition, straining an analogy, and failure of an analogy.

Lack of Definition

Stories are told by a teller, from a point of view, in order to serve a purpose. It is important that we are very clear about why we are constructing scenarios, and how we want them to be used. This is especially true when constructing scenarios based on analogy. Analogical reasoning is, as I have suggested, extremely pragmatic in that its soundness is dependent on context and interests. Everything is analogous to everything else, in some respect, from some perspective. It is important for us to be clear about why we are privileging some respects and perspectives.

It is clear at the outset that our investigation presupposes that global warming is a problem. This suggests (following Burke, 1945) that there is a lack of fit between an agent, an action, a goal, a setting, or an instrument. Perhaps our actions do not conduce to our goals, or our goals are not consistent with our setting. There are other ways as well in which trouble could arise.

In this case the nature of the problem is suggested by the metaphor that is commonly used to characterize global warming: "the greenhouse effect." As the earth warms it will be as though we are trapped inside of a greenhouse. This way of setting the problem also suggests some solutions: prevent the greenhouse from being constructed, figure out how to let the heat escape, develop some way of cooling the greenhouse, or learn to live in greenhouse conditions.

There is nothing wrong with viewing global warming as a problem, and metaphorically identifying it with the problem of being trapped in a greenhouse. What is important is that we recognize this as one way of thinking about global warming, and that we are convinced that this is a good way of thinking about it (for additional discussion of the use of this metaphor, see Glantz, this volume).

This is especially important when our problem-definition is established by a metaphor. Recent research suggests that metaphors deeply affect the way we perceive, think, and act (Lakoff and Johnson, 1980; Johnson, 1987; Lakoff, 1987). Metaphors can be helpful in guiding our thinking, but they can also obscure and obfuscate alternative ways of looking at a situation. Sometimes we begin with an apt metaphor, and then later hear it as a literal description. We come to think that reality is exhausted by the content of our metaphor.

An example of an area in which a bad metaphor has misled discussion is fiscal policy (McCloskey, 1986). We tend to identify the national budget with a family budget, and to view the significance of a deficit as being the same in both cases. Yet the consequences of a government deficit, and the alternatives available for responding to it, are very different than in the case of families.

A scenario embodies a particular way of viewing a situation. It is important that we characterize the situation carefully, and that we are very clear about what we want to do with our scenarios.

Straining an Analogy

A good analogy will do a lot for us, but there are some things it cannot do while retaining its virtue. As we have seen, constructing scenarios on the basis of analogy involves projecting beyond the points of one-to-one correspondence. Sometimes the projection is a matter of "filling in" in a fairly natural way. Sometimes it involves extending the story in ways that are more imaginative, but still grounded in the analogy. Other times, however, we find ourselves spinning important parts of the story out of whole cloth. When this occurs, the analogy is being strained.

Consider the following example of a strained analogy (see Scriven, 1976, 213). During the Watergate Hearings, John Erlichman was asked why burglarizing the office of Daniel Ellsberg's psychiatrist was regarded by the Nixon White House as an appropriate tactic in their efforts to silence Ellsberg. Erlichman responded with an analogy. Suppose that you have discovered that there is a map in a safety deposit box, showing the location of an atomic bomb that will explode tomorrow in downtown Washington. Wouldn't it be right for you to break into the bank vault? Similarly, Erlichman suggested, it was appropriate to break into the office of Ellsberg's psychiatrist.

There are some points of analogy between these two cases. In both cases the actor will suffer bad consequences unless the action is undertaken. In both cases there seem to be no other alternatives open to the actor to prevent these consequences. Despite these points of analogy there are important differences in these two cases. In the atom bomb case the stakes are very much higher than in the Ellsberg case. Moreover, in the atom bomb case, the bad consequences would be caused by an act of gratuitous violence, and would be suffered by innocent people. In the Ellsberg case the bad consequences would be suffered by those of questionable innocence, and would be caused by acts of political dissent. These differences are so significant that it is reasonable to say that the analogy between these two cases has been strained.

Sometimes it is hard to tell when an analogy is being strained. When we are focusing on global warming, for example, our reading of other cases is conditioned by our interests. Disanalogies tend to be ignored in favor of points of analogy. This is virtually unavoidable, and indeed, without such selective reading, the case scenario approach would be impossible. Analogy, after all, is not identity. Still, interesting cases are very complex and extremely rich. One and the same case can be analogous to many different cases. It is important that, in our eagerness to find interesting analogies, we do not ignore important points of disanalogy.

Failure of Analogy

Sometimes we are tempted to force analogies between what we know and what we are confronted with. It has been persuasively argued that succumbing to this temptation is often an important reason why social policy goes wrong. During the Vietnam War, for example, American decisionmakers tended to see the Vietnamese as analogous to the Nazis in the threat they posed to American interests and in their willingness to wage a wider war. Consequently a negotiated settlement with the Vietnamese was seen as capitulation to an insatiable aggressor: It would be another "Munich" (Neustadt and May, 1987).

In the area of human response to climate change, analogies may fail for a number of different reasons. I will discuss three.

Proposed historical or cross-cultural analogies sometimes fail due to significant differences in technological possibilities. It has been suggested that the Medieval Warm Epoch (c. 800–1200 A.D.) is analogous to the climate regime that we will face in case of global warming (Bolin et al., 1986, 288). Although this proposed analogy may be useful for some purposes, it is doubtful that we will be able to learn much about possible human responses to global warming from an account of how people responded in the Medieval Warm Epoch. The technological possibilities for response that we have today are much different than those that were available a millenium ago.

A second reason why analogies can fail is due to differences in political and social organization. Our society has institutions devoted to responding to emergencies. A society that failed to have such institutions, or one that was much more highly organized in this respect, might be importantly disanalogous to us in ways that would matter.

A third reason why analogies fail is due to different informational positions. Suppose that some society, virtually like ours in many important respects, faced some environmental changes similar to those that we will face in the near future. If their information about these changes was radically better or worse than ours, analogies between their responses and ours could be wholly inappropriate.

CONCLUSION

Good scenarios about various subjects are found in a broad range of different literatures: novels, histories, anthropological, geological, and biological accounts. These scenarios are based on a wide variety of sources. Some of the most useful scenarios of our time have been produced by writers as diverse as Darwin and Brecht. The case scenario approach constructs scenarios on the basis of analogies. It is one way of grappling for a glimpse of the future.

As I have described this approach it may sound more like science fiction than science. In some ways it is like science fiction: Science fiction involves scenario construction, and good science fiction is plausible and insightful. Scenarios can vary, however, in their level of abstraction and detail. A good scenario concerning global warming may be more like the "small" stories embedded in everyday discourse than like a Russian novel.

Still, we should not object to the close relationship between the CSA and literature. For the affinities between them do not undermine the CSA's scientific standing. Scientific thought, from its very origins, has often moved in metaphor. Newtonian space, for example, has been thought of as being like a container, and Newtonian time as flowing like a river. With the advent of relativity theory the metaphors and analogies changed. The notion of "curved" space-time is constructed by analogy with curves on two-dimensional planes. Stories about trains and travelers have been important in explanations of relativity theory. In general, scientific discourse is replete with metaphors and analogies that we would be hard put to eliminate (Gentner, 1982; Hesse, 1966; Hoffman, 1980; Jones, 1982). Indeed, scientific writing may be closer to story-telling than many have thought. Perhaps reports of scientific research are suppressed narratives, with many of the same features as other narratives in other genres. Perhaps some of the significance of climate models rests implicitly on their evoking stories about what it would be like to live in worlds with different climates.

These speculations may be extreme. Ultimately, the viability of the case scenario approach does not rest on the truth of the scenarios. What matters in assessing the CSA, or any method, is whether or not it contributes to our understanding of the phenomena under investigation.

ACKNOWLEDGMENTS

This chapter originated in conversations with Michael Glantz (NCAR) and has benefited in all stages of its preparation from discussions with him. I have also been helped by Steve Fuller (University of Colorado), David Hawkins (University of Colorado), Richard W. Katz (NCAR), Maria Krenz (NCAR), Linda Mearns (NCAR), James W. Nickel (University of Colorado), Tom Wigley (University of East Anglia), and especially Barbara Brown (NCAR) and Martin Hollis (University of East Anglia).

REFERENCES

Associated Press, 1987: Law against predicting future causes confusion. *Daily Camera*, 28 February, A6.

Bolin, B., R.D. Döös, J. Jäger, and R. Warrick (Eds.), 1986: *The Greenhouse Effect, Climatic Change, and Ecosystems.* SCOPE 29. New York, NY: John Wiley and Sons.

Bruner, J., 1987: Life as narrative. *Social Research, 54*, 11–32.

Burke, K., 1945: *The Grammar of Motives.* New York, NY: Prentice-Hall.

Chomsky, N., 1965: *Aspects of the Theory of Syntax.* Cambridge, MA: MIT Press.

Dreyfus, H., and S. Dreyfus, 1985: *Mind over Machine: The Power of Human Intuition and Expertise in the Era of the Computer.* New York, NY: MacMillan.

Fiske, D.W., and R.A. Shweder, 1986: *Metatheory in Social Science: Pluralisms and Subjectivities.* Chicago, IL: University of Chicago Press.

Gentner, D., 1982: Are scientific analogues metaphors? In D.S. Miall (Ed.), *Metaphor: Problems and Perspectives.* Atlantic Highlands, NJ: Humanities Press, 106–32.

Goodman, N., 1970: Seven strictures on similarity. In *Problems and Projects.* Indianapolis, IN: Hackett Publishing Company, 437–46.

Hesse, M.B., 1966: *Models and Analogies in Science*. Notre Dame, IN: Notre Dame University Press.

Hoffman, R.R., 1980: Metaphor in science. In R. Honeck, and R.R. Hoffman, (Eds.), *Cognition and Figurative Language*. Hillsdale, NJ: Lawrence Erlbaum Associates, Inc., 393–423.

Huss, W.R., and E.J. Horton, 1987: Alternative methods for developing business scenarios. *Technological Forecasting and Social Change, 31*, 219–38.

Johnson, M., 1987: *The Body in the Mind*. Chicago, IL: The University of Chicago Press.

Jones, R.S., 1982: *Physics as Metaphor*. New York, NY: New American Library.

Krieger, M.H., 1987: The possibility of doom. *Technology and Society, 9*, 181–90.

Lakoff, G., 1987: *Women, Fire, and Dangerous Things: What Categories Reveal About the Mind*. Chicago, IL: University of Chicago Press.

Lakoff, G., and M. Johnson, 1980: *Metaphors We Live By*. Chicago, IL: University of Chicago Press.

Martino, J.P., 1983: *Technological Forecasting for Decision Making*. Second Edition. New York, NY: Elsevier Science Publishing Co.

McCloskey, D.N., 1986: *The Rhetoric of Economics*. Madison, WI: University of Wisconsin Press.

McNown, R., 1986: On the uses of econometric models: A guide for policy-makers. *Policy Sciences, 19*, 359–80.

Neustadt, R.E., and E.R. May, 1986: *Thinking in Time: The Uses of History for Decision-Makers*. New York, NY: Free Press.

Passmore, J., 1974: *Man's Responsibility for Nature: Ecological Problems and Western Traditions*. New York, NY: Charles Scribner's Sons.

Popper, K., 1957: *The Poverty of Historicism*. New York, NY: Harper and Row.

Ricoeur, P., 1984: *Time and Narrative*. Chicago, IL: University of Chicago Press.

Roth, P., 1987: *Meaning and Method in the Social Sciences*. Ithaca, NY: Cornell University Press.

Scriven, M., 1976: *Reasoning*. New York, NY: McGraw-Hill Book Company.

Skorstad, J., B. Falkenhainer, and D. Gentner, 1987: Analogical processing: A simulation and empirical corroboration. In *Proceedings of the AAAI-87, The Sixth National Conference on Artificial Intelligence*, 369.

Traub, J., 1979: Futurology: The rise of the predicting profession. *Saturday Review*, December, 24–32.

Tversky, A., and D. Kahneman, 1986: Judgements under uncertainty: Heuristics and biases. In H.R. Arkes, and K.R. Hammond (Eds.), *Judgement and Decisionmaking*. Cambridge, UK: Cambridge University Press, 38–55.

Wigley, T.M.L., P.D. Jones, and P.M. Kelly, 1986: Empirical climate studies. In B. Bolin, D. Döös, J. Jäger, and R. Warrick (Eds.), *The Greenhouse Effect, Climatic Change, and Ecosystems*. SCOPE 29. New York, NY: John Wiley and Sons, 271–322.

Winch, P., 1958: *The Idea of a Social Science*. London, UK: Routledge and Kegan Paul.

5

Statistics of Climate Change: Implications for Scenario Development

Richard W. Katz

INTRODUCTION

If anything is "constant" about climate, it is its ceaseless variation over all temporal and spatial scales. For example, Figure 1 shows the time series of annual total precipitation and of May total precipitation at Denver, Colorado, over a 100-year period. In looking at these plots, the first impression is of a process dominated by "noise" rather than by any strong "signal." When questions of possible future climate change arise, their resolution is confounded with this inherent variability. If the climate is always varying, what can possibly be said about the nature of future climate over, say, the next 10 or 50 years? Alternatively, looking back in time, what climate changes can be identified as having occurred in the recent past and can these changes be attributed to specific causes?

For the purposes of climate impact assessment, it is convenient to approach these issues from the perspective of a decision-maker. This individual is faced with selecting actions to be taken the consequences of which depend, in part, on the climate over some future time horizon. What particular assumptions should

Richard W. Katz is a scientist in the Environmental and Societal Impacts Group at the National Center for Atmospheric Research, Boulder, Colorado. He has a Ph.D. in statistics and works on applications of statistics to environmental problems. These applications have included the statistical analysis of climate experiments and the decision-analytic assessment of the economic value of information about weather and climate.

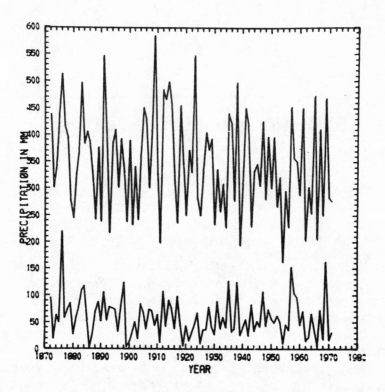

Figure 1. Annual total precipitation (upper line) and May total precipitation (lower line), Denver, Colorado, 1872–
1971. (Source: Katz and Glantz, 1979, 134.)

this individual make about how the likelihood of occurrence of
future climate events might differ from their observed relative frequency of occurrence in the past?

In this chapter, the nature of the information about future
climate that   would be needed for impact studies is first considered. It is clear that information is required, not only about any
possible changes in mean climate, but also about possible changes
in the variability of climate, in order to determine how the probabilities of individual climate events would change. Keeping these
needs in mind, the adequacy of the output from climate experiments based on general circulation models (GCMs) is assessed.
Moreover, the prospects for resolving the so-called "first-detection"

problem, concerned with identifying and attributing any climate change that might have taken place by subjecting the recent observational record of climate to statistical analysis, are discussed. Explanation is given as to why it is inherently more difficult to make statistical inferences about changes in climate variability than about changes in climate means. Ultimately, a decisionmaker needs to have available the probabilities of various climate events in order to select an optimal action. The problem of how best to estimate these climatological probabilities is examined from a decision-analytic viewpoint.

The case scenario approach (Jamieson, this volume) does not directly address these issues; instead, it simply assesses the societal impacts of specific climate events that have actually occurred in the recent past. Nevertheless, in drawing an analogy to the impacts of future climate changes, a projection must be made. The validity of this projection depends on, among other things, the correspondence between the actual climate event about which the case study is centered and anticipated future climate changes. Consequently, the issues to be examined in this chapter are of direct relevance to the case scenario approach. The concept of a "scenario," as referred to in this chapter, is concerned only with the climate process (i.e., as generally used in the climate literature). It should be kept in mind that these entities are not necessarily "scenarios" in the more general sense adopted by Jamieson (this volume).

MODEL-BASED SCENARIOS OF CLIMATE CHANGE

Much of the impetus for the growing concern about possible ongoing and future changes in global climate can be traced to the results of climate experiments based on GCMs (e.g., WMO, 1986). These models have three spatial dimensions and allow for the dynamic evolution of the atmosphere over time (see Dickinson, 1986, for a general review of the use of GCMs for climate experiments). The question naturally arises as to how much confidence should be attached to such model-based outcomes. Further, is the output of these models in a form that is appropriate for use in impact studies?

In many situations, the potential impact of changes in climate variability may be as great as or greater than the impact of any changes in climate means. Consequently, information is needed about how the variability of climate would change. Such information is especially necessary to determine how the likelihood of extreme climate events would change. Moreover, information about changes in climate is needed on small enough regional scales to be meaningful for societal impact considerations. Finally, although temperature receives the most attention in GCM experiments (especially those concerned with increases in carbon dioxide), any effects on precipitation would probably have more substantial societal impacts in many circumstances.

Part of the difficulty in assessing the performance of GCMs is due to certain statistical issues that arise. On the one hand, GCMs do have the advantage of being able to be employed to perform controlled experiments (e.g., comparing "control runs" representing the current climate with "experiment runs" representing the climate under a doubling in carbon dioxide), making the interpretation of any results more straightforward than those based on observational studies. On the other hand, climate simulations should exhibit variability just as actual climate observations do. The complexity of GCMs implies that only a limited number of computer runs can be made. Hence, questions of statistical significance enter into the evaluation of the outcomes of GCM climate experiments. Because the experiments are relatively short and because of the relatively high degree of climate variability over a wide range of temporal and spatial scales (even when simulated by a GCM), many observed differences between GCM experiment and control runs are not large enough to reject the hypothesis that they are simply attributable to chance variation. This situation is especially discouraging, given that it is common for GCM control runs to produce differences from observed data that are highly statistically significant.

For the case of a doubling in carbon dioxide, only when the results of GCM experiments are averaged over relatively large regions can changes in mean temperature be identified as statistically significant. However, little is known about how the variability of temperature would change as the mean changes (e.g., as induced by increases in carbon dioxide and other trace gases). At this point, GCM control runs do not even possess a degree of variability that is

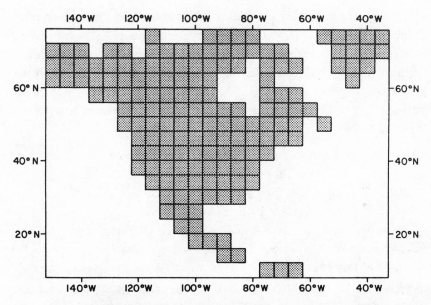

Figure 2. Spatial distribution over North America of grid points for one version of Oregon State University atmospheric GCM. (Source: Katz, 1982, 1451.)

similar in magnitude to the variability of the real climate, a necessary condition in order to have any confidence in results concerning changes in variability produced by GCM climate experiments.

GCMs cannot currently provide meaningful information about changes in temperature on small enough spatial scales to be useful in impact studies. One reason for this deficiency of GCMs is their lack of spatial resolution. Figure 2 shows the spatial resolution over North America of one version of the Oregon State University atmospheric GCM. The geographic boundaries are drawn coarsely in an attempt to faithfully represent the nature of their appearance in the model. Considering the speculation about the possible extent of flooding in Florida from a sea-level rise associated with a carbon dioxide-induced warming (Schneider and Chen, 1980), it is amusing to note that the state of Florida essentially does not exist in this model.

The physical validity of regional-scale climate characteristics generated by GCMs is also lacking (Dickinson, 1986). It is not even clear what GCM output at individual grid points represents: in some respects it is comparable to real atmospheric measurements taken at a single site, whereas in other respects it is more like spatial averages of real atmospheric measurements. Termed the "climate inversion" problem, some attempts have been made to relate GCM grid point output to local weather (Kim et al., 1984; Wilks, 1986). However, the viability of this approach has yet to be established.

Precipitation is relatively more variable than temperature. Moreover, GCM control runs produce precipitation sequences the statistical characteristics of which are not necessarily in close agreement with observed precipitation data. GCM results for a doubling of carbon dioxide suggest a slight increase in global precipitation (Schlesinger and Mitchell, 1985). However, no conclusive results are available regarding effects on precipitation on any relevant spatial scales (e.g., as induced by increases in carbon dioxide and other trace gases). In spite of the expected net increase in precipitation, there might well be relatively large regions in which decreases in precipitation occur. Katz (1983) discusses methods for making statistical inferences about changes in precipitation based on GCM climate experiments.

In summary, GCM climate experiments do not currently produce information (e.g., scenarios of future climate) in a form that is at all useful for impact studies. This fact should not be surprising because GCMs were developed with the intention of aiding in basic research about the atmosphere, rather than with the needs of climate impact research in mind. It is not at all clear that the prospects are any more favorable in the near future for GCMs to be able to better meet these requirements. More detailed reviews of the adequacy of GCM-based climate scenarios have been provided by, for example, Gates (1985), Wigley et al. (1986), and Lamb (1987). More detailed discussions of the statistical issues involved in interpreting the outcomes of GCM climate experiments have been given by, for example, Chervin and Schneider (1976), Laurmann and Gates (1977), Katz (1982), Livezey (1985), Zwiers and Thiebaux (1987), and Zwiers (1987).

FIRST DETECTION OF CLIMATE CHANGE

Another way of attempting to provide information about what the future climate will be like involves monitoring the observed climate. The so-called "first-detection" problem is concerned with how to detect any changes in climate that might currently be taking place and how to identify the sources causing these changes. In principle, the detection of change in the observed climate record should be a relatively straightforward statistical question. However, the attribution of any detected change is a much more difficult, perhaps unresolvable, issue. Unfortunately, without resolution of the attribution problem, the projection of observed climate changes (e.g., trends) into the future, as is necessary to produce scenarios for use in climate impact studies, cannot be completely justifiable.

Much research has been devoted to the first-detection problem, especially in the context of the increasing concentration of carbon dioxide in the atmosphere (DOE, 1982; MacCracken and Luther, 1985; Wigley et al., 1986; Barnett and Schlesinger, 1987). In practice, a number of complications arise that make the resolution of the statistical detection component of this problem less straightforward. First, as with GCM climate experiments, the relatively high degree of variability typical of climate data implies that statistical tests may not be very powerful (i.e., climate changes that are large enough to be practically significant may not be large enough to be correctly identified as statistically significant). Statistical methods for detecting climate change have been proposed, for example, by Wigley and Jones (1981), Bell (1982), and Epstein (1982).

A second complication, the effects of which are somewhat more subtle, involves the "multiplicity" problem (e.g., Tukey, 1977; Livezey and Chen, 1983; Solow, 1987). This issue arises because climate is monitored on a sequential basis, and over the years overlapping sets of observations will be repeatedly subjected to statistical analysis. This situation complicates the interpretation of tests of statistical significance, making it more difficult to resolve whether an observed change in climate is real or simply attributable to chance variation. The exact nature of this problem can perhaps best be illustrated by means of an example that involves paraphrasing a birthday card featuring the Peanuts characters. On the front of the card, Lucy is shown tossing a coin and says "Heads I get you a present, tails I get you a card." On the inside of the card, Lucy happily exclaims "It came out tails." But Snoopy points out that "She had to toss the coin ten times."

Yet another complication concerns the issue of "transient response" (e.g., Schneider and Thompson, 1981). Most scenarios of future climate change induced by carbon dioxide are based on steady-state (equilibrium) models; generally comparing control runs using pre-industrial levels of carbon dioxide with experiment runs using a constant but higher level of carbon dioxide (usually a doubling in concentration). In the real world, however, the atmospheric concentration of carbon dioxide has been steadily increasing over time. Moreover, the response of the atmosphere to any increase in carbon dioxide should gradually take place over a period of at least several decades, in part because of the relatively slow evolution of ocean processes. Keeping all these complications in mind, it is not surprising that Wigley et al. (1986, 309) conclude that "unequivocal, statistically convincing detection of effects of carbon dioxide and trace-gas levels on climate is not yet possible."

Unlike GCMs, controlled experiments cannot be performed using the real atmosphere. Hence, any conclusions about the existence of cause-and-effect relationships cannot be based on statistical analysis of the observed climate alone. In particular, it is not justified to conclude simply on the basis of an observed warming that greenhouse-gas forcing is to blame. Other factors, which have not been held constant and whose effects on climate have not been isolated, may be partly or entirely responsible instead. For a general discussion of the dangers of drawing any conclusions about cause-and-effect relationships from observational studies, see Freedman et al. (1978).

The so-called "fingerprint" approach has been proposed as one way to provide evidence that would be more convincing for identifying the source of a climate change (MacCracken and Moses, 1982; MacCracken and Luther, 1985). As predicted by GCM experiments, the effects on climate of increases in the atmospheric concentration of carbon dioxide should have a unique "signature," which might be identified in the past observations. In particular, although the surface air temperature should be warmer, the stratosphere should be cooler when greenhouse gases are the cause. But the acknowledged inadequacies of GCMs (mentioned earlier) imply that this fingerprint cannot be precisely specified. Consequently, the prospects for the attribution of any observed climate change do not appear to be very favorable. Barnett and Schlesinger (1987), for example, failed to find confirmation of the carbon dioxide fingerprint in the observed climate record. Despite this lack of confirmation, such a signature may well exist, perhaps in a somewhat different form than that for which they searched.

In summary, the projection of future climate on the basis of currently observed trends is not necessarily justified. This issue raises a dilemma for decisionmakers. Should they retain the "stationarity" hypothesis that the climate is not changing in any permanent fashion and estimate the probabilities of occurrence of future climate events using the observed frequencies of occurrence of these events in the historical record? Or should these probability estimates be based only on the relatively recent historical record or on the extrapolation of a trend or cycle or on the outcome of a GCM experiment? More will be said about this dilemma in a later section.

INFERENCES ABOUT CLIMATE VARIABILITY

In this section, the difficulties in making inferences about changes in climate variability, alluded to earlier, are examined in more detail. The problem of estimating changes in the variance is inherently more difficult than that of estimating changes in the mean, and is complicated by certain statistical characteristics of real or simulated climate time series. It is more difficult because of its second-order nature. That is, making inferences about means is a first-order problem in the sense that statistical procedures are based on *variances of means*, whereas making inferences about variances is a second-order problem in the sense that statistical procedures are based on *variances of variances*.

Among other things, this second-order nature implies that the estimates of climate variability will be relatively less precise than the corresponding estimates of climate means. In particular, it will be more difficult to detect changes in climate variability that are of practical significance than to detect such changes in climate means. The estimates of climate variability also will be more sensitive to the particular statistical characteristics of climate time series, meaning that conventional statistical tests may be more likely than usual to provide erroneous conclusions about possible changes in variance. Katz (1988a) considers in detail this question of how to make inferences about climate variability.

Because climate time series are generally autocorrelated (i.e., correlated over time), the conventional definition of variance has some drawbacks. For one thing, the concept of variance is not as meaningful when the observations are correlated. For another, if autocorrelation is present, the results of tests of significance for variances are distorted in a manner analogous to its effects on tests of significance for means. Specifically, climate variables are generally positively autocorrelated (i.e., weather spells have a tendency to "persist"), implying that climate means are more variable than if the observations being averaged were actually uncorrelated. Katz (1988a) has advocated the use of a specific procedure that adjusts for autocorrelation.

A second problem arises because climate variables at best have distributions that are only approximately normal in shape. Tests for changes in means are not particularly dependent on the normality assumption, whereas standard tests for changes in variance are extremely sensitive to departures from the normality assumption. In fact, Box (1953) has suggested that one particular test for equality of variances could be better used as a test for normal distributions. In another context, he has commented that making the standard test for equality of variances is "like putting to sea in a rowing boat to find out whether conditions are sufficiently calm for an ocean liner to leave port!" (Box, 1953, 333). Katz (1988a) has applied a procedure that adjusts for both autocorrelation and non-normal distributions to output generated by GCM control runs. Wilson and Mitchell (1987) have applied this procedure to compare GCM-simulated climate variability with observed climate variability.

Besides the difficulties just mentioned, the problem of making inferences about climate variability has not received sufficient attention for various reasons. In particular, it could be argued that the importance of the problem to climate impact studies is not fully appreciated. In an attempt to combat this neglect, Mearns et al. (1984) illustrated, by concrete examples, why this issue needs to be addressed. They demonstrated the nonlinear nature of the relationship between the mean climate and the probabilities of extreme events. Among other things, it is evident that changes in the likelihood of occurrence of a specific climate event depend as much on changes in the variance as on changes in the mean.

Finally, it should be mentioned that a statistical theory for extreme values is available that is parallel, if not completely analogous, to that for means. This theory has apparently been largely neglected in climate applications. In attempting to apply such a theory to climate processes, some practical obstacles might arise. Nevertheless, it is clear that extremes (including the occurrence of record-breaking values) are statistics that are important in specifying scenarios of future climate that are realistic for societal impact studies. Katz (1988b) has discussed in more detail the statistical theory of extreme values and records and its applicability to climate.

DECISIONMAKING UNDER UNCERTAINTY

As mentioned earlier, the question of how best to estimate the future probability of occurrence of a climate event from the available historical climate data base does not have a simple answer. Should a record as long as possible be relied on or should only the relatively recent record be used? The first option would clearly produce the most precise probability estimates if the climate were not changing, whereas the second option would have the advantage of adapting to a possible ongoing climate change. One approach to the resolution of such an issue is to view it as a problem in economics, one that falls within the realm of the theory of decisionmaking.

This theory is concerned with the problem of how best to make decisions given the information available (generally imperfect) about future events of interest. Formally, a decisionmaker (e.g., a farmer) is faced with the task of selecting an action to be taken from a set of possible actions (e.g., plant one type of crop or plant another type of crop). The economic consequences of these actions are dependent on future weather or climate events (e.g., drought or no drought) about which the decisionmaker has available some sort of information specifying their likelihood of occurrence. The decisionmaker is assumed to rely on some criteria for selecting the action to be taken. One such criterion is to choose the action having the greatest expected economic return (or more generally, maximum expected utility). Such expected returns are determined by weighting the possible consequences by the probabilities of occurrence of the corresponding events.

A review of decision analysis in the context of information about weather and climate has been made by Winkler and Murphy (1985). This decision-analytic approach has been employed to determine the optimal actions to be prescribed for decisionmakers who have imperfect weather or climate forecasts available. For example, Brown et al. (1986) have studied the problem of a farmer's decision concerning whether to plant spring wheat or to let the land lie fallow given a forecast of growing season precipitation. Such a study could be extended to examine how a farmer ought to adapt to a climate change in which the probabilities of certain precipitation events are gradually increasing or decreasing.

The same type of methodological approach could be applied to the issue of how best to estimate the likelihood of climate events. In theory, a decisionmaker could choose among different estimates of the probability of a climate event on the basis of which estimation procedure leads to maximizing expected return. Although there have been calls for revising climate normals and other statistics to reflect the shorter, most recent time period in light of possible climate change (Todorov, 1985), several examples suggest that this approach is not necessarily best.

The West African Sahel has experienced a long run of dry years starting about 1970 (Lamb, 1982), whereas rainfall was abnormally high during the immediately preceding time period of the 1950s and 1960s (Figure 3). Consequently, if in 1970 the rainfall normals had been based on a short time period (say, 1950–1970) rather than the entire record, these normals would have differed to an even greater extent from the rainfall actually observed during the 1970s and 1980s. Another example is the allocation of water in the Colorado River Basin in the western United States. On the basis of a relatively short period of streamflow records, water rights were allocated between the Upper and Lower Basin states in 1922. A subsequent study (Stockton and Jacoby, 1976) reconstructed a longer historical record of streamflow using tree rings. This study established that the streamflow data on which the water compact is based constitutes a period of unusually high flow (see Brown, Figure 2, in this volume).

Each of these examples suggests a fundamental quandary concerning the best way to estimate the likelihood of climate

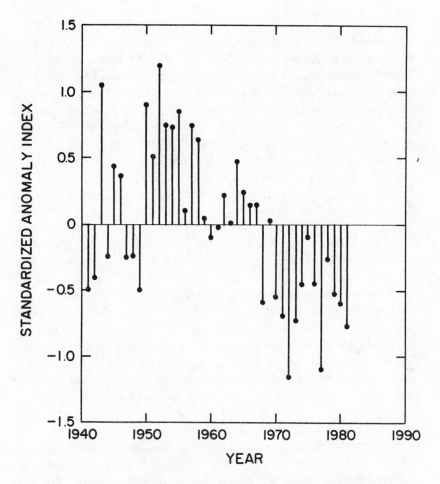

Figure 3. Time series of Standardized Anomaly Index for West
African Sahel based on April–October rainfall during
1941–1981. (Source: Katz and Glantz, 1986, 768).

events. On the one hand, if the climate is indeed changing, using
a long prior record to estimate these probabilities would provide
precise, but misleading, statistics. On the other hand, using a relatively short prior record would likely miss some of the natural
climate variability and provide at best an imprecise estimate. The
trade-off between these two approaches is especially delicate if the
climate event occurs relatively rarely.

IMPLICATIONS

Several fundamental issues need to be resolved before impact studies should rely on scenarios of future climate based on GCM climate experiments or based on the extrapolation of observed climate change. Are GCMs capable of producing information about climate change in a form that is useful for impact studies? Can the cause of an observed climate change be unequivocably identified, as necessary to justify any extrapolation into the future? Facing these unresolved issues, does a decisionmaker have any strategy available that is preferable to simply relying on the observed relative frequency of occurrence of a weather or climate event in the historical record as a basis for estimating the likelihood of occurrence of the same event in the future?

It is difficult to conceive of a plausible scenario in which these issues will turn out to be resolved in the near future. No revolutionary improvements, just minor refinements, in GCMs should be expected, meaning that essentially the same limitations will remain. Although it is sure that some climate changes that have taken place in the recent past will eventually be identified, the attribution problem will remain, at least in some respects, not completely resolvable. In view of unanticipated, new discoveries of sources of uncertainty that are sure to occur, it would be unrealistic to expect that decisionmakers soon will have available information specifying in a probabilistic form the nature of future climate change over time horizons important for climate impact assessments.

The validity of the case scenario approach (Jamieson, this volume) is not solely dependent on whether the climate event considered is analogous in some respects to an anticipated future climate change. Instead, the value of this approach rests more in providing information about how society has dealt with climate events in the past, regardless of whether the specific event being examined is at all analogous to events of interest in the future. Given the drawbacks associated with both the case scenario approach and with various methods of projecting future climate change, it is clear that these different approaches need to be employed in a complementary fashion to produce the most reliable climate impact assessments.

ACKNOWLEDGMENTS

I thank B.G. Brown, M.H. Glantz, and T.M.L. Wigley for their comments on a draft version of this manuscript.

REFERENCES

Barnett, T.P., and M.E. Schlesinger, 1987: Detecting changes in global climate induced by greenhouse gases. *Journal of Geophysical Research, 92,* 14772–80.

Bell, T.L., 1982: Optimal weighting of data to detect climatic change: Application to the carbon dioxide problem. *Journal of Geophysical Research, 87,* 11161–70.

Box, G.E.P., 1953: Non-normality and tests on variances. *Biometrika, 40,* 318–35.

Brown, B.G., R.W. Katz, and A.H. Murphy, 1986: On the economic value of seasonal-precipitation forecasts: The fallowing/planting problem. *Bulletin of the American Meteorological Society, 67,* 833–41.

Chervin, R.M., and S.H. Schneider, 1976: On determining the statistical significance of climate experiments with general circulation models. *Journal of the Atmospheric Sciences, 33,* 405–12.

Dickinson, R.E., 1986: How will climate change? The climate system and modelling of future climate. In B. Bolin, B.R. Döös, J. Jäger, and R.A. Warrick (Eds.), *The Greenhouse Effect, Climatic Change and Ecosystems.* Scope 29. Chichester, UK: Wiley, 206–70.

DOE (Department of Energy), 1982: *Proceedings of the Workshop on First Detection of Carbon Dioxide Effects.* Washington, DC: U.S. Department of Energy.

Epstein, E.S., 1982: Detecting climate change. *Journal of Applied Meteorology, 21,* 1172–82.

Freedman, D., R. Pisani, and R. Purves, 1978: *Statistics.* New York: Norton.

Gates, W.L., 1985: The use of general circulation models in the analysis of ecosystem impacts of climatic change. *Climatic Change, 7,* 267–84.

Katz, R.W., 1982: Statistical evaluation of climate experiments with general circulation models: A parametric time series modeling approach. *Journal of the Atmospheric Sciences, 39*, 1446–55.

Katz, R.W., 1983: Statistical procedures for making inferences about precipitation changes simulated by an atmospheric general circulation model. *Journal of the Atmospheric Sciences, 40*, 2193–201.

Katz, R.W., 1988a: Statistical procedures for making inferences about climate variability. *Journal of Climate* (in press).

Katz, R.W., 1988b: Statistics and decision making for extreme meteorological events. In M. Glantz and E. Antal (Eds.), *Identifying and Coping with Extreme Meteorological Events.* Budapest: Hungarian Academy of Sciences (in press).

Katz, R.W., and M.H. Glantz, 1979: Weather modification for food production: Panacea or placebo? *Journal of Soil and Water Conservation, 34*, 132–4.

Katz, R.W., and M.H. Glantz, 1986: Anatomy of a rainfall index. *Monthly Weather Review, 114*, 764–71.

Kim, J.-W., J.-T. Chang, N.L. Baker, D.S. Wilks, and W.L. Gates, 1984: The statistical problem of climate inversion: Determination of the relationship between local and large-scale climate. *Monthly Weather Review, 112*, 2069–77.

Lamb, P.J., 1982: Persistence of Subsaharan drought. *Nature, 299*, 46–8.

Lamb, P.J., 1987: On the development of regional climatic scenarios for policy-oriented climatic-impact assessment. *Bulletin of the American Meteorological Society, 68*, 1116–23.

Laurmann, J.A., and W.L. Gates, 1977: Statistical considerations in the evaluation of climate experiments with atmospheric general circulation models. *Journal of the Atmospheric Sciences, 34*, 1187–99.

Livezey, R.E., 1985: Statistical analysis of GCM climate simulation, sensitivity, and prediction experiments. *Journal of the Atmospheric Sciences, 42*, 1139–49.

Livezey, R.E., and W.Y. Chen, 1983: Statistical field significance and its determination by Monte Carlo techniques. *Monthly Weather Review, 111*, 46–59.

MacCracken, M.C., and F.M. Luther (Eds.), 1985: *Detecting the Climatic Effects of Increasing Carbon Dioxide.* Washington, DC: U.S. Department of Energy.

MacCracken, M.C., and H. Moses, 1982: The first detection of carbon dioxide effects: Workshop summary, 8–10 June 1981, Harpers Ferry, WV. *Bulletin of the American Meteorological Society, 63,* 1164–78.

Mearns, L.O., R.W. Katz, and S.H. Schneider, 1984: Extreme high-temperature events: Changes in their probabilities with changes in mean temperature. *Journal of Climate and Applied Meteorology, 23,* 1601–13.

Schlesinger, M.E., and J.F.B. Mitchell, 1985: Model projections of the equilibrium climatic response to increased carbon dioxide. In M.C. MacCracken and F.M. Luther (Eds.), *The Potential Climatic Effects of Increasing Carbon Dioxide.* Washington, DC: U.S. Department of Energy, 81–147.

Schneider, S.H., and R.S. Chen, 1980: Carbon dioxide flooding: Physical factors and climatic impact. *Annual Review of Energy, 5,* 107–40.

Schneider, S.H., and S.L. Thompson, 1981: Atmospheric CO_2 and climate: Importance of the transient response. *Journal of Geophysical Research, 86,* 3135–47.

Solow, A.R., 1987: Testing for climate change: An application of the two-phase regression model. *Journal of Climate and Applied Meteorology, 26,* 1401–5.

Stockton, C.W., and G.C. Jacoby, Jr., 1976: *Long-term Surface-water Supply and Streamflow Trends in the Upper Colorado River Basin.* Lake Powell Research Project Bulletin 18. Washington, DC: National Science Foundation.

Todorov, A.V., 1985: Sahel: The changing rainfall regime and the "normals" used for its assessment. *Journal of Climate and Applied Meteorology, 24,* 97–107.

Tukey, J.W., 1977: Some thoughts on clinical trials, especially problems of multiplicity. *Science, 198,* 679–84.

Wigley, T.M.L., and P.D. Jones, 1981: Detecting CO_2-induced climatic change. *Nature, 292,* 205-8.

Wigley, T.M.L., P.D. Jones, and P.M. Kelly, 1986: Empirical climate studies: Warm world scenarios and the detection of climatic change induced by radiatively active gases. In B. Bolin, B.R. Döös, J. Jäger, and R.A. Warrick (Eds.), *The Greenhouse Effect, Climatic Change and Ecosystems.* Scope 29. Chichester, UK: Wiley, 271–322.

Wilson, C.A., and J.F.B. Mitchell, 1987: Simulated climate and CO_2-induced climate change over western Europe. *Climatic Change, 10*, 11–42.

Wilks, D.S., 1986: *Specification of Local Surface Weather Elements from Large-Scale General Circulation Model Information, with Application to Agricultural Impact Assessment.* SCIL Report 86-1, Corvallis, OR: Oregon State University.

Winkler, R.L., and A.H. Murphy, 1985: Decision analysis. In A.H. Murphy and R.W. Katz (Eds.), *Probability, Statistics, and Decision Making in the Atmospheric Sciences.* Boulder, CO: Westview Press, 493–524.

WMO (World Meteorological Organization), 1986: *Report of the International Conference on the Assessment of the Role of Carbon Dioxide and of Other Greenhouse Gases in Climate Variations and Associated Impacts*, Villach, Austria, 9–15 October 1985, WMO No. 661. Geneva: World Meteorological Organization.

Zwiers, F.W., 1987: Statistical considerations for climate experiments. Part II: Multivariate tests. *Journal of Climate and Applied Meteorology, 26*, 477–87.

Zwiers, F.W., and H.J. Thiebaux, 1987: Statistical considerations for climate experiments. Part I: Scalar tests. *Journal of Climate and Applied Meteorology, 26*, 464–76.

6

Impact Assessment by Analogy: Comparing the Impacts of the Ogallala Aquifer Depletion and CO_2-Induced Climate Change

Michael H. Glantz and Jesse H. Ausubel

EDITOR'S NOTE

This discussion paper was prepared in 1984 and a short-ened version of it was published in *Environmental Conservation*, *11* (1984) 123–31. It was a first attempt to apply this approach of impact assessment by analogy to the global warming issue. It served as the stimulus to the EPA-supported project on forecast-ing by analogy the societal responses to the regional impacts of a global warming. It was based on scientific information available in 1983. The findings of the scientific community on this issue today are essentially the same as those of a few years ago. Therefore the references to this methodological chapter have not been updated. Today, however, there may be greater agreement about a likely occurrence of a global warming as a result of increased loading of the atmosphere of carbon dioxide and other trace gases.

Michael H. Glantz is head of the Environmental and Societal Im-pacts Group at the National Center for Atmospheric Research. He received his Ph.D. in political science from the University of Penn-sylvania. His main areas of research relate to the interactions be-tween climate and society with a special focus on drought.

Jesse H. Ausubel is director of the Program Office of the National Academy of Engineering. Mr. Ausubel became a resident fellow of the National Academy of Sciences in science and public policy in 1977. From 1981 to 1983 he served as a National Research Council staff officer principally responsible for studies of the greenhouse ef-fect. His interests generally revolve around long-term interactions of technology and environment.

INTRODUCTION

This chapter compares two long-term, gradual, cumulative environmental changes with potentially severe societal consequences for the American Great Plains: the depletion of the Ogallala Aquifer and a possible global warming induced by increasing carbon dioxide (CO_2) in the atmosphere. While these two environmental issues (one probable and one potential) have been addressed separately in the past, there are compelling reasons to consider them together. Both have implications for the amount of water available for sustained agricultural production in this region. Linking such issues could enable us to learn, on the basis of how society responds to one issue, how it might respond to others. The linkage also shows the need to identify the combined regional effects and societal responses to these problems. As Pittock and Salinger (1982) have suggested for the CO_2 issue, "It is the regional effects of such a global warming which will largely determine the social and economic consequences." Assessing these environmental issues together suggests that researchers and policymakers should formalize their use of "casual" analogies by undertaking more rigorous comparisons and making explicit their assumptions about specific analogies that they use (Martino, 1975).

The chapter is divided into four parts. The first section discusses the logic and possible value of reasoning by analogy. The second section describes the two cases, identifies similarities and differences, and suggests a possible convergence between the Ogallala and CO_2 issues. The third section examines categories of potential societal responses for these issues. A final section offers conclusions and recommendations.

REASONING BY ANALOGY

Reasoning by analogy is by no means a newly discovered endeavor but has been one of the traditional tools of the art of discourse. Mathematical models might be regarded as a formalized version of analogy, in which extended, elaborate comparisons are often drawn between logico-mathematical structures and the behavior of natural, hypothetical, or human systems. The purpose of this chapter is more modest; we claim merely that certain issues are amenable to comparison, both in general and in detail, that the elements of such comparisons should be made explicit, and that the similarity of properties may be enlightening and useful for assessment and could possibly have some predictive value.

Economy of Effort

The use of analogy in the study of the impacts of climate on society was proposed by participants in the Working Group on Impacts at the World Climate Conference (WMO, 1979). There were at least two reasons for suggesting such an approach. First, it is important in climate-related impact assessment research to absorb the lessons of the past in order to ensure progress as one climate impact study follows another (Glantz et al., 1982). Large impact assessment studies continue to be undertaken on the CO_2 issue, and it is important to assure that those undertaking these efforts be made aware of, and exploit, experiences gained during the past decade and a half in climate impact assessment as well as in other areas of research that may have a bearing on the CO_2 issue. There are undoubtedly similarities in structure, function, and methods of research between current CO_2 efforts and recent efforts to examine, for example, the potential consequences of climatic changes brought about by stratospheric supersonic flight or chlorofluorocarbon releases to the atmosphere or the consequences of prolonged drought in various parts of sub-Saharan Africa.

Secondly, there are several environmental issues that may have strong similarities with regard to societal effects—for example, CO_2, acid rain, depletion of the ozone layer, soil erosion, and depletion of groundwater. Efforts must be made to share research experiences so that in each case the impact assessment will not have to be undertaken as if it were the first time one had faced this genre of environmental problem.

To these two reasons for using analogy may be added a third. A comprehensive assessment can be extremely expensive. By contrast, reviews of past, similar experiences are generally quite inexpensive. They enable us to avoid duplication and to begin further studies from a higher level of sophistication and with a greater awareness of the successes as well as the pitfalls of climate-related impact analysis.

A workshop on the effect of a CO_2-induced global warming of the atmosphere recommended comparisons with previous warm periods in history.

Appropriate case studies might include the late Roman Empire (300-600 AD), Iceland in the Middle Ages (1100-1400 AD), Western and Central Europe (14th Century), North America during initial European colonization (1500-1650), or the U.S. High Plains from 1870 to the present (U.S. DOE, 1980).

However, as acknowledged by the report, differences between periods may outweigh their similarities, and thus raise doubt about the value of such historical analogies.

Comparing two contemporary environmental changes may raise fewer a priori questions. To provide an example, we propose to compare the regional effects of and responses to groundwater depletion in the Ogallala Aquifer region with those resulting from a climatic change induced by CO_2 (and trace gases) loading of the atmosphere. First we will identify similarities in the two issues and show that they share enough important characteristics to justify further discussion. On the basis of this comparison, we will try to determine if responses of decisionmakers, from farmers to national policymakers, to the one issue might be similar to responses to the other.

THE OGALLALA AQUIFER DEPLETION

The Ogallala Aquifer is a geological formation of water-bearing porous rocks which underlies parts of eight states in the American Great Plains, stretching about 800 miles from north to south and about 400 miles from east to west (Figure 1).

The depth of the aquifer from the surface of the land, its rate of natural recharge, and its saturated thickness vary from region to region (Gutentag and Weeks, 1980). The aquifer provides a major source of water for agricultural, municipal, and industrial development in a large section of the Great Plains. It is particularly valuable in the dry climate of that area, where the rainfall is highly variable, with runs of wet or dry years sometimes occurring.

Exploitation of the aquifer has varied through time as well as from place to place. It began at the turn of the century in the southern part of the High Plains but did not become extensive there or in other parts of the Great Plains until farmers were faced with an extended drought at a time when technological developments had reduced the costs of tapping the groundwater (Bittinger

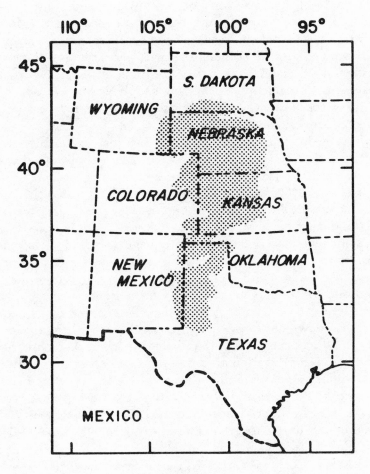

Figure 1.

and Green, 1980; Walsh, 1980). The first major impetus to exploitation of the aquifer was the drought of the 1930s, when many
dryland farmers resorted to the use of groundwater to reduce their
vulnerability to uncertain seasonal and interannual rainfall. Once
drilled, the wells tended to remain in operation, even after the
drought was over.

Since World War II, reliance on groundwater in the region
has increased steadily, stimulated by drought in the 1950s, as well
as by new irrigation technology, cheap energy for pumping, and

higher prices in the 1970s for food, feed grains, and cotton. Tens of thousands of wells now dot the region above the aquifer. A large, new agricultural economy has developed because irrigation farming provides much greater economic returns per unit area than was possible with dryland farming. In addition, an extensive feedlot industry has developed in this region, supplying by 1977 about 38 percent of the national total production of grain-fed beef (High Plains Associates, 1982).

The hydrologic impacts of this sharp increase in the use of water from the aquifer have been major. The withdrawal of water has greatly surpassed the aquifer's rate of natural recharge (e.g., Wyatt et al., 1976; Weeks, 1978). Because the saturated thickness and rates of withdrawal and recharge vary spatially, and because the lateral movement of the water in the aquifer is exceedingly slow, some places overlying the aquifer have already depleted their water supply as a source of irrigation; in others rising costs of energy have made it either uneconomic to rely on groundwater or have prompted farmers to conserve. On the High Plains of Texas, for example, where the aquifer is the main source of irrigation water, drawdown has been viewed as critical since the early 1950s (Firey, 1960). Other parts of the Plains, especially Nebraska, have more favorable saturation thickness and recharge rates and are not so vulnerable (Weeks and Gutentag, 1981). Reinforcing this difference is that "present laws concerning ground water vary from no statewide regulatory controls in Texas to full authority of the State Engineer to control ground water extractions in New Mexico" (High Plains Associates, 1982, I-4).

In 1976, Congress expressed its concern about the water resource situation in the High Plains by authorizing a $6 million, five-year study (Section 193 of Public Law 94-587):

> In order to assure an adequate supply of food to the nation and to promote the economic vitality of the High Plains Region, the Secretary of Commerce ... is authorized and directed to study the depletion of the natural resources of those regions ... presently utilizing the declining water resources of the Ogallala aquifer, and to develop plans, to increase water supplies in the areas and report thereon to Congress. ... In formulating these plans, the Secretary is directed ... to examine the

feasibility of various alternatives to provide adequate water supplies to the area ... to assure the continued economic growth and vitality of the region ... (High Plains Associates, 1982, I-5-6).

The year 2020 was selected as the date toward which to project. In addition to this study, a $5.5 million study was commissioned for the U.S. Geological Survey "to provide (1) hydrologic information needed to evaluate the effects of continued groundwater development; and (2) computer models to predict aquifer response to changes in groundwater development" (Weeks and Gutentag, 1981).

The High Plains Study Council, composed of the governors of the six key states involved in the study, presented recommendations to the U.S. Congress in early 1983. The final report of the High Plains Study showed favorable conditions for agriculture until 2020 in most areas of the aquifer. By 2020, however, most of the marginal lands will have been put into use and there will be little additional land after that time to bring into production in order to offset the loss of the land removed from irrigation because of the depletion of the aquifer elsewhere in the region.

CARBON DIOXIDE AND CLIMATIC CHANGE

Carbon dioxide is a trace gas that is important in regulating the climate of the earth. Human activities are increasing the concentration of carbon dioxide (as well as other radiatively active trace gases such as the chlorofluorocarbons (CFCs), methane, and oxides of nitrogen) in the atmosphere. With respect to CO_2 increases these activities include deforestation and various other land uses that lead to combustion of wood and oxidation of organic materials in soils, but by far the most important contribution is from the burning of coal, oil, and gas. With respect to the other trace gases such activities include the manufacture of CFCs, agricultural practices (methane) and increased use of fertilizers (oxides of nitrogen). Projections of continuing growth in population and economic activity imply increasing levels of CO_2, if a substantial share of global energy is extracted from fossil fuels, especially from the abundant supplies of coal. Such projections predict a doubling of pre-industrial levels of atmospheric CO_2 within the next 50–100 years (Nordhaus, 1979; IIASA, 1981; NRC, 1977, 1982, 1983).

While there is little doubt that carbon dioxide traps long-wave radiation, thus warming the atmosphere near the surface of the earth, there is substantial uncertainty about what the climatic consequences of increased CO_2 would be. Many numerical models of the atmosphere suggest an increase in global average annual surface temperature of about 2–3°C with a doubling of CO_2 (NAS, 1982; Clark, 1982). The earth would then be as warm as it appears to have been during any period in the last 100,000 years (Flohn, 1981). The temperature increase would not be distributed uniformly among regions: if there is a 2°C warming near the equator, there might be a 6–10°C warming (annual average) near the poles. Very little can be said with confidence, however, about redistributions of rainfall that might occur at the regional level. Moreover, estimates have been based on equilibrium responses of climatic models to very large fixed increases in CO_2.

Numerical models of the climate system suggest that, given an increase of CO_2 in the atmosphere, precipitation will increase in the tropics, whereas the midlatitude rainbelt will shift northward. With higher evaporation rate due to higher temperatures in these areas in North America that presently provide water for irrigated agriculture, these areas will grow drier (Manabe et al., 1981).

The modeling results, which must be qualified by substantial uncertainty, are supported by findings in paleoclimatology, whose data are likewise tentative. As can be seen in Figure 2, reconstruction of the climate of the Altithermal (about 4,000–8,000 years ago), a period for which paleoclimatological evidence suggests that warmer atmospheric temperatures may have occurred on a hemispheric scale, shows that a drier band existed in the midlatitudes of the Northern Hemisphere (Kellogg, 1977; Butzer, 1980).

Scientists have also plotted temperature and precipitation patterns that occurred in the Northern Hemisphere during the ten warmest Arctic winters and summers in the twentieth century (Williams, 1980; Jäger and Kellogg, 1983), as well as departures for annually averaged conditions only (Wigley et al., 1980). A short period of instrumental records makes it difficult to define quantitative scenarios for longer periods (i.e., decadal), so we looked at ensemble averages of individual years. If the characteristics of these ensembles (i.e., sets) of individual years prove to be realistic indicators of long-term changes, then these particular observed

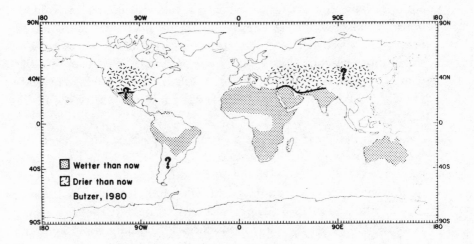

Figure 2.

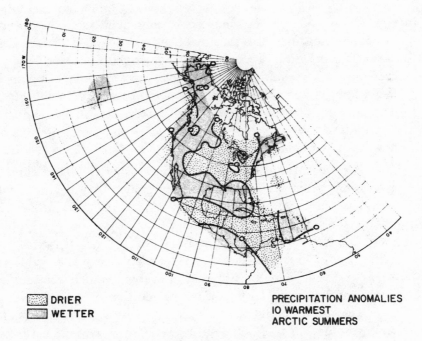

Figure 3.

climatic data could suggest what the regional effects of a CO_2-induced global warming might be. Figure 3 depicts the summer precipitation anomalies that occurred during the ten warmest Arctic summer seasons (Jäger and Kellogg, 1983). For example, the chart suggests drier conditions in the U.S. Great Plains correlating with warmer Arctic summers and, by inference, with a warmer earth.

It is not clear whether anthropogenic effects have been detected in present climatic statistics. However, if model estimates are reasonably correct, those effects should appear unambiguously in the next 5 to 15 years (e.g., Madden and Ramanathan, 1980; Wigley and Jones, 1981; Thompson and Schneider, 1982). As CO_2 continues to increase, the consequences for climate should emerge more distinctly, and the consequences in agriculture, water supply systems, and many aspects of human life and the environment are likely to manifest themselves with increasing frequency and intensity. Of course, some of the regional consequences of the changes in climate and increases in atmospheric CO_2 could be beneficial as well as harmful. Indeed, CO_2 is essential for plant life, and increasing CO_2 could have a positive effect on food supplies by increasing photosynthetic rates for some plants (Wittwer, 1982). The numerous possible beneficial or detrimental effects and their distribution over regions, industries, and population groups are now being explored.

Although the consequences of increasing CO_2 and climatic change are unknown, numerous policy groups have begun to take an interest in the question (e.g., U.S. Senate, 1980; Council on Environmental Quality, 1981; National Commission on Air Quality, 1980; U.S. House of Representatives, 1981). The Department of Energy has conducted a multiyear research program and evaluation of the problem, and the National Research Council has recently completed a review (NRC, 1983). In addition, CO_2 is attracting the attention of an increasing number of physical scientists (for bibliographies on the CO_2 issue, see Olson et al., 1980; Chilton et al., 1981; NCPO, 1981), as well as social scientists and humanists. The media too has shown a heightened interest in the global warming issue.

SIMILARITIES, DIFFERENCES, AND CONVERGENCE

Do the CO_2 and Ogallala issues have enough in common to justify an analogy? They share several important characteristics, most notably a long time scale. Both issues are expected to become serious within the next 50 years (as the Ogallala becomes largely depleted in some areas and the climate presumably departs from the range within which it has remained for the past tens of thousands of years) and to become increasingly critical through the next century. Both issues also become *gradually* more serious. The drawdown of water and the burning of fossil fuels are essentially continuous and have been steadily increasing; they are not comparable to the sudden collapse of a dam or the dumping of toxic wastes, in which the potential problem is connected to a small number of identifiable, discrete events. Moreover, both issues are cumulative and difficult or costly to reverse. Each foot of drawdown of the Ogallala brings the total depletion closer. Similarly, a portion of each year's CO_2 emissions adds to previous accumulation, and the ability of the earth to withdraw carbon from the atmosphere is not growing as rapidly as our capacity to add to it. Implementation of policies to recharge the Ogallala or reduce atmospheric CO_2 concentration can be effective only over a long period, so that if policy action is postponed for many years, the only option may be to accept the environmental changes and to adapt.

Another parallel is that both issues are affected by our tendency to opt for short-term gains from the use of large amounts of water or energy without considering possible long-term costs. Finally, cheap fuel can aggravate both issues; if it is expensive to buy fossil fuels or pump water, the problems may grow at a slower pace.

In some characteristics, the two issues show weaker similarities or differences. Both the CO_2 increase and the Ogallala depletion have features of a potential "tragedy of the commons." A tragedy of the commons occurs when a resource which is freely available to all, but the management of which is the responsibility of no one, is very heavily used and is ultimately exhausted or degraded (Hardin, 1968; Ausubel, 1980). It is worth noting that the histories of various local, regional, and subnational commons (e.g., grazing lands, river systems, fisheries) offer excellent lessons in the variety of responses by human institutions to changes in resource supply and demand. In the case of CO_2, all countries share the atmosphere and its capacity for disposal of waste; if many nations emit large amounts of CO_2, they, along with others, may end up with climates quite different from the ones they have today. The Ogallala, like CO_2, is an issue which becomes serious if many users behave in the same way.

However, the consequences are not necessarily distributed or shared in similar fashion. Nations which burn the most coal may or may not suffer the most adverse consequences, although climate studies do suggest that the USSR, China, and the United States, the countries with the largest coal reserves, will probably all experience some drying. Their actions could well impose large costs on some others.

In contrast, there is much strong local cause and effect with the Ogallala. Although its depletion could have secondary but nonetheless dire effects on food-importing members of the international community, the Ogallala is, in fact, a national or a regional problem. The heaviest users of the aquifer are quite likely to suffer the most immediate adverse consequences. The flow within the aquifer is not so free or fast that the consequences will spread evenly or rapidly. CO_2 also presents greater possibilities for benefit, for example, due to increased photosynthetic productivity (Wittwer, 1982) than the Ogallala depletion, though judicious farmers might gain from the adversity of others. Both issues do require widespread action to make most responses effective. The nature of economic markets means that even careful behavior by an individual farmer or nation to gain insulation from the issue will meet with limited success.

While the degree of "commonality" in the CO_2 issue is greater, so is the degree of uncertainty. It is highly speculative how much CO_2 will be emitted, how the climate will change, what the regional impacts on society and the environment will be, and in what context of national and global events the climate-induced changes will take place. In contrast, rates of drawdown of the Ogallala Aquifer, degrees of depletion, depth from surface, saturated thickness, and recharge rates are to a large degree measurable or calculable, although predictions into the future, here too, are fraught with uncertainty. The effects of drawing water from the aquifer are relatively direct and identifiable, while it may be arguable for a long time what might be the impacts of climate change on society and whether those impacts are attributable to CO_2.*

In one important aspect, with regard to their ultimate societal consequences, the two issues may converge. If, as inferred, one of the effects of increasing the level of CO_2 will be to increase the frequency, duration, and severity of droughts in the Great Plains, this in turn will lead to reduced recharge of and increased demand on the Ogallala Aquifer.

It is worth noting that there is a major difference in the ways the "drying out" of the Great Plains might occur. The Ogallala issue is essentially driven by increasing or unabating water *demand*. A CO_2 drying out of the Plains would be essentially driven by a decreasing water *supply*. However, since both lead to reduced water availability throughout this region, the convergence or reinforcement of the two could create an extremely difficult resource management problem.

COMPARING SOCIETAL RESPONSES

In order to compare possible societal responses to the depletion of the Ogallala and to CO_2-induced climatic change, it is necessary to define categories of response and give examples. Discussion of such responses has been under way for several years in the natural hazards literature (e.g., Burton et al., 1978). Cultural

---

* It is interesting to note that efforts to evaluate the costs of impacts of climatic change and the Ogallala depletion have gone in a similar and perhaps challengeable direction: the Climate Impact Assessment Program (CIAP) study (CIAP, 1975) and parts of the High Plains Associates study (1982) (among others) employed economic models that were not designed to estimate costs of altered agricultural activities in the long term. The CIAP study arrived at global cost estimates by aggregating and extrapolating questionable, geographically specific, partial equilibrium models for prices, crop yields, and temperatures. High Plains Associates employed an econometric model to project decades into the future, though such models are designed for short-term phenomena (up to a few years) and are dubious even then (High Plains Associates, 1981; Robinson, 1981).

anthropologist Michael Thompson (1980) has suggested recently a categorization of individual and group strategies in terms of relative emphasis on management of resources (supply) versus management of needs (demand). The response categories employed below come from the CO_2 impacts literature. There are several reasons to take the categories from the CO_2 literature. Most importantly, they appear to be applicable to the Ogallala case. Secondly, it is convenient to begin with categories which appear to have already been effective with one of the issues.

Responses to CO_2-induced impacts have been categorized in several ways. Meyer-Abich (1980), for example, wrote about prevention, compensation, and adaptation; Corbett (1980) wrote of prevention, mitigation, and compensation; Schelling (1980) discussed prevention and adaptation; Kellogg and Schware (1981) discussed at length averting and mitigating strategies; Lave and Ausubel (1979) presented in the same article adaptation and prevention, adaptation and adjustment, amelioration and mitigation; finally, Lave and Ausubel noted in the same report additional categories of available social reactions to: (1) prevent, finesse, or control the rate of disturbances (i.e., causes) or (2) remedy, ameliorate, or adapt to the effects of the disturbances. These conflicting, often overlapping, meanings associated with the same strategic category when discussed by different authors (and sometimes the same author in the same paper) clearly indicate the need to develop a more consistent set of terms to describe possible responses to a proposed CO_2-induced regional climate change. We will use the terms prevention, compensation, and adaptation, as defined by Meyer-Abich (1980) and illustrated extensively by Robinson and Ausubel (1981).

The three responses are broad stretches on a continuum which ranges from dealing exclusively with the causes of the environmental change to dealing exclusively with the consequences of the environmental change. *Prevention* refers to strategies which attack the problem at its origin, that is, strategies which reduce the production of CO_2 or reduce the consumption of water from the Ogallala. *Compensation* refers to strategies which allow the production of CO_2 or extensive use of Ogallala water but try to compensate for it; for example, by reforestation which would absorb CO_2 or by importation of water to recharge the aquifer. *Adaptation* refers to strategies which allow the environmental changes to occur (a change in climate, reduced water supply) and involve becoming attuned to the new regime. Examples of the three responses may well involve factors which could be ascribed to more than one category. They are classified here according to the chapter authors' perceptions about whether they are *predominantly* preventive, compensative, or adaptive.

Prevention

Prevention with respect to the Ogallala issue translates into an attempt to conserve present water levels in the aquifer. Conservation might, for example, require limiting withdrawals to an amount equalling the natural recharge, an impractical measure that would essentially eliminate use. Conservation could also include: farming practices that are conducive to the retention of soil moisture (such as stubble mulching, minimum tillage, fallow, and crop rotation); cultivation of crops that consume little or no irrigation water (such as gopherweed, kenaf, Jerusalem artichoke); shifting from high-water-use irrigated crops to relatively lower water-use crops (for example, from corn to wheat, or from sugar beets to sunflowers); the use of irrigation only as a buffer during periods of severe moisture stress; acceptance of lower yields for conventional crops by using less water and fertilizer; reduction of the number of cattle dependent on feed; and so on.

Institutional tactics would include tighter regulations or higher prices for water withdrawals. Lending institutions, such as the Federal Land Bank, could encourage conservation through loan practices that would discourage farmers from developing new irrigation facilities or expanding existing ones.

Prevention would be the most difficult (and drastic) option for the Texas High Plains area because of the extremely low recharge rate and relatively low saturated thickness of the aquifer in that region, and because of the region's dependence on Ogallala water. This strategy might be implemented (in theory) if, for example, it was believed that the value to society of the groundwater in the future would be much greater than it is at present. Prevention would require policymakers to repeatedly weigh present short-term economic benefits against future long-term benefits. Prevention seems an unlikely response for this reason and because there may be few farmers today who would willingly end their dependence on water from the Ogallala and revert to dryland farming and ranching. Many of the farmers in the High Plains of Texas, for example, do not really have a choice concerning their dependence on irrigation. Their land was purchased at a price reflecting irrigated land values, and their annual payments preclude any options that diminishes gross returns. To other farmers it would mean a return to dependence on variable rainfall without irrigation as a buffer.

With respect to CO_2, prevention means evolving a system which uses less fossil fuel. On the one hand, conservation might be used to reduce energy demand. On the other hand, energy sources which produce less CO_2 could be emphasized. Development of hydroelectric power could be subsidized; greater encouragement could be given to solar technologies for heating, generation of electricity, and production of fluid fuels for transportation; biomass fuels, which balance creation and absorption of CO_2, could be substituted for certain fossil fuels. Alternatively, one could rapidly expand both conventional nuclear power and more advanced forms, such as breeder technology. An additional strategy would be to encourage exploration for and exploitation of natural gas which emits less CO_2 per unit energy than do other fossil fuels.

At least in theory, a variety of institutionally oriented tactics could be directed toward prevention of CO_2 emissions as well. For example, higher taxes might be introduced for CO_2-emitting fuels; carbon residuals permits could be required; standards for ambient CO_2 could be established; and principles of assigning legal liability for damages on account of climatic change could be accepted. Tactics outside the energy economy might also be effective. For example, slowing down expansion of agriculture into forested lands and improving land use practices could reduce emissions.

Several experts have argued that prevention strategies, whether based on law, regulation, price mechanisms, or rationing, are unlikely with respect to CO_2 (Glantz, 1979; Meyer-Abich, 1980; Kellogg and Schware, 1981; Lave, 1981). Shifts away from fossil fuel would require international cooperation, but some nations are apparently willing to gamble on the possibility of benefits from CO_2-induced climate change. The energy alternatives continue to appear costly and/or risky themselves. Because of the long lead time needed for building new energy systems, significant decisions would probably be required well before statistical evidence about climatic change is convincing and well before signs of widespread degradation become attributable to CO_2 increases. Finally, the long time span and gradual nature of the issue do not match the political process, which tends to be myopic and focused on problems that appear to require more immediate political attention (Glantz, 1980). On the other hand, strong arguments can be made for the benefits of shifting to renewable energy sources and improving efficiency of water use. Conservationist views of the aquifer and the atmosphere as sacrosanct domains may also provide impetus for adoption of prevention strategies.

Compensation

In the Ogallala case, a compensation response would be a commitment to maintain the level or rate of development of the economic activities which depend on the aquifer's water supply. One way to maintain the supply would be to recharge the aquifer artificially. This could be accomplished by importing water into the region and by land management practices designed to increase natural rates of recharge. (Such strategies are sometimes referred to as technological or engineering fixes.)

A second approach would be to maintain the present level and style of agriculture in the region, but not necessarily with the water in the Ogallala itself. The most obvious strategy would again be to import water. Interstate water importation schemes for various parts of the United States have been suggested for several decades with varying degrees of seriousness. Studies of interstate transfers have also been included as part of the High Plains Study; the U.S. Army Corps of Engineers received $800,000 to complete a feasibility study for transfers suggested in Figure 4 (Colorado Department of Agriculture, 1981). However, political opposition to interbasin water transfers is often very strong. For example, in the charge given to the Army Corps of Engineers, both the Mississippi and the Columbia Rivers were placed off limits as possible sources of water to be transferred to the High Plains regions to offset the depletion of the Ogallala. Another strategy frequently suggested to compensate for diminished water supply, resulting either from the Ogallala Aquifer depletion or from CO_2-induced climate change, is weather modification (precipitation enhancement). This strategy also evokes strong political opposition (often local) from competing

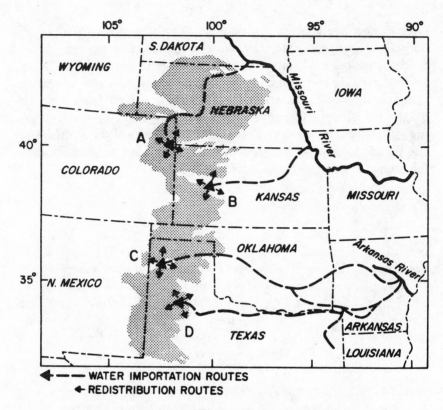

Figure 4.

interest groups, not to mention skepticism from within the scientific community as to its feasibility.

Several compensation strategies have been proposed with respect to CO_2. The one most frequently mentioned is reforestation: decreasing atmospheric carbon dioxide by planting millions of trees which would transfer carbon to the biosphere. Strategies have also been proposed which are focused on different reservoirs and processes in the carbon cycle. It is theoretically possible to absorb carbon dioxide into soil carbon banks by growing short-lived plants for conversion to humus, which would be stored in artificial peat bogs. Biological transfer to the deep oceans is also a theoretical possibility; supplying phosphates and nitrates to surface waters could fertilize growth of those marine organisms that produce carbonate shell structures, and these would eventually sink and either dissolve in the deep ocean or settle safely on the ocean floor. Physical transfer to the deep ocean is conceivable as well; pipelines might, for example, gather CO_2 from power plants and deliver it to points in the oceans where currents would carry it down to deeper layers where it would remain for centuries. Another compensation strategy would involve using solar- or nuclear-generated electricity to extract carbon from the atmosphere and convert it to a liquid hydrocarbon. As in the case of the Ogallala, there are proposals for "climate management" in which other factors in the climate system such as clouds and albedo (reflectivity of a surface) would be manipulated to compensate for CO_2-induced climatic changes, if the physical and political obstacles can be overcome. Finally, large-scale water schemes, ranging from interbasin transfers to iceberg importation, have been mentioned as compensation efforts for CO_2 increases in the atmosphere.

Doubts are often expressed about the effectiveness and practicality of compensation strategies for CO_2. Lack of mechanisms for international cooperation and pressures for use of the biosphere for food, firewood, and fiber limit opportunities. And once the level of atmospheric CO_2 becomes very high, the strategies might not be feasible on a worthwhile scale. Yet, such compensation strategies should not be ruled out; the scale and diffusion of many of today's technologies were scarcely imagined a generation ago, and it is in about a generation that the compensation strategies might become necessary alternatives.

Adaptation

If one chooses neither to prevent an issue from arising by suspending the cause, nor to compensate with countermeasures to suspend the undesirable effects, the remaining choice is to allow the effects to take place and let society adapt. Of course, adaptation may involve anticipatory as well as ex post facto actions.

Several adaptive responses to the Ogallala depletion are already being pursued or considered for implementation either by individual farmers and communities or by states and federal agencies. For example, on a local or state basis, once increasingly adverse effects of the Ogallala depletion begin to be felt, policymakers may encourage economic diversification to minimize dependence on the aquifer. At the farm level adaptation could mean acceptance of lower well yields, lower crop yields, shifting to different crops (often of lower value), more efficient irrigation practices, less irrigated acreage, lower overall agricultural production, and ultimately relocation or a change to economic activities not necessarily related to agriculture.

Unequal distribution of costs may be a reason that adaptation is a likely response. The effects of mining the Ogallala will initially be local because the rate of flow of the groundwater is exceedingly slow, and not all places in the Great Plains will suffer the same consequences at the same time from similar rates of withdrawal. In Texas, the Plains region is highly dependent on aquifer water, and depletion would have a major impact on local communities, as well as on the state's economy. For Colorado, on the other hand, where there is much less dependence on Ogallala water, it was suggested by that state's High Plains Study Team that the economic effect of depleting its portion of the aquifer would be of the same magnitude as closing a large military installation or factory (Colorado Department of Agriculture, 1981). Nebraska, with land that could be put into production (the Sand Hills) and with a relatively high recharge rate and deep saturated thickness, is in a more favorable position than other states above the aquifer with respect to future availability of groundwater. These geographic disparities and variations in effects of mining the aquifer suggest local adaptive responses, rather than a comprehensive interstate compact governing rates of use of Ogallala water.

Suggested adaptive responses to the CO_2 issue in the Great Plains, cited in the CO_2 literature, encompass, in fact, several of the preventive, compensatory, and adaptive responses to the Ogallala depletion, thereby showing the need for more consistent use of these terms. One description of adaptive response to a CO_2-induced warming in the Great Plains goes as follows:

> Methods to mitigate Plains irrigation abandonment include centralized planning of groundwater and surface water use beginning immediately; maintenance of present-day underground water levels; careful rationing of water to back up a change in the present irrigation areas to effect a mixed irrigated/non-irrigated agriculture. Additional water is brought from non-Plains basins, even as far away as the Great Lakes and

Canadian North. A small percentage of the increased personal income in the United States is used to ensure greater support for Plains food production, both dry and irrigated. To maintain wheat in the Plains, several techniques are used: planting of drier/hotter tolerant varieties of wheat and sorghum; no-till practices; increased herbicides and pesticides; crop-growing geared to long-range weather forecasts, cycles, and cumulative soil moisture (S.M.I.) indices. ... Occasional failures are balanced by increased wheat growing in what is at present the Corn Belt and in the East. Businesses and farmers are given incentives to move to the increasingly useful Canadian North.

To mitigate desertification in the Great Plains the following are introduced, in addition to improved wheat practices: careful management of native, drought-resistant grasses with production capacity in each area geared to soil-moisture levels; institution of timely and effective soil conservation practices throughout the Plains, with no setbacks like those of the mid-1970s; a soil-bank reserve program; shelter-belt planting and windbreaks; new tree species suitable for the Southern plains [sic]; and snow supply management (Lave and Ausubel, 1980, 119).

While adaptive responses would also occur in relation to forestry, fisheries, and so forth, Meyer-Abich (1980) argues that qualitatively the basic forms of adaptation to climatic change are migration, reeducation, and industrialization. Change in agricultural productivity, which is one of the basic results of climatic change, is followed by corresponding adjustments in population density (migration) or by increasing agricultural or other economic activities (reeducation and industrialization).

For a variety of reasons, Meyer-Abich also concludes that adaptation, whether innovative or passive, is most likely to be the predominant societal response. Adaptation has these advantages:

- does not require an agreement on long-term goals but is rather flexible when goals and values are changing;
- does not require long-term international cooperation but allows a maximum of self-determination in evaluating costs and benefits·

- allows the appropriation of positive externalities of climatic changes, if there are any;
- allows one to confine oneself to the least marginal action at present;
- allows deferment of expenses most distantly into the future; and
- is the line of least resistance with respect to present patterns of interest and incentives.

These categories of response to the societal implications of a CO_2 contribution to a global warming show how people might react to such an environmental situation but give no indication of the reasons and motivations behind their responses. Many authors, in several disciplines, have sought to investigate why individuals and groups make decisions regarding environmental change the way they do (e.g., Kluckhohn and Strodtbeck, 1961; Saarinen, 1966; Dawes, 1975; Douglas, 1978; Douglas and Wildavsky, 1982). This is an important area of research related to but beyond the scope of our chapter.

CONCLUSIONS AND RECOMMENDATIONS

We have attempted to show that environmental issues might be compared for the purpose of learning from how society responds to one issue about how it might respond to others. To make such a comparison as concrete as possible, we took two contemporary environmental issues centering on changes that could adversely affect the net moisture that would be available to crops, i.e., the water balance, and have pointed out similarities and differences between these two environmental changes. We have selected a specific region, the American Great Plains, as the focus for comparison. One of the environmental issues, a CO_2-induced climatic change, still remains more hypothetical than actual. The other issue, depletion of the Ogallala Aquifer, is already a serious problem, with growing adverse consequences projected to occur in the next century. Potentially serious effects of both changes could occur at about the same time, in the middle of the 21st century.

Three points emerge from our comparison.

1. *Responses detailed for one issue may offer analogues for responses to other issues.* In particular, because studies of possible responses to the Ogallala depletion are more advanced, they

may shed light on responses to a hypothetical but increasingly expected CO_2-induced climatic change. Portions of the High Plains Study (High Plains Associates, 1982), which analyzed the Ogallala depletion for its impacts on the national, regional, state, and local economies, could provide useful first approximations of how farmers and other decisionmakers might respond to a CO_2-induced change in the regional water balance. More thorough review of the High Plains Study would determine which components of the study might contribute to the identification of potential CO_2 socioeconomic impacts (see Wilhite, this volume).

2. *Categories of response can benefit from more consistent definition and may be applicable to more than one issue.* We have attempted to reduce the confusion that surrounds categories of responses to the impacts of a CO_2-induced climatic change. Having chosen three categories taken from the CO_2 impacts literature— prevention, compensation, and adaptation—we placed in those categories specific suggested responses to the potential impacts of a CO_2-induced warming. These categories (which are not necessarily mutually exclusive) also provide a useful framework for actual and possible responses to the Ogallala Aquifer depletion.

It is interesting to note that while there is richer detail in the assessments of the Ogallala issue, there appears to have been more consideration of conceptual frameworks for the CO_2 issue. It may be timely to take these two issues, and other long-term, gradual, cumulative environmental changes that have thus far been expressed only as casual analogies with CO_2 such as acid rain and air pollution (*CO_2 Newsletter*, 1982) and soil erosion (Schneider, 1981) and to develop an overall framework in which possible responses could be identified and evaluated.

3. *The CO_2 and Ogallala issues should be considered jointly.* To date, these two issues have been assessed separately in the scientific as well as policymaking communities, as contemporary but unrelated environmental changes that could affect agricultural production in the American Great Plains. The separation is partly a result of the jurisdiction between the Department of Energy, directing CO_2 research, and the Department of Commerce, involved with regional economic development. Yet, in addition to the potential research benefits of comparing these issues, it is important to look at them together because *their effects on society may converge*, if present trends in exploitation of the aquifer continue and

if speculation about the CO_2 issue and its impacts on this particular region prove to be correct. Impacts that have been suggested separately for these environmental changes could combine to make a difficult situation a much more desperate one. *Moreover, policies judged expensive for one issue might seem more affordable as responses to a combination of both (and perhaps other) issues.* Finally, by combining the assessment of these two issues, allusions to the long-term stability of regional climate (such as the following) would receive the critical scrutiny that they deserve:

> Key resources of land and climate, well suited to large agricultural enterprises, *remain* but the Region is faced with a simultaneous decline in water and the energy resources to support such enterprises (High Plains Associates, 1982, preface) (italics added).

In sum, the development of credible comparisons and analogies appears to be a fruitful as well as inexpensive means of advancing climate-related impact assessment research for a variety of important environmental issues.

ACKNOWLEDGMENTS

We would like to express our appreciation to the Weyerhaeuser Foundation for their encouragement and financial support for the Great Plains Planning Workshop held in 1981 at NCAR in Boulder, Colorado, and for research activities related to the topic of assessing climate-related impacts by analogy. We would like to thank the workshop participants who were instrumental in redirecting the focus of our research and to the many colleagues who read drafts of the manuscript and provided critical reviews and suggestions. We would also like to thank Dr. Walter Orr Roberts, President Emeritus of the University Corporation for Atmospheric Research (UCAR) in Boulder, Colorado, for his strong support. Finally, special thanks are extended to Maria Krenz for her editorial help during the many stages of this chapter and for her assistance in organizing the workshop.

REFERENCES

Ausubel, J.H., 1980: Economics in the air: An introduction to economic issues of the atmosphere and climate. In J.H. Ausubel and A.K. Biswas (Eds.), *Climatic Constraints and Human Activities*. Oxford, UK: Pergamon Press, 13–60.

Ausubel, J.H., and A.K. Biswas (Eds.), 1980: *Climatic Constraints and Human Activities*. Oxford, UK: Pergamon Press.

Bittinger, W.D., and E.B. Green, 1980: *You Never Miss the Water Till . . . .* Littleton, CO: Water Resources Publications.

Burton, I., R.W. Kates, and G.F. White, 1978: *The Environment as Hazard*. New York, NY: Oxford University Press.

Butzer, K.W., 1980: Adaptation to global environmental change. *Professional Geographer, 32*, 269–78.

Chen, R.S., 1981: Interdisciplinary research and integration: The case of CO_2 and climate. *Climatic Change, 3*, 429–47.

Chilton, B.D., L.J. Allison, and S.S. Talmage, 1981: *Global Aspects of Carbon Dioxide: An Annotated Bibliography*. Oak Ridge, TN: Oak Ridge National Laboratory.

CIAP (Climate Impact Assessment Program), 1975: *Economic and Social Measures of Biological and Climatic Change*. Monograph 6, DOT-TST-75-20. Washington, DC: Department of Transportation.

Clark, W. (Ed)., 1982: *Carbon Dioxide Review: 1982*. New York, NY: Oxford University Press.

CO_2 Newsletter, 1982: Editorial. January–February. Teton Village, WY: William N. Barbat Associates, 2.

Colorado Department of Agriculture, 1981: Increase water supply. *Ogallala Aquifer Newsletter*, November. Denver, CO: Colorado Department of Agriculture, 9.

Corbett, J.G., 1980: Paper presented at the AAAS/DOE Workshop on Environmental and Societal Consequences of a Possible CO_2-Induced Climatic Change. Annapolis, MD, 2–6 April 1979. See Report of Panel IV, Social and Institutional Responses, in *Proceedings*. Springfield, VA: National Technical Information Service, 79–103.

Council on Environmental Quality, 1980: *Global Energy Futures and the Carbon Dioxide Problem.* Washington, DC: U.S. Government Printing Office.

Dawes, R.M., 1975: Formal models of dilemmas in social decisionmaking. In M.F. Kaplan and S. Schwartz (Eds.), *Human Judgment and Decision Processes.* New York, NY: Academic Press, 87–108.

Douglas, M., 1978: Cultural bias. Occasional paper No. 35. London, UK: Royal Anthropological Institute of Great Britain and Ireland.

Douglas, M., and A. Wildavsky, 1982: *Risk and Culture: An Essay on the Selection of Technological and Environmental Dangers.* Berkeley, CA: University of California Press.

Firey, M., 1960: *Man, Mind and Land: A Theory of Resource Use.* Glencoe, IL: Free Press.

Flohn, H., 1981: *Possible Climatic Consequences of a Man-Made Global Warming.* Laxenburg, Austria: IIASA.

Glantz, M.H., 1979: A political view of CO_2. *Nature, 280,* 189–90.

Glantz, M.H., 1980: Some political considerations in dealing with drought. In N. Rosenberg (Ed.), *Drought in the Great Plains: Research on Impacts and Strategies.* Littleton, CO: Water Resources Publications, 60–85.

Glantz, M.H., J. Robinson, and M.E. Krenz, 1982: Climate-related impact studies: A review of past experiences. In W. Clark (Ed.), *Carbon Dioxide Review: 1982.* New York, NY: Oxford University Press, 57–93.

Gutentag, E.D., and J.B. Weeks, 1980: *Hydrological Investigation Atlas HA-642.* Washington, DC: U.S. Geological Survey.

Hardin, G., 1968: The tragedy of the commons. *Science, 162,* 1243–8.

High Plains Associates, 1981: *Congressional Briefing on the Six-State High Plains-Ogallala Aquifer Regional Resources Study.* Austin, TX: Camp Dresser and McKee, Inc.

High Plains Associates, 1982: *The Six-State High Plains-Ogallala Aquifer Regional Resources Study.* Preliminary Draft. Austin, TX: Camp Dresser and McKee, Inc.

IIASA (International Institute for Applied Systems Analysis), 1981: *Energy in a Finite World: A Global Systems Analysis*. Cambridge, MA: Ballinger.

Jäger, J., and W.W. Kellogg, 1983: Anomalies in temperature and rainfall during warm Arctic seasons. *Climatic Change, 5,* 39–60.

Kellogg, W.W., 1977: *Effects of Human Activities on Global Climate*. Tech Note 156. Geneva, Switzerland: World Meteorological Organization.

Kellogg, W.W., and R. Schware, 1981: *Climate Change and Society*. Boulder, CO: Westview Press.

Kluckhohn, F.R., and F.L. Strodtbeck, 1961: *Variations in Value Orientations*. Evanston, IL: Row, Peterson & Co.

Lave, L.B., 1981: *Mitigating Strategies for CO_2 Problems*. CP-81-14. Laxenburg, Austria: IIASA.

Lave, L.B., and J. Ausubel, 1980: Issues associated with analysis of economic and geopolitical consequences of a potential CO_2-induced climatic change. Report of Panel V. In Proceeding of the AAAS/DOE workshop on Environmental and Societal Consequences of a Possible CO_2-Induced Climatic Change, Annapolis, Maryland, 2–6 April 1979. Springfield, VA: National Technical Information Service, 104–21.

Madden, R.A., and V. Ramanathan, 1980: Detecting climate change due to increasing carbon dioxide. *Science, 209,* 763–8.

Manabe, S., R.T. Wetherald, and R.J. Stouffer, 1981: Summer dryness due to an increase of atmospheric CO_2. *Climatic Change, 3,* 347–86.

Martino, J., 1975: *Technological Forecasting for Decisionmaking*. New York, NY: Elsevier Science Publishing Co.

Meyer-Abich, K.M., 1980: Chalk on the white wall? On the transformation of climatological facts into political facts. In J.H. Ausubel and A.K. Biswas (Eds.), *Climatic Constraints and Human Activities*. Oxford, UK: Pergamon Press, 61–74.

NAS (National Academy of Sciences), 1977: *Energy and Climate*. Washington, DC: National Academy Press.

NAS (National Academy of Sciences), 1982: *Carbon Dioxide and Climate: A Second Assessment.* Washington, DC: National Academy Press.

National Commission on Air Quality, 1980: *Summary Report of the NCAQ Carbon Dioxide Workshop, St. Petersburg, Florida. 30–31 October 1980.* Washington, DC: National Commission on Air Quality.

NCPO (National Climate Program Office), 1981: *Research Issues and Supporting Research of the National Program on Carbon Dioxide, Environment and Society.* Fiscal Year 1980. Springfield, VA: National Technical Information Service.

Nordhaus, W.D., 1979: *The Efficient Use of Energy Resources.* New Haven, CT: Yale University Press.

NRC (National Research Council), 1983: *Changing Climate.* Report of the Carbon Dioxide Assessment Committee, Board on Atmospheric Science and Climate. Washington, DC: National Academy Press.

Olson, J.S., L.H. Allison, and B.N. Collier, 1980: *Carbon Dioxide and Climate: A Selected Bibliography.* Volumes 1–3. Oak Ridge, TN: Oak Ridge National Laboratory.

Pittock, A.B., and M.J. Salinger, 1982: Towards regional scenarios for a CO_2-warmed earth. *Climatic Change, 4,* 23–40.

Robinson, J., 1981: *Climate Impacts and Global Models.* Laxenburg, Austria: IIASA.

Robinson, J., and J.H. Ausubel, 1981: *A Framework for Scenario Generation for CO_2 Gaming.* Laxenburg, Austria: IIASA.

Saarinen, T.F., 1966: *Perceptions of the Drought Hazard on the Great Plains.* Research Paper No. 106. Chicago, IL: Department of Geography, University of Chicago.

Schelling, T.C., 1980: Letter Memo from the Ad Hoc Study Panel of the Climate Board on Economic and Social Aspects of Carbon Dioxide Increase to Dr. Philip Handler, President, National Academy of Sciences, 18 April 1980.

Schneider, S.H., 1981: Research on potential environmental and societal impacts. *Hearing before the Subcommittee on Investigation and Oversight of the Committee on Science and Technology on CO_2 and Climate: The Greenhouse Effect.*

97th Congress. Washington, DC: U.S. Government Printing Office, 32.

Thompson, M., 1980: *The Social Landscape of Poverty.* WP-80-174. Laxenburg, Austria: IIASA.

Thompson, S., and S. Schneider, 1982: Carbon dioxide and climate: Has a signal been observed yet? *Nature, 295,* 645–6.

U.S. DOE (Department of Energy), 1980: *Proceeding of the Workshop on Environmental and Societal Consequences of a Possible CO_2-Induced Climate Change.* 2–6 April 1979, Annapolis, MD. Springfield, VA: National Technical Information Service.

U.S. House of Representatives, 1981: *Hearing before the Subcommittee on Investigation and Oversight of the Committee on Science and Technology on CO_2 and Climate: The Greenhouse Effect.* 97th Congress. Washington, DC: U.S. Government Printing Office.

U.S. Senate, 1980: *Hearing on Carbon Dioxide Buildup in the Atmosphere.* 96th Congress, Publication No. 96–107. Washington, DC: U.S. Government Printing Office.

Walsh, J., 1980: What to do when the well runs dry. *Science, 210,* 754–6.

Weeks, J.B., 1978: *Plan of Study for the High Plains Regional Aquifer: Systems Analysis in Parts of Colorado, Kansas, Nebraska, New Mexico, Oklahoma, South Dakota, Texas and Wyoming.* Water Resources Investigations 78–80. Lakewood, CO: U.S. Geological Survey.

Weeks, J.B., and E.D. Gutentag, 1981: *Hydrological Investigations Atlas.* HA-648. Washington, DC: U.S. Geological Survey.

Wigley, T.M.L., and P.D. Jones, 1981: Detecting CO_2-induced climatic change, *Nature, 292,* 205–8.

Wigley, T.M.L., P.D. Jones, and P.M. Kelly, 1980: Scenario for a warm, high-CO_2 world. *Nature, 283,* 17–21.

Williams, J., 1980: Anomalies in temperature and rainfall during warm Arctic seasons as a guide to the formulation of climate scenarios. *Climatic Change, 2,* 249–66.

Wittwer, S.H., 1982: Carbon dioxide and crop productivity. *New Scientist, 95,* 233–4.

WMO (World Meteorological Organization), 1979: *Proceedings of the World Climate Conference*. Geneva, Switzerland: WMO.

Wyatt, A.W., A.E. Bell, and S. Morrison, 1976: *Analytical Study of the Ogallala Aquifer in Farmer Country, Texas*. Report No. 205. Austin, TX: Texas Water Development Board.

7

Great Lakes Levels and Climate Change: Impacts, Responses, and Futures

Stewart J. Cohen

PROLOGUE

In November 1985, newspapers in the Great Lakes region reported that water levels were so high that a severe storm would cause considerable damage to shoreline facilities. By September 1987, high water level problems had eased on all lakes. In fact, Lake Ontario's level was 13 cm *below* its long-term average, thus causing problems of a different kind for property owners and recreational boaters in the Kingston–Cornwall area of the St. Lawrence River. The rapid change occurred because of dry weather and high outflows through control works at Cornwall (Environment Canada news release, 16 September 1987).

The above illustrates that the Great Lakes Basin is an oscillating system because of climate variability, and that any resource management initiative must have the capacity to handle such oscillations. Paleoclimatic research suggests that the fluctuation experienced between November 1985 and September 1987 was not unprecedented. In fact, geologic and archaeologic evidence indicate that lake levels over the last 7,000 years have been considerably higher and lower than any recorded this century (Larsen,

Stewart J. Cohen is a professional staff member of the Canadian Climate Centre, a component of the Atmospheric Environment Service, Toronto, Canada. He has a Ph.D. in geography from the University of Illinois and has taught at York University in Toronto. In 1984 Dr. Cohen received a fellowship from the Natural Science and Engineering Research Council (Canada) to study the potential impacts of global climatic change in the Great Lakes Region.

1985). Consequently, if control measures are introduced that cannot respond to lake-level variability, these measures could have the potential to amplify any negative effects.

Water resources in any region of the world are influenced by a number of environmental factors. One of these factors, climate, is a highly variable component of nature. In recent centuries and decades, seasonal and interannual climate variations have resulted in short-term extreme events such as droughts and floods, longer-term fluctuations in streamflows and lake levels, and changes in snowpack and glacial extent. These extreme events and long-term changes have had significant impacts on society, including economic fluctuations, property damage, and loss of life. Consequently, the assessment of climate impacts on water resources involves not only climatology and hydrology, but many other disciplines as well, including the social sciences, engineering, biological sciences, and economics.

The past can teach us a great deal about how climate, water resources, and society interact. However, projections of the future must include a new element, not experienced by previous generations. That element is the "greenhouse effect," a possible warming of the earth's climate due to continuous increases in atmospheric concentrations of CO_2 and other trace gases. There is considerable uncertainty regarding many aspects of the projected warming, but the possible magnitude of this change to the global environment could lead to changes in the hydrologic cycle that would be too great to ignore. Such changes could affect water resources planning and management in many regions.

The following discussion is divided into five parts: causes of recent lake-level fluctuations in the Great Lakes Basin; impacts of these fluctuations; responses to these fluctuations; scenarios of future climate change (i.e., greenhouse effect) and their possible impacts; and implications for future water resources management.

LAKE-LEVEL FLUCTUATIONS AND RECENT CLIMATE

The Great Lakes Basin contains the world's largest system of freshwater lakes (Figure 1). The basin has a drainage area of approximately 766,000 km^2, of which 32 percent is water surface. Water flows from Lake Superior through the regulated St. Mary's River near Sault Ste. Marie. Lakes Michigan and Huron

Figure 1. The Great Lakes of North America, with grid points from two general circulation models.

are at the same elevation, 7 meters below Superior, so they are sometimes referred to as "Michigan–Huron." The outflow of these three lakes proceeds through the unregulated St. Clair River and Lake St. Clair into Lake Erie, which drains into Lake Ontario via the Welland Canal, Niagara River, and the well-known Niagara Falls. Finally, water leaves the Great Lakes Basin via the regulated St. Lawrence River at Cornwall, where it mixes with water from neighboring basins as it proceeds toward the Gulf of St. Lawrence, and the Atlantic Ocean.

Monitoring and management of the basin's water resources involves agencies from various levels of government in the United States and Canada. However, an important component is the International Joint Commission (IJC), a binational organization established by treaty in 1909. The basin's discharge is regulated by control works near Sault Ste. Marie and Cornwall (see Figure 1). There are diversions bringing water into Lake Superior from the north, and taking water out at Chicago. Discharge at all of these control points is regulated under the direction of the IJC in consultation with the two countries. There is also a shipping canal (Welland Canal) on the Ontario side of the Niagara River, which permits navigation between Lakes Ontario and Erie, and the New York State Barge Canal, which provides a similar service between Lake Erie and western New York state.

The region has a midlatitude continental climate, with four distinct temperature seasons. Monthly precipitation does not vary a great deal between winter and summer, except in the west, which has a drier winter. The annual total is 700–1,000 mm, depending on location. Winter (December–March) is cold enough to produce a stable snow cover on the land, and ice on the lakes, which effectively shuts down the commercial shipping industry for about three months. However, some open water persists throughout the winter (Saulesleja, 1986), and this can modify the regional climate, particularly by generating "lake effect" snowfall in the lee areas to the east and south of each lake (Eichenlaub, 1979).

Lake-level fluctuations can occur on various time scales. Over a period of several hours to days, wind combined with barometric pressure differences can cause unusual local water level changes, particularly at the extreme ends of the shallower lakes, such as Lake Erie. When the wind is blowing along the axis of the lake, a rapid difference in levels develops between one end of the lake and the other. These extreme event storm surges have led to lake-level differences as large as five meters between the windward and leeward sides. A seiche occurs after the wind dies down and the lake level returns to equilibrium as water moves from one end of the lake to the other.

On a seasonal basis, there are water level fluctuations which reflect the normal hydrological cycle. The peak in spring and summer occurs because of high runoff (mostly snowmelt) and low evapotranspiration. This seasonal fluctuation is relatively small, averaging 38 cm on Lake Ontario, 37 cm on Lake Erie, and 30–31 cm on Lakes Michigan–Huron and Superior.

Our main interest, however, is in longer-term fluctuations, and this requires consideration of climatic variation and its influence on Net Basin Supply (NBS). This can be modeled as follows:

$$NBS = P_{lake} - E_{lake} + R + \text{diversions} - \text{consumptive use}$$

where P_{lake} is over-lake precipitation, E_{lake} is open-water evaporation, and R is overland runoff. At present, the net effect of the Long Lac, Ogoki, Chicago, and Welland Canal diversions is a minor increase of less than 1 percent of "normal" NBS at Cornwall. For individual lakes, small increases or decreases occur. The resulting net effect on the levels of Lakes Superior, Michigan–Huron, Erie, and Ontario are +2.0 cm, –0.6 cm, –10 cm, and +3.2 cm, respectively. Consumptive use is estimated at 106–180 m^3/sec, equivalent to 1.4–2.4 percent of "normal" NBS (Cohen, 1986a). Groundwater variations are assumed to be negligible in the long term. Thus, the major elements influencing historical variations in NBS are P_{lake}, E_{lake} and R. Specific effects of IJC control actions on lake levels will be discussed later.

Climatic variability has had a significant impact on mean annual NBS, which is equivalent to mean annual discharge at Cornwall. The 1959–85 "normal" annual NBS is approximately 7,500 m^3/sec, considerably higher than the 1861–1985 mean shown in Figure 2. Pre-1959 data were obtained from nearby Iroquois, and represent preregulation flow. However, the recent high NBS is due to a long period of above-average precipitation, rather than regulation.

Annual NBS has varied widely throughout this century, but has shown considerable fluctuation in the last 30 years (Figure 2). Low lake-level records for this century were set on some lakes in 1963–65, while record highs have been observed in 1973–74. These were exceeded at all lakes except Ontario in 1986. The high levels have caused an increased incidence of flooding, erosion, and damage to shoreline properties. Fortunately, the dry 1986–87 winter allowed lake levels to recede somewhat as of this writing (November 1987). The level at Lake Superior is now below its long-term average. A similar situation exists at Lake Ontario, though in this case, it is a result of dry weather combined with increased releases at Cornwall, authorized by the IJC.

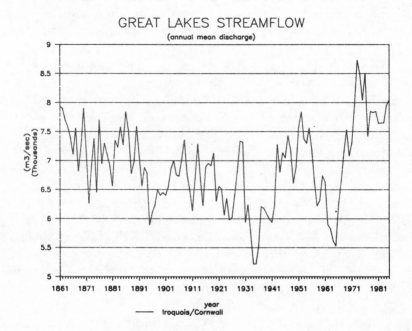

Figure 2. St. Lawrence discharge measured at Iroquois (pre-1959) and Cornwall (1959–present). (Source: Environment Canada, 1986.)

The recent high lake levels and NBS were largely a result of above-average precipitation (Figure 3), especially over Lakes Superior and Huron, as well as relatively cool temperatures which reduced E_{lake} losses (Quinn, 1981, and personal communication). The low NBS experienced in 1963–65 occurred primarily because of below-average precipitation (Cohen, 1986b). However, the importance of E_{lake} must be kept in mind, since the open water surface represents such a large portion of the basin. Under present "normal" climate conditions, E_{lake} losses are almost as large as each of the main inputs, P_{lake} and R (Cohen, 1986c). The declining lake levels in 1987 have been due to below-average precipitation, as well as above-average air and lake temperatures which have increased E_{lake} losses.

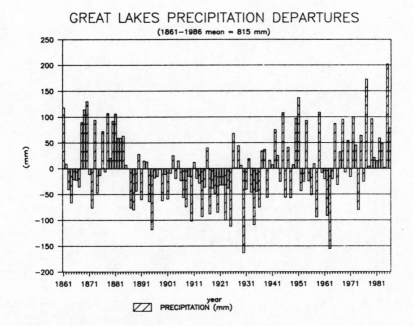

Figure 3. Great Lakes Basin precipitation. (Source: Quinn, 1981; personal communication, November 1985 and January 1987.)

IMPACTS OF CLIMATE AND LAKE-LEVEL FLUCTUATIONS

With a population of 29 million on the U.S. side and 8 million on the Canadian side, the Great Lakes is a highly industrialized international basin; a major producer of hydroelectric power, crops, and wood products; and an important transportation corridor. Some of these activities (e.g., shipping) are tied to the lakes themselves. Others (e.g., winter recreation) are not dependent on the lakes directly, but are sensitive to variations in climate. In addition, certain climate-sensitive activities (e.g., agriculture) have the potential to become major consumers of water, so variations in climate that affect water demand could indirectly affect lake levels.

A model framework of these and other linkages between climate, lake levels and NBS, and the Great Lakes region's economy, is presented in Figure 4. Starting from the left side of the figure, a variation in climate is identified either by a change to a new

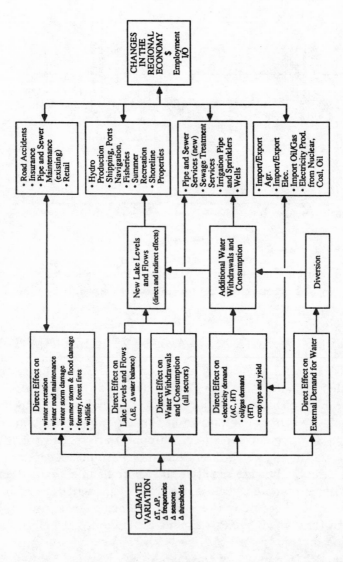

Figure 4. Interconnected components of climate impacts and societal responses within the Great Lakes region. Δ = change, T = temperature, P = precipitation, E = evaporation, AC = air conditioning, HT = space heating, I/O = inputs/outputs. (Source: Cohen, 1986c.)

mean condition (e.g., warmer temperatures), or by changes in frequencies of extreme events, lengths of season, or the exceedance of critical thresholds (e.g., temperature below 0°C). This variation could directly influence land-based activities, basin hydrology, regional water demand, energy demand for heating and air conditioning, and perhaps demand for future exports of various commodities (including water), as well as changes in demand for imported goods. Changes in water demand could alter lake levels and NBS, and lead to changes in water distribution services. The new lake levels and streamflows could affect several water-based activities, including hydroelectric production.

A number of scenarios of "climatic impact" could be drawn from a scenario of long-term climatic change. Cohen (1986c) describes possible impacts on the energy sector of a warmer, drier climate, with various ripple effects being felt by water distribution services due to changes in water demand for cooling purposes at fossil fuel and nuclear power stations. Another example within this climate scenario is an increase in demand for irrigation water by the agriculture sector. If either or both sectors were to experience significant increases in water demand, reductions in NBS and lake levels could occur, with subsequent ripple effects being felt by hydroelectric utilities and other water-based activities.

The early 1960s was a period of low NBS and lake levels (Figures 2 and 3) caused by extensive drought. Lakes Michigan–Huron and St. Clair recorded their lowest lake levels this century. A number of examples of specific impacts are listed in Allsopp et al. (1981). Hydroelectric utilities experienced losses of 19–26 percent at facilities on the Niagara and St. Lawrence Rivers. Insufficient water depths at connecting channels resulted in reduced cargo loads per trip, and a total carrying loss of 8.6 million tons for the 1964 shipping season. A number of wells dried up, resulting in water being transported by truck. During the summer of 1963, drought in southern Ontario caused yield reductions in soybeans and grain corn of 25 percent and 15 percent, respectively, from the 1961–65 average. Recreational boaters and harbors suffered from lack of sufficient water depths, and aesthetically unpleasant mud flats were exposed along the shoreline (personal communication, Doug Cuthbert, 1 December 1987). There were also reports of problems at water intakes and municipal sewage outlets along the Great Lakes shoreline.

Since the drought ended in 1965, the region has received precipitation greater than the 1861–1986 average in 17 out of 22 years. This resulted in record high NBS and lake levels being recorded in the early 1970s, which were subsequently exceeded in 1986 at all lakes except Ontario. During both episodes, damage to shoreline property was extensive, particularly on Lakes Erie and Michigan–Huron. In 1973, damage and protection costs on the U.S. side exceeded US$200 million (IJC, 1981a). Losses in Canada were nearly CAN$17 million (IJC, 1981a).

During the 1970s, shoreline erosion ensued and trees bordering marsh areas were damaged or killed. Excessive moisture also caused significant damage to many crops in southern Ontario, the worst cases being the 55 percent reduction in white bean yields in 1977 and a similar loss in winter wheat production in 1978. On the positive side, hydroelectric facilities were producing electricity at full capacity (Allsopp et al., 1981). Meanwhile, several regions were actually experiencing losses due to local drought. One example is the 15 percent loss in provincial grain corn yield in 1978 due to low precipitation which occurred in a zone extending from north and east of Toronto all the way to Windsor, Ontario (Allsopp et al., 1981).

As of this writing (November 1987), I have been unable to obtain a precise assessment of damages from the 1985–87 high-levels episode. There were reports of damage from Chicago's Lakeshore Drive to Lake Erie's shoreline cottages during storm surges in December 1985 and 1986. Preliminary estimates of damage in several Ontario counties exceed CAN$16 million (Stewart and Yee, 1987). If not for the relatively low frequency of severe storms, damage would have been much worse during this episode.

GOVERNMENT RESPONSES

Four words describe the main thrust of government responses to the three recent lake-level crises: regulation, protection (United States only), research, and communication. The latter has taken on greater importance during the most recent episode.

1963–65 Low Lake Levels

The low-level crisis of 1963–65 was of great concern to hydro-electric power producers, shipping companies, recreational boaters, riparians, and municipalities. The IJC responded in three ways. First, to improve conditions in Lakes Michigan–Huron, St. Clair, and Erie, the IJC ordered increased outflows from Superior. However, due to extremely low water supplies, only minimal assistance could be provided. Second, Lake Ontario levels were maintained slightly higher than they otherwise would have been by reducing outflows at the St. Lawrence River control works. Third, a study of Great Lakes levels was initiated to determine the best alternatives for lake-level regulation and control of shoreline erosion.

Meanwhile, shoreline property in certain areas was being converted from forestry and agriculture to recreational, industrial, and residential uses. Major shoreline urban centers were expanding. Some of this property was developed during the low-level period. Since lakeshore land was highly valued, development was carried out down to the lowest possible elevation, even for that historically low period (IGLLB, 1974). It would appear that local governments and the general public thought that lower levels represented a new "normal."

1973 High Lake Levels

The high levels of 1973 caused extensive flood damage to the newly developed shoreline properties. The IJC responded by reducing Lake Superior outflow to Lakes Michigan–Huron, and increasing outflow from Lake Ontario. This was a deviation from the existing regulation plan, known as Plan 1958-D. After several months, resulting changes in levels at Superior, Michigan–Huron, and Ontario were +20 cm, −12.5 cm, and −30 cm, respectively, compared to what they would have been under Plan 1958-D. The response at Lake Ontario appeared to have a significant effect on Lake Ontario levels. However, the levels at Michigan–Huron were 60–90 cm above average, so the effectiveness of the response at Superior was relatively low (IGLLB, 1974).

In the United States, the Army Corps of Engineers in cooperation with a number of municipalities, undertook flood protection projects at all lakes except Superior. These projects included the construction of dikes, rock cribs, and sand cribs.

The IJC study, initiated in 1964 and completed in 1974, recommended that a new regulation plan be established for Lake Superior. It was the least costly of six alternatives being considered, and involved maintaining Lake Superior at a higher minimum level by modifying the operation of the control works at St. Mary's River. Lake Ontario's plan would remain the same. This new plan, Plan 1977, was instituted in 1979 and had a small effect (approximately 1 cm) on the levels of the upper lakes (personal communication, Doug Cuthbert, 1 December 1987). Shoreline zoning and setback of structures were also recommended, to avoid repeating the mistakes of the 1960s. However, implementation of shoreline management plans was beyond IJC's mandate. Regulation and dredging of the St. Clair and Detroit Rivers, and Lakes Michigan–Huron were rejected as being too costly, while further study would be required for Erie. As a follow-up, the IJC commissioned the Lake Erie Regulation Study in 1977. Completed in 1981 (IJC, 1981a), this study concluded that dredging or controlled diversion at the Niagara River was not justified. In all the above cases, the benefits to shoreline properties were not as great as the costs of the plans.

Another area of concern identified in the 1974 report was that steadily increasing consumptive use of water in the basin could affect mean lake levels. Though there was no specific recommendation, the IJC commissioned the Great Lakes Diversions and Consumptive Uses study in 1977. The original report (IJC, 1981b) projected that consumptive use would increase from 170 m^3/sec (1975 estimate) to 280 m^3/sec by 2000 and 720 m^3/sec by 2035 (approximately 10 percent of NBS), largely because of anticipated growth in demand for cooling water by nuclear and fossil-fueled power plants. An update (IJC, 1985) lowered the projected increase to 238 m^3/sec by 2000, because of new uncertainties in projecting economic growth and technological change.

1985–87 High Lake Levels

During this period, lake levels were marginally higher than those recorded in 1973, except at Ontario. Lakes Michigan–Huron and Erie levels were 60–100 cm above average. NBS was 15 percent above average. Public outcry was tremendous due to damage to shoreline property.

As a result, the response to this crisis was much broader and involved a greater number of agencies and activities. As before, the first response by the IJC in April 1985 was to reduce Lake Superior outflow, this time by 30 percent below Plan 1977 flow. As of November 1985, the IJC action raised Lake Superior levels by 10 cm and lowered Michigan–Huron, St. Clair, and Erie by 7 cm, 4 cm, and 3 cm, respectively (personal communication, Doug Cuthbert, 1 December 1987). In addition, Ontario Hydro, the provincially owned utility, reduced the outflow of the Ogoki diversion by 110 m^3/sec (1.3 percent of 1985 NBS), although the effect of this on Lake Superior levels was only 0.25 cm. In December 1985, the IJC ordered increased outflows from Lake Ontario due to continued high precipitation and high inflows from Lake Erie. At no time did the IJC order increased outflows at the Chicago diversion (Pratt, 1985).

The response of the U.S. Army Corps of Engineers was similar to its actions of 1973, i.e., flood protection projects at Lakes Erie and St. Clair, and the Detroit River. State agencies from Michigan and Ohio, as well as local governments, cooperated in these projects (Pratt, 1985).

In a November 1985 U.S. Army briefing (Pratt, 1985), nothing new could be suggested to reduce the high levels. The combined effect of emergency actions during 1985 was only a 3–7 cm reduction, while the lakes were up to 100 cm above average. Subsequently, calculations of the effect for 1986–87 showed that Lake Superior's levels were raised by 24 cm, while levels at Lakes Michigan–Huron, St. Clair, and Erie were lowered by 9, 9, and 6 cm, respectively (Stewart and Yee, 1987). Fortunately, dry weather from November 1986 to July 1987 reduced supplies to the upper lakes, thereby lowering these levels considerably. By October 1987, Lakes Michigan–Huron and Erie were less than 30 cm above the long-term average.

The Canadian government response was to establish a special forecast service as well as a water level communications center. The "water level forecast desk" was established at the Environment Canada weather office at Pearson International Airport near Toronto. Special forecasts of wind speed and direction, waves, and water levels (including near shore) were produced. Warnings and watches were issued for damage-prone shoreline areas, using criteria developed by the Ontario Ministry of Natural Resources

(OMNR) in consultation with regional conservation authorities. These forecasts were also provided to municipalities. The criteria were revised downward in 1987 due to the drop in lake levels that year.

In March 1986, Environment Canada also created the Great Lakes Water Level Communications Centre at the Canada Centre for Inland Waters in Burlington, near Hamilton, Ontario. Its purpose was to disseminate public warnings of the approach of severe storms that could cause flooding and erosion damage. The center received warnings and watches from the forecast desk at Toronto and operated 24 hours a day during storm events. It was a primary source of information to the public, the media, politicians and government agencies on the causes of the high levels. The center responded to telephone inquiries and media requests for information, provided speakers on the subject for public meetings, and acted as a focal point for information on the issue (Stewart and Yee, 1987). Ongoing government studies and activities, including those of the IJC, were publicized. An information document, using interviews with Great Lakes specialists, was prepared for insertion into 21 area newspapers. The center also conducted a survey of shoreline erosion damage.

In 1986, the IJC began its third study of Great Lakes levels. It is due to be completed in 1991. The short-term goals are to propose measures that could be taken to alleviate the problems due to the current high-level conditions. Of greater importance, perhaps, are the study's long-term goals. Its mandate includes examination of present shoreline management practices and land uses, and determination of the costs and benefits of alternatives such as new zoning regulations. Taking a long-term view of the future, one of the factors to be considered is possible climatic warming resulting from the greenhouse effect.

POSSIBLE IMPACTS OF A WARMER CLIMATE

Recent literature has provided considerable documentation on the probable causes (and uncertainties) of future climatic warming, and its possible impacts on society (Clark, 1982; MacCracken and Luther, 1985; Kates et al., 1985; WMO, 1986; Titus, 1986; Bolin et al., 1986, NAS, 1987). Included in this global effort are a number of studies of regional scale impacts. One of these was a pilot study of the Ontario portion of the Great Lakes region, initiated by Environment Canada in 1984 (Cohen and Allsopp, 1988). There was particular interest in impacts on hydroelectric power production, shipping, agriculture, energy and water demand, and recreation. Possible changes in NBS and lake levels for the entire basin were also investigated. The working group consisted of researchers from government, universities, and the private sector.

The pilot study used a scenario derived from the Goddard Institute for Space Studies (GISS) model projection of a greenhouse warming (Hansen et al., 1983). The grid points in Figure 1 represent data points from that scenario. Details are available in Cohen (1986c). The GISS scenario of warmer temperatures and slightly higher annual precipitation would result in reduced NBS and lower lake levels (Southam and Dumont, 1985; Cohen, 1986c). This result is similar to those of a majority of warming scenarios (Cohen, 1987; personal communication, Frank Quinn, 29 October 1987), including several derived from the Geophysical Fluid Dynamics Laboratory (GFDL) model projection (Manabe and Stouffer, 1980). Data points are shown in Figure 1.

Regional economic impacts of the GISS scenario would appear to be mixed, but mostly negative. There would be economic losses for hydroelectric power producers and commercial shipping. Energy demand would probably be reduced because of the shorter winter, though the effects of the longer summer on electricity use could not be discounted entirely. Agriculture in Ontario would experience wider interannual variations in productivity, unless irrigation facilities were to expand so as to reduce losses during dry years. Summer recreation would benefit from longer seasons, while winter recreation in southern areas of the region would lose its economic viability because of lack of reliable snow for skiing.

In a scenario of lower NBS, there is a potential for conflicts over water within the region. Consumptive use is already increasing because of urbanization, industrialization, and increased demands for electric power from fossil-fuel and nuclear power plants (IJC, 1985). If additional demands for irrigation water were superimposed on existing *in situ* demands (e.g., shipping), the IJC scenario of increased use and a climatic warming scenario of lower NBS (Bruce, 1984; Cohen, 1986c, 1987), then there is potential for conflicts similar to those that exist in other interstate and international basins where water shortages have previously occurred (Gleick, 1987).

Can we find an analogue that could provide some insight into possible responses to projected lower lake levels and NBS in the Great Lakes? First, let us consider the historical cases described in the previous section.

NBS and lake levels during the 1963–65 episode were similar to those projected in a majority of greenhouse warming scenarios. The episode was of relatively short duration. However, it is possible that a lengthy dispute between the states of Wisconsin and Illinois over water allocations had its roots during this period (personal communication, Stanley Changnon, 31 August 1987). At issue was whether the city of Chicago could account differently for water from two sources, Lake Michigan and groundwater aquifers. The dispute was eventually settled by the U.S. Supreme Court. The operating rules for the Chicago diversion established by this ruling, including annual withdrawals equivalent to 90 m^3/sec, are still in effect in spite of recent high lake levels.

Beyond the Wisconsin–Illinois dispute, the above low-levels episode did not include any attempt to allocate water among various riparian interests and, as such, provides no insights into how such allocations might be determined in a future greenhouse climate. However, research in progress (personal communication, Stanley Changnon, 31 August 1987) may shed some light into this matter, since the investigators intend to look at how past policy decisions may have been affected by prevailing climatic and lake-level conditions.

The 1973 and 1985–87 high-levels episodes were also of short duration. In addition, they represent an event that is opposite to the one projected in a future warmer climate. Despite these problems, could an analogy be based on these high-level episodes? For instance, it is important to recognize that the most recent of these has led to some serious reconsideration of traditional engineering approaches to management. In its present study, the IJC is examining the possibility of using policy instruments (e.g., zoning) to reduce flood damage along shorelines. However, in order to use high levels as an analogy for a future low-levels episode, one must ask the following: would institutions and the general public respond to high and low lake levels in similar fashion? The most probable answer is no.

In a high-levels episode, the greatest damage is experienced by shoreline property owners. This group may have few response options besides structural protection (e.g., walls, dikes, etc.), particularly those in major urban areas (e.g., Chicago). Meanwhile, other water users benefit from high levels (as long as they remain within the range of recent experience), and if greater demands for consumptive use would be required by agriculture or industry, few would object to increased water allocations. If low levels would prevail as the new "normal," everyone would lose, including utilities, shipping, riparians, recreational boaters, and land-based activities requiring lake water (e.g., agriculture and municipal services).

In the long term, an allocation system would have to be established, including amounts and timing of water withdrawals and return flows. Perhaps a schedule of water prices would be part of this. In case of disputes, the IJC might be asked to act as arbiter between countries, much as the U.S. Supreme Court settled interstate disputes in the 1960s and 1970s. At the moment, the U.S. and Canadian governments do not have a precedent for determining water allocations in the Great Lakes in times of low supplies. In a future warmer climate, they might have to set one.

Another possible response would be new proposals for dredging navigation channels to maintain conditions suitable for commercial shipping. Dredging tends to result in lower lake levels in the upstream lake. Historically, a dredging project has also included "compensation," which is the term used to describe a means of offsetting the effects of channel enlargements by placing ungated works in a river to restrict its flow (Korkigian, 1963). For example, when the St. Clair River between Lakes Huron and St. Clair was deepened, it was proposed that several submerged sills be constructed at the head of the St. Clair River in order to compensate for the resultant lowering of Lake Huron. Sills are horizontal rock barriers located on the river bottom. These sills would have reduced the rate of outflow, thereby maintaining levels somewhat higher than they would have been without the deeper channel downstream. When higher lake levels returned in late 1965, construction was postponed. In the greenhouse scenario, it is possible that these or other compensation works would be reconsidered in order to keep the lakes at a higher level (personal communication, Frank Quinn, 29 October 1987).

The IJC and both federal governments would also have to pay more attention to proposed diversion schemes in this scenario. Such schemes, particularly at the continental scale, have generally involved the export of Canadian water to the southwest United States. One of these, the so-called "Grand Canal" project (Bourassa, 1985), would use the Great Lakes as part of a continent-wide diversion of water from the James Bay area to the southwest United States. Some have argued that under certain economic and management conditions, such exports could be considered (Laycock, 1987). Proponents describe the benefits to be gained from the additional water, including improved water quality in the Great Lakes (see Laycock, 1987) and controlled lake levels suitable for shipping and hydroelectric power production.

There has been strong resistance in Canada and in the Great Lakes region to such plans because of their high cost, potential environmental damages, and fears that this would lead to reduced regional control of the resources in favor of "continental" interests. An example of this is the Great Lakes Charter, signed in 1985 by the governments of eight states and two provinces located in the region. The charter calls for and promotes regional unity against any large diversions that might harm the lakes and current uses of the resource (Harris, 1985). However, what would happen to this opposition to diversions, including the Grand Canal project, if warmer, drier conditions and low lake levels were to become the "norm?"

It is possible that local and regional agencies would promote increased water use efficiency and water conservation through new pricing arrangements, permit systems, technological improvements, and other means. This would enhance regional resilience to droughts and ensure that the region would have a voice in setting basin management policy. In addition, dredging may also occur on a limited scale. However, diversion remains a wild card (Glantz and Ausubel, 1984, also in this volume), and we will have to wait until the next lengthy period of low NBS and lake levels before we get a better indication of how regional, federal, and international agencies might respond to greenhouse warming.

THE FUTURE(S): FLEXIBILITY IS A KEY

Climatic variability has influenced the people of the North American Great Lakes region directly, as well as indirectly via fluctuations in lake levels and NBS. Impacts of past events and projected impacts of modeled scenarios of future climate are not severe (compared with well-known climate hazards such as the Sahelian drought), but they are regionally significant. A number of policy responses are possible, but it is difficult to identify the most likely actions in a greenhouse climate with low lake levels because there are no adequate historical analogues. The 1963–65 low-level episode was too brief. High-level events could not be used as analogues because the problem is different, and so are the likely responses.

Another complicating factor is that water resources management policies in the Great Lakes have traditionally developed in an ad hoc manner as specific issues occurred (Changnon, 1987). Future responses will depend, in part, on whether this style of basin management will change, and there is some evidence that it might.

The ongoing IJC study represents a paradigmatic shift in attitudes toward basin management. The high levels of 1985–87 were at the threshold of society's historical framework, and the feeling among participants of a recent IJC discussion session, held in January 1987 in Ontario (IJC, 1987), was that the human system would have to accommodate the natural system. The consensus was that effective remedial measures would most likely require a strategy combining both structural and nonstructural measures. This could not be done successfully without establishing a solid link between "good science" and policymakers' needs, greater participation in the decisionmaking process by key stakeholders (government, utilities, shoreline property owners, etc.), and increased efforts at educating the public. The commissioners concluded that management would have to expand its horizons and consider alternative futures that may be different from historical experience. Flexibility would be a key element. Thus, nonstructural solutions (e.g., zoning) may become important in future resource management.

Much more research is needed to resolve the uncertainties both in atmospheric modeling and the assessment of impacts due to climatic change and variability. Case studies of future scenarios, such as the one described here, provide an opportunity to identify future research needs, as well as providing good preliminary estimates that can sensitize governments to the potential problems. In addition, we must begin to think about how to model the various linkages between society and the natural environment and subsequent feedbacks that a societal response may impose on the entire system (e.g., see Titus, 1986; Chen and Parry, 1987; Parry et al., 1987). This "second generation" of studies represents an exciting challenge for researchers, policy advisors, and managers.

EPILOGUE: JULY 1988

During the first six months of 1988, much of North America, including the Great Lakes region, experienced below average precipitation. As a result of the ensuing drought, the levels of Lakes Michigan–Huron, St. Clair, and Erie declined slightly at a time of year when their levels would normally be rising (Environment Canada news release, 15 June 1988). Lake levels at the above lakes during June 1988 were 43–46 cm below June 1987 levels. Lakes Superior and Ontario experienced declines of 23 and 10 cm, respectively (*Toronto Star* report, 12 July 1988).

Meanwhile, the neighboring Mississippi River was near its lowest levels of the century. This created significant problems for commercial shipping, prompting urgent requests for increased outflows from the Chicago diversion at Lake Michigan into the Illinois River, which drains into the Mississippi River (*Toronto Star* report, 25 June 1988). The government of Illinois, a state which is in both the Great Lakes and Mississippi basins, was in favor. In response, the government of Ontario began discussion of a "Water Transfer Control Act" which would prohibit transfers of water from any of Ontario's major basins without the permission of the province's minister of natural resources (*Toronto Star* report, 30 June 1988). It is ironic that the proposed diversion to aid Mississippi shipping would probably lead to lower Great Lakes levels, and, consequently, losses for Great Lakes shipping (and other interests).

There were also a number of comments made regarding the role of water transfers in the proposed free trade agreement between Canada and the United States. Although Canadian officials insisted that there was no connection (*Toronto Star* reports, 8 and

12 July 1988), a U.S. representative from Iowa disagreed, referring to comments made by a U.S. trade official (*Toronto Star* report, 9 July 1988). Opponents of the increased diversion point out that if the greenhouse effect creates a warmer, drier climate with reduced moisture supplies, such proposals represent "the first indication of how the U.S. will instinctively search for new water supplies from the Great Lakes and Canada instead of tackling the problem locally" (*Toronto Star* report, 9 July 1988).

Politically, Lake Michigan is solely within the United States. Hydrologically, it is part of a larger international basin. No matter what decision is taken during this drought period, there will certainly be jurisdictional disputes because of the present lack of formal structure to basin management. What is needed, perhaps more than anything else, is an international agency with *broad decisionmaking authority*. Perhaps the IJC should be given a broader mandate than it now has, but that would require agreement of the various user groups (commercial, recreational, riparian, etc.) as well as local, regional, and national interests. Eventually, the citizens of the Great Lakes, through their elected representatives in both countries, will have to decide where the "buck should stop."

ACKNOWLEDGMENTS

My thanks to Stan Changnon, Doug Cuthbert, Zane Goodwin, Ralph Moulton, John O'Reilly, Stu McNair and Frank Quinn for their assistance in obtaining information, and to Doug Cuthbert, Andrej Saulesleja, and Linda Mortsch for their comments on this manuscript. The views expressed herein are my own and are not necessarily those of Environment Canada.

REFERENCES

Allsopp, T.R., P.A. Lachapelle, and F.A. Richardson, 1981: *Climatic Variability and Its Impact on the Province of Ontario*. Environment Service Report CLI 4-81. Downsview, Ontario: Canadian Climate Centre.

Bolin, B., B.R. Döös, J. Jäger, and R.A. Warrick (Eds.), 1986: *The Greenhouse Effect, Climatic Change, and Ecosystems*. SCOPE 29. Chicester, U.K.: John Wiley & Sons.

164

Bourassa, R., 1985: *Power from the North*. Scarborough, Canada: Prentice–Hall.

Bruce, J.P., 1984: Great Lakes levels and flows: Past and future. *Journal of Great Lakes Research, 10*, 126–34.

Changnon, S.A., Jr., 1987: Great Lakes policies and hydrospheric and atmospheric research needs. *Journal of Water Resources Planning and Management, 113*, 274–82.

Chen, R.S., and M.L. Parry (Eds.), 1987: *Policy-Oriented Impact Assessment of Climatic Variations*. RR-87-7. Laxenburg, Austria: International Institute for Applied Systems Analysis.

Clark, W.C. (Ed.), 1982: *Carbon Dioxide Review: 1982*. New York, NY: Oxford University Press.

Cohen, S.J., 1986a: Climatic change, population growth, and their effects on Great Lakes water supplies. *The Professional Geographer, 38*, 317–23.

Cohen, S.J., 1986b: The effects of climate change on the Great Lakes. In J.G. Titus (Ed.), *Effects of Changes in Stratospheric Ozone and Global Climate, 3*. Washington, DC: United Nations Environment Programme and U.S. Environmental Protection Agency, 163–84.

Cohen, S.J., 1986c: Impacts of CO_2-induced climatic change on water resources in the Great Lakes Basin. *Climatic Change, 8*, 135–53.

Cohen, S.J., 1987: Sensitivity of water resources in the Great Lakes region to changes in temperature, precipitation, humidity and wind speed. In S.I. Solomon, M. Beran, and W. Hogg (Eds.), *The Influence of Climate Change and Climatic Variability on the Hydrologic Regime and Water Resources*. Publication No. 168. Wallingford, Berkshire, UK: International Association of Hydrological Sciences, 489–500.

Cohen, S.J., and T.R. Allsopp, 1988: The potential impacts of a scenario of CO_2-induced climatic change on Ontario, Canada. *Journal of Climate*, in press.

Eichenlaub, V., 1979: *The Weather and Climate of the Great Lakes Region*. South Bend, IN: University of Notre Dame Press.

Environment Canada, 1986: *Historical Streamflow Summary, Ontario, to 1985.* Ottawa, Ontario: Inland Waters Directorate.

Glantz, M.H., and J.H. Ausubel, 1984: The Ogallala Aquifer and carbon dioxide: Comparison and convergence. *Environmental Conservation, 11,* 123–31.

Gleick, P.H., 1987: *The Implications of Global Climatic Changes for International Security.* Publication No. ERG-87-3. Berkeley, CA: Energy and Resources Group, University of California.

Hansen, J., G. Russell, D. Rind, P. Stone, A. Lacis, S. Lebedeff, R. Ruedy, and L. Travis, 1983: Efficient three-dimensional global models for climate studies: Models I and II. *Monthly Weather Review, 111,* 609–62.

Harris, E., 1985: Great Lakes Charter. *Focus on Great Lakes Water Quality, 10,* 13.

IGLLB (International Great Lakes Levels Board), 1974: *Regulation of Great Lakes Water Levels: A Summary Report.* Washington, DC and Ottawa, Ontario: International Joint Commission.

IJC (International Joint Commission), 1981a: *Lake Erie Water Level Study.* Report to the IJC by the International Lake Erie Regulation Study Board. Washington, DC and Ottawa, Ontario: International Joint Commission.

IJC (International Joint Commission), 1981b: *Great Lakes Diversions and Consumptive Uses.* Report to the IJC by the International Great Lakes Diversions and Consumptive Uses Study Board. Washington, DC and Ottawa, Ontario: International Joint Commission.

IJC (International Joint Commission), 1985: *Great Lakes Diversions and Consumptive Uses.* Report to the Governments of the United States and Canada. Washington, DC and Ottawa, Ontario: International Joint Commission.

IJC (International Joint Commission), 1987: *Design Exploration Discussions Regarding the Great Lakes Levels Reference.* Draft Report, 11 February 1987, discussion session held 13–14 January 1987. Alton, Ontario: International Joint Commission.

Kates, R.W., J.H. Ausubel, and M. Berberian (Eds.), 1985: *Climate Impact Assessment: Studies of the Interaction of Climate and Society.* New York, NY: John Wiley & Sons.

Korkigian, I.M., 1963: Channel changes in the St. Clair River since 1933. Proceedings of the American Society of Civil Engineers. *Journal of the Waterways and Harbors Division, 89*, WW2, May.

Larsen, C.E., 1985: Lake level, uplift, and outlet incision: The Nipissing and Algoma Great Lakes. In P.F. Karrow and P.E. Calkin (Eds.), *Quaternary Evolution of the Great Lakes.* Special Paper 30. Ottawa, Ontario: Geological Association of Canada.

Laycock, A.H., 1987: Free trade and water export. *Canadian Water Resources Journal, 12*, i–x.

MacCracken, M.C., and F.M. Luther (Eds.), 1985: *Projecting the Climatic Effects of Increasing Carbon Dioxide.* DOE/ER-0237. Washington, DC: U.S. Department of Energy.

Manabe, S., and R.J. Stouffer, 1980: Sensitivity of a global climate model to an increase of CO_2 concentration in the atmosphere. *Journal of Geophysical Research, 85*, C10, 5529–54.

NAS (National Academy of Sciences), 1987: *Current Issues in Atmospheric Change.* Washington, DC: National Academy Press.

Parry, M.L., T.R. Carter, and N.T. Konijn (Eds.), 1987: *The Impact of Climatic Variations on Agriculture.* Volume 1: *Assessments in Cool Temperate and Cold Regions.* Volume 2: *Assessments in Semi-Arid Regions.* Dordrecht, The Netherlands: Reidel.

Pratt, J., 1985: Great Lakes levels and flows conditions and related activities. Briefing, 15 November 1985. Chicago, IL: U.S. Army Corps of Engineers, North Central Division.

Quinn, F.H., 1981: Secular changes in annual and seasonal Great Lakes precipitation, 1854–1979, and their implications for Great Lakes water resources studies. *Water Resources Research, 17*, 1619–24.

Saulesleja, A. (Ed.), 1986: *Great Lakes Climatological Atlas.* Ottawa, Ontario: Ministry of Supply and Services Canada.

Southam, C., and S. Dumont, 1985: *Impacts of Climate Change on Great Lakes Levels and Outflows.* Unpublished report available from Inland Waters/Lands Directorate. Burlington, Ontario: Environment Canada.

Stewart, C.J., and P. Yee, 1987: *Interim Report on the 1987 High Water Levels of the Great Lakes, January to June 1987.* Available from Inland Waters/Lands Directorate. Burlington, Ontario: Environment Canada.

Titus, J.C. (Ed.), 1986: *Effects of Changes in Stratospheric Ozone and Global Climate.* Washington, DC: United Nations Environment Programme and U.S. Environmental Protection Agency.

WMO (World Meteorological Organization), 1986: *Report of the International Conference on the Assessment of the Role of Carbon Dioxide and of Other Greenhouse Gases in Climate Variations and Associated Impacts.* 9–15 October 1985, Villach, Austria. WMO No. 661. Geneva, Switzerland: World Meteorological Organization.

8

The Rising Level of the Great Salt Lake: An Analogue of Societal Adjustment to Climate Change

Peter M. Morrisette

INTRODUCTION

How will society respond and adjust to a greenhouse-induced climatic change? This is a difficult question to pursue. It involves predicting the future with respect to both the effects and impacts of a greenhouse-induced climatic change as well as predicting societal response. This task is made particularly difficult by the uncertainties that surround the greenhouse question. There is, for example, uncertainty about the rate at which greenhouse gases are increasing, the impacts of greenhouse gases on the atmosphere, and how the global environment might respond to increased greenhouse gases. In addition, there will likely be difficulty in planning for the long-term effects of greenhouse-induced climatic change given the uncertainty about other potential changes in the climate and environment, and more important, in social institutions. Schelling (1983, 453), in fact, argues that the more perplexing uncertainties do not involve physical aspects of weather and climate, rather they are centered around the question of "what the world will look like."

Clearly, given these uncertainties, a reliable prediction of how society will respond to a greenhouse-induced climatic change is

Peter M. Morrisette is a postdoctoral fellow at the National Center for Atmospheric Research working with the Environmental and Societal Impacts Group. He has written several articles on the response of decisionmakers during the early and mid-1980s to the rising level of Utah's Great Salt Lake. His current research is focused on policy responses to ozone depletion and global warming.

neither practical nor possible. However, it is possible to look at actual cases of how society responds to the problems and impacts of fluctuations in climate. While these actual cases are not blueprints to how society will respond to the more complicated problems and issues that are likely to be characteristic of greenhouse-induced climatic change, they may provide useful insight.

The rising level of Utah's Great Salt Lake is such a case. The Great Salt Lake has risen 12 ft since 1982, reaching a new historic high level in 1986, and resulting in over $300 million in damages and incurred costs. This unprecedented rise in the lake level is directly related to increased precipitation in northern Utah, and it has forced resource managers and policymakers to address the issue of variability in lake levels and in climate. This chapter analyzes the response of decisionmakers to the rising level of the Great Salt Lake in the context of how society adjusts to climatic stress. It argues that decisionmakers have long assumed that the lake was relatively stable. As a result of this perception, decision-makers have favored short-term structural measures for protecting threatened facilities and controlling the level of the lake, rather than more flexible long-term strategies for adapting to variability in lake levels and avoiding future impacts. The assumption has been that normal conditions will return, and thus long-term adjustment is not necessary.

THE ANALOGY

The rising level of the Great Salt Lake is not a physical analogue to greenhouse-induced climatic change. The unusually high precipitation of the past several years is clearly an anomalous event; however, it is not unprecedented in the extended record for the lake's drainage basin (Karl and Young, 1986). Thus it is not at all clear that the climate has changed, rather the recent heavy precipitation is probably more indicative of the variable nature of the climate in the Great Basin. Furthermore, it is possible that a greenhouse-induced climatic change will actually result in a decrease in precipitation (or perhaps in higher evaporation rates) for the Great Basin, and thus result in lower rather than higher lake levels (see Kellogg and Zhao, 1988 for a review of potential precipitation and soil moisture changes from greenhouse-induced warming).

The analogy presented in this chapter between the rising level of the Great Salt Lake and a greenhouse-induced climatic change is one of how society responds to climatic stress. More specifically, it is concerned with how society responds to fluctuation and change in the environment resulting from a change or shift in climatic conditions. Between 1982 and 1986 resource managers and policymakers in Utah were confronted with an unprecedented rise in the level of the Great Salt Lake. Decisionmakers are now forced to deal with the task of adjusting to a lake level that is significantly more variable than previously thought. Similarly, with a greenhouse-induced climatic change, resource managers will be confronted with the need to adjust to extreme conditions not previously experienced or anticipated.

Furthermore, despite the high awareness and visibility surrounding the greenhouse problem, it is likely that the initial impacts of greenhouse-induced climatic change will not be readily perceived. That is, resource managers will likely perceive the initial impacts as anomalous events that may necessitate a short-term adjustment but will likely not require any long-term change in strategy. Only after a change in climatic conditions has persisted and impacts accumulated are resource managers likely to consider the long-term aspects of the problem and its possible greenhouse connections. Thus the argument in this analogy is that on a local or regional scale, response and adjustment to the impacts of a greenhouse-induced climatic change may not be all that different from how society presently adjusts to problems of climatic variability and stress—that is, adapting to impacts on a short-term incremental basis.

HOW SOCIETY ADJUSTS TO CLIMATIC STRESS

In adjusting to climatic stress, society has three general options—(1) prevent impacts before they occur, (2) implement countervailing or compensating measures to offset impacts, or (3) adapt to impacts as they occur (Mann, 1983; Schelling, 1983; Glantz and Ausubel, this volume). Mann (1983) explains that both preventive and curative or compensating strategies impose high costs on society early in the adjustment process. Major costs are imposed in a relatively short period of time before serious impacts are incurred. On the other hand, adaptive strategies assume that existing technologies and societal institutions can deal with whatever changes occur. Costs are spread over a longer time period, and individuals and groups are for the most part allowed to determine the means of adjustment (Mann, 1983). Each of these strategies imposes costs; they differ, however, in how they deal with those costs.

Policies and strategies designed to prevent or compensate for the impacts of climatic stress will likely involve major change and innovation in existing policies (Schelling, 1984; Mann, 1983). Such change requires convincing policymakers of both the seriousness of the problem and the urgent need for action. Even if policymakers are convinced of the seriousness of the problem, new and innovative policy will still be difficult to formulate and implement given the uncertainty, and the potential high costs and risks associated with such policy changes. Thus adapting to impacts as they occur is most often the favored response. It allows society to postpone costs into the future by emphasizing the short term, and therefore it is the path of least resistance (Glantz and Ausubel, this volume).

Furthermore, policies designed to adjust to the impacts of climatic stress and change as they occur can be formulated more slowly and evolve incrementally, and thus would be less constrained by existing parameters of policymaking (Mann, 1983). In fact, the traditional approach to environmental problems is incremental or "muddling through" (Glantz, 1979; Lindblom, 1959). Such an approach usually relies on past experience and small change in existing policies. Furthermore, decisionmakers tend to rely on traditional measures or strategies for solving a problem. Such solutions are often satisfactory in the short run, but they are far from the ideal solution in the long run (Brooks, 1986). They do little to prepare society for future stress.

Even when confronted with the reality of the need for long-term adjustment, society is resistant to changing existing use patterns and management practices. Instead of adapting to environmental change and building flexibility into the system of response and adjustment, society often relies on technical fixes and rigid short-term structural solutions to reduce or eliminate the negative impacts of variability and maintain the operational status quo (Brooks, 1986; Holling, 1986; Morrisette, 1987, 1988). In the case of the rising level of the Great Salt Lake, this study argues that such an approach is fostered by the belief that normal (i.e., stable) conditions will prevail in the long run, and as a result long-term (preventive or compensating) adjustment is not seen as necessary.

THE LAKE-LEVEL RISE PROBLEM

The Great Salt Lake is a terminal saline lake located in an enclosed drainage basin in northern Utah (Figure 1). The modern-day lake is a remnant of Pleistocene Lake Bonneville. The normal level of the lake, defined as the historic average (based on a 140-year record), is approximately 4,200 ft (Arnow, 1984). Under assumed normal or equilibrium conditions, the level of the lake should be maintained by a relative balance between inflow (precipitation and runoff) and outflow (evaporation). However, the level of the Great Salt Lake during the period of historic record has been anything but stable. The historic record for the Great Salt Lake is shown in Figure 2. The secular trends evident in this historic record are the result of imbalances in the lake's equilibrium conditions. In June 1986, the Great Salt Lake reached an historic high of 4,211.85 ft, surpassing the old mark of 4,211.6 ft reached in 1873. The historic record low level for the lake of 4,191.35 ft occurred in 1963. In the past 24 years the lake has risen 20.5 ft. The most dramatic rise, however, has occurred since the fall of 1982. Between the fall of 1982 and the summer of 1986 the lake recorded a net increase of 12 ft. This included seasonal rises of 5.1 ft (the most extreme on record), 5.0 ft, 2.05 ft, and 3.5 ft in 1983, 1984, 1985, and 1986 respectively. The lake peaked at a record high level of 4,211.85 ft in June of 1986, and again reached that level in 1987. With the onset of drier conditions and the implementation of the west desert pumping project, the lake declined to 4,209.45 ft in November of 1987.

Recent research on the climate of northern Utah has indicated that there is a direct relationship between precipitation in the Great Salt Lake drainage basin and lake-level rise (Kay and Diaz, 1985; Karl and Young, 1986). Furthermore, the climate record for northern Utah indicates that the Great Salt Lake Basin has been in a wet period for some years, with the last several years being among the wettest on record (Figure 3). Using data from ten index stations in the Great Salt Lake Basin, annual precipitation for the period 1980–1986 has been 129, 112, 150, 180, 133, 107, 129 percent

174

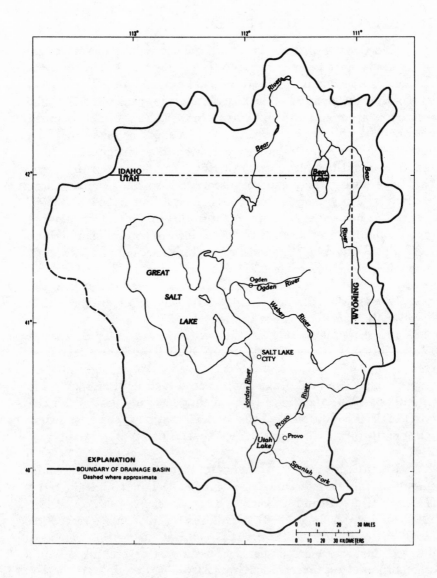

Figure 1. Great Salt Lake drainage basin (Arnow, 1984).

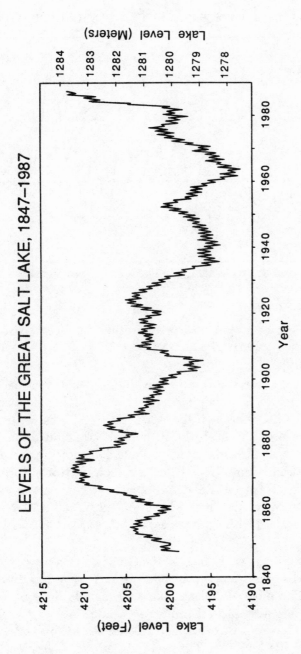

Figure 2. Historic lake-level record for the Great Salt Lake.

GREAT SALT LAKE BASIN PRECIPITATION, 1951-86

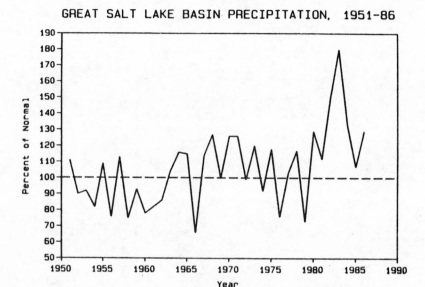

Figure 3. Percent of normal precipitation in Great Salt Lake Basin, 1951–86.

of normal respectively (normal is the 1951–1980 30-year average). Karl and Young (1986), using a record for the Salt Lake drainage from 1864 to 1984, identified the wet spell of 1981–84 as the most extreme on record. However, the authors also concluded that while the current wet spell is highly anomalous, it is not unprecedented in the long-term climate record for the region. Not surprisingly, runoff into the Great Salt Lake in the past several years has shown a similar pattern to that of precipitation—265, 330, 190, and 300 percent of normal for the water years 1983, 1984, 1985, and 1986, respectively.

Since 1982, the rising level of the Great Salt Lake has resulted in a diverse set of impacts to both public and private facilities and activities. The most significant impacts include flooding of and damage to lakeshore mineral industries, major highways and railroads which pass near or over the lake, and recreation and wildlife areas around the lake (see Figure 4). In addition, residential ar-

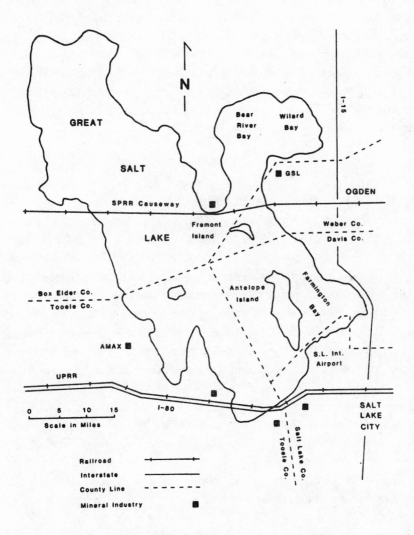

Figure 4. Great Salt Lake and vicinity and facilities threatened by rising lake levels.

eas around Salt Lake City and Salt Lake International Airport are threatened by rising lake levels. Since 1982, damages and capital investments from lake flooding have exceeded $300 million (U.S. Army Corps of Engineers, 1986). In an attempt to protect threatened facilities, state and local authorities have diked and raised large sections of major highways such as Interstate 80, as well as diking other public facilities such as waste water treatment plants. In addition, private interests such as the railroad and mineral companies have undertaken extensive measures to protect their threatened facilities (Utah Division of Comprehensive Emergency Management, 1985).

RESPONSE AND ADJUSTMENT TO THE LAKE-LEVEL RISE PROBLEM

In general, the State of Utah's response to the lake-level rise problem has been one of wait-and-see and incremental adjustments. The emphasis has been on short-term "structural fixes" rather than long-term adaptive strategies. As the lake began to rise rapidly in the fall of 1982, resource managers became concerned with the potential problems from high lake levels. However, it was widely believed that the lake would soon recede. The problem here was that resource managers and policymakers had long viewed the Great Salt Lake as a relatively stable system (see, for example, State Road Commission of Utah, 1958). Past experience with the lake and efforts in the 1970s at modeling the lake's hydrologic system identified a level of 4,202 ft as the most likely high level of the lake for long-term planning purposes (Utah Division of Great Salt Lake, 1976; Utah Division of Water Resource, 1984). Resource managers and policymakers felt that sudden and persistent lake-level rise was not possible. Thus existing resource management strategies were not capable of dealing with the rapid rise in the level of the Great Salt Lake between 1982 and 1986. A chronology of important policy and management decisions concerning the Great Salt Lake is provided in Table 1.

Table 1
Chronology of major social and political response
to Great Salt Lake level change 1950s to 1987

Date	Lake Level	Social and Political Response
1950s–1960s	Declining lake level Historic low of 4,191.35 ft in 1963, followed by slowly rising lake	Development encroaches on lake's floodplain. Plans are proposed for developing the lake as an economic and recreational resource.
1973	4,200.45 ft Rising lake	Legislature establishes policy-advisory committee to prepare long-term management plan for the lake.
1975	4,201.55 ft Rising lake	(1) U.S. Supreme Court granted ownership of the lakebed and shoreland to the state of Utah. (2) House Bill No. 23 established Division of Great Salt Lake (DGSL) and directed it to develop a comprehensive plan for the lake.
1976	4,202.25 ft Rising lake prior to 1976; declining lake after 1976	(1) Comprehensive plan released outlining management goals including lake-level control and floodplain management. (2) Ongoing planning process outlines options for controlling lake levels. Hydrological models suggest narrow range of potential lake levels.
1979	4,199.90 ft Declining lake	Legislature eliminates DGSL. Division of State Lands and Forestry (DSLF) assumes management responsibility for the lake. Comprehensive plan is never implemented. Legislature mandates that the lake be maintained below 4,202 ft.
1983	4,204.70 ft Rising lake with record rise of 5.1 ft for a single season	(1) In January the DSLF released a contingency plan for the lake outlining management options and predicting a 1983 peak of 4,203 ft.

Table 1, continued

Date	Lake Level	Social and Political Response
		(2) Legislature directed DSLF to develop strategies to maintain the lake below 4,202 ft rescinding its 1979 mandate.
		(3) State and local agencies and private interests are forced to implement emergency mitigation efforts.
		(4) Damage from flooding statewide (including damages from lake-level rise) totaled $478 million. Federal disaster declared.
1984	4,209.25 ft Rising lake	(1) Emergency mitigation efforts continue and a second federal disaster is declared. State investigates an array of lake-level control strategies.
		(2) Southern Pacific Railroad causeway is breached lowering the south arm of the lake by one foot.
		(3) Federal government rules that existing authorities do not permit their involvement in large-scale flood mitigation efforts.
1985	4,209.90 ft Rising lake	(1) Scientific workshop held to assess future lake-level and climate trends.
		(2) Delineation of the lake's shoreland as a Beneficial Development Area (BDA) is proposed by Division of Comprehensive Emergency Management and FEMA.
		(3) Senate Bill 97 allocates $96 million for lake flooding mitigation efforts. Final allocation of funds was to be made during a special session of the legislature in June. Lake-level rise problem moderates and the special session is canceled. Only $20 million is actually spent on mitigating the lake-level rise problem.
		(4) Planning on lake-level control options continues.

Table 1, continued

Date	Lake Level	Social and Political Response
1986	4,211.85 ft New historic record high level; lake predicted to continue rising	(1) With rising lake levels the legislature is forced to schedule a special session in May to resolve the issue. (2) After lengthy debate, the legislature funds the $60 million west desert pumping project which had been endorsed by the governor and the state's resource management agencies. (3) Local governments implement some zoning restrictions; however, the BDA is not implemented.
1987	4,211.85 ft Same level as previous year; with drier conditions prospects look good for declining lake levels.	West desert pumping becomes operational in the late spring; however, with moderating lake level conditions the project becomes a political liability for its supporters.

The Assumption of Stability

Schemes for managing the Great Salt Lake can be traced back to the 1930s. However, it was not until the 1950s and 1960s that efforts at lake management and development began to receive serious attention (see Wasatch Front Regional Council, 1973). The 1950s and 1960s was a period of both relatively low lake levels (the lake reached its historic low in 1963), and a period of extensive development and growth in northern Utah. Thus the shoreland around the lake was experiencing great development pressure. The goal of these early attempts at lake management was the development of the lake and its shoreland. Lake-level fluctuation was not considered a problem. In fact, a report issued by the State Road Commission of Utah (1958) predicted that the lake would never again reach the 4,205 foot level. The principal goal of these early management efforts was to develop a large freshwater lake or embayment along the east shore of the lake. It was thought that a freshwater embayment would provide important recreational and commercial opportunities. It should be noted that the idea of an east shore embayment would surface again in the 1980s as a solution to the problem of high lake levels.

The first effort at long-term coordinated planning for the lake began in the early 1970s with the establishment of a comprehensive planning process for the Great Salt Lake. This comprehensive planning process began formally in late 1973 when the state legislature established a policy-advisory committee to prepare a long-term management and development plan for the Great Salt Lake (Christensen and Searle, 1974). In 1975, the Utah State legislature passed House Bill No. 23, which established the Division of Great Salt Lake and the Great Salt Lake Board within the Utah Department of Natural Resources, and empowered them to develop and maintain a comprehensive plan for the Great Salt Lake (Utah Code 65–8a). Prior to the formation of the Division of Great Salt Lake, management of the lake occurred in an uncoordinated, piecemeal manner under the direction of various agencies of the Department of Natural Resources. Two factors gave rise to the comprehensive planning process and the formation of the Division of Great Salt Lake: (1) the recognition that a coordinated, comprehensive approach was necessary for the long-term development and management of the lake, and (2) the granting of ownership of the lakebed and shoreland to the State of Utah by the U.S. Supreme Court in 1975 (Burnham, 1980; Jones et al., 1976). In addition, the lake had been rising since 1963, and in 1973 rose above the 4,200 foot level for only the second time since 1930; thus there was some concern about potential damage resulting from a rising lake level.

The comprehensive plan developed by the Division of Great Salt Lake was intended as a general framework on which to base decisions for the management and use of the Great Salt Lake. The plan included several important policy statements regarding lake levels. In particular, the plan called for identification of the Great Salt Lake's floodplain and its recognition as a hazard zone. The plan also called for investigation of "the physical, economic, and political feasibility of procedures which would be used to control the level of the Great Salt Lake," and stated that until such measures were adopted "the comprehensive plan and the ongoing planning process will recognize varying lake levels resulting from nature's wet-year and dry-year cycles" (Utah Division of Great Salt Lake, 1976, 8).

However, despite these seemingly innovative policy statements regarding floodplain delineation and the recognition of lake-level fluctuations, the comprehensive plan suggested that lakeshore users plan for lake-level fluctuations of between only 4,192 and 4,202 ft. This represented a range of potential lake-level fluctuations between the historic low level and a level slightly above the historic average. However, this narrow range of potential lake-level fluctuations was supported by both the past experience of resource managers, most of whom had never seen the lake above 4,202 ft, and more important, by computer models of the lake's hydrologic system which suggested that a lake level above 4,202 ft was highly improbable (Jones et al., 1976; Utah Department of Development Services, 1977; Stauffer, 1980). These computer models, however, used only a limited part of the historic lake-level record; in some cases only the last 30 to 40 years of record. Shorter data sets were used because they were believed to be more reliable and to provide a more accurate assessment of the modern lake and future lake-level conditions.

Thus it seemed reasonable for resource managers to believe, based on past experience and the most recent studies, that a lake level above 4,202 ft was highly unlikely. However, the 4,202 foot level was also notable because it was the level at which significant and widespread impacts would occur. The 4,202 foot level marked a transition between less developed and more developed shoreland. Thus to claim that the lake would fluctuate between 4,192 and 4,202 ft was not a threatening statement implying the need for shoreline protection. This range of lake levels essentially represented what resource managers and lakeshore users had experienced as the lake's "natural" or "traditional" floodplain. However, to suggest or imply that the lake would not exceed this level was clearly wishful thinking.

Despite its shortcomings, the 1976 comprehensive plan was a relatively complete assessment of both short- and long-term management goals for the lake. However, the plan was never implemented. No floodplain or hazard zones were identified, and no lake-level control strategies were adopted. In 1979, the Division of Great Salt Lake was dissolved. The elimination of the division was due to two factors: (1) the belief that it was no longer needed (the comprehensive plan had been prepared and the lake level had been declining since 1976), and (2) political conflicts within state government. More important, however, the elimination of the division was indicative of the state's lack of concern over the lake. While the duties and goals of the division were transferred to another division within the Department of Natural Resources (Division of State Lands and Forestry), the impetus behind the comprehensive management process of the early and mid-1970s was undoubtedly lost.

Decisionmakers Opt to "Muddle Through" the Current Crisis

In 1982, despite the planning efforts of the previous ten years, the State of Utah was completely unprepared to deal with a rapid and persistent rise in the level of the Great Salt Lake. The events of 1983 and 1984, two consecutive years in which the lake rose five feet, were simply not thought possible. Thus state and local governments were forced to implement emergency flood mitigation measures. Included among these measures were the raising and diking of sections of Interstate Highway 80, and the diking of critical public facilities such as wastewater treatment plants and drain canals. The state was also forced to abandon lakeshore recreation areas and waterfowl management areas. In addition, the mineral and railroad companies were spending millions of dollars building dikes to protect their facilities.

As the lake-level rise problem persisted, the State of Utah has reluctantly been forced to deal with it. Major lake-level control and flood mitigation projects were beyond the means of local governments, and the federal government refused to assume any major responsibility (Federal Emergency Management Agency, 1986; U.S. Army Corps of Engineers, 1986). Great Salt Lake flooding, however, was included in federal disaster declarations for Utah in both 1983 and 1984, and thus the state received some disaster funding and assistance through FEMA. In addition, the state was receiving some funds for diking from the U.S. Army Corps of Engineers. In general, however, the state has been forced to assume the financial responsibility for the lake-level rise problem. The state has reviewed a variety of structural lake-level control and flood mitigation strategies (see Utah Division of Water Resources, 1986). However, these alternatives were controversial; the legislature was uncertain whether lake-level control was either necessary or the responsibility of the state. As a result, the state opted to take a wait-and-see approach. Decisionmakers were hoping that the lake would soon recede and therefore were not willing to take any action that was not immediately necessary.

The first measure taken by the state to control the level of the Great Salt Lake was the breaching of the Southern Pacific Railroad Causeway in the summer of 1984 (Utah Division of Comprehensive Emergency Management, 1985). The railroad causeway divides the lake into a smaller northern arm and a larger southern arm. Because most of the inflow occurs in the southern arm of the lake, the effect of the causeway has been to create higher brine concentrations in the northern arm of the lake and a higher lake level in the southern arm. The causeway breach had been suggested as a lake-level control strategy in the 1976 comprehensive plan; however, it was not implemented at that time. Initially there was strong opposition to the breach on the part of mineral companies who pumped brine from the north arm, and by the legislature who questioned its need and cost. However, opposition to the breach softened as the lake continued to rise. The causeway breach, however, had only a one-time effect on the level of the lake. It lowered the south arm of the lake one foot at a cost of $3.7 million. Thus the state was still forced to deal with the problem of rising lake levels.

In 1985, after two years of unprecedented lake-level rise and nearly two hundred million dollars in damages, the legislature authorized $96 million for flood mitigation and lake-level control measures (Utah Code 73–10e). However, the lake-level rise problem moderated substantially during the spring of 1985, peaking far below its predicted level. Thus only $20 million was actually spent (State of Utah Office of Legislative Research and General Counsel, 1986). With an apparent moderating in the lake-level rise, concern among politicians for the problem nearly disappeared, and the legislature backed away from taking any further action regarding the lake level. Decisionmakers had opted to "muddle through;" they were not willing to take any action other than what was immediately necessary to mitigate an impending crisis, and in 1985 there was no apparent crisis.

The Choice of a Rigid Solution

In 1986, the state legislature did not plan to address the issue of lake-level control; however, when it became apparent that the lake was going to rise substantially due to extremely heavy late winter and spring precipitation, the governor and the legislature were forced to respond. Past adjustments had bought time in the hope that the lake would soon recede. However, the lake continued to rise, and damages became both more substantial and widespread. A special session of the Utah Legislature was scheduled in May of 1986 to address the questions of lake-level control and flood protection, and again the controversy resurfaced over whether the state should take action, and if so, what action.

State resource management agencies had for several years been considering an array of structural lake-level control and flood mitigation measures. Included among these were plans to increase upstream storage on the Bear River (the lake's principal tributary), the diversion of water from the Bear River into the Snake River in Idaho, the widespread construction of protective dikes along the shore of the lake, two different plans to create east shore embayments, and a proposal to pump water from the lake into a large evaporation pond in the desert west of the lake (Utah Division of Water Resources, 1986).

After considerable debate and compromise, the legislature chose to implement the west desert pumping project which had been endorsed by the governor and the state's resource management agencies as the most cost-effective project. However, many in the legislature favored the creation of a large east shore embayment. The project, referred to as inter-island diking, involved the construction of deep water dikes which would connect the south shore of the lake to Antelope Island, Fremont Island and the north shore of the lake (see Figure 4). This proposal was favored by the heavily developed and politically powerful east shore counties because it offered the east shore of the lake both direct protection from rising lake levels and the possibility of economic and recreational development. However, the project did nothing to protect the south and west shores of the lake, and in fact left these unprotected areas more vulnerable to future lake-level fluctuations. The impetus behind inter-island diking was not lake-level management and flood mitigation, rather it was lake development. Support for the project was also indicative of a wider belief among politicians that any lake-level management project should produce economic benefits in addition to lake-level control. Many in the legislature found it difficult to support a project such as west desert pumping which had the single purpose of wasting water. Furthermore, many legislators still questioned the need for state action and involvement. The compromise reached by the legislature allowed for the funding and construction of the west desert pumping project and provided funding for final design and technical reports on the inter-island diking alternative.

The idea behind the west desert pumping project is to pump water from the lake into a large evaporation pond that would be created in the desert west of the lake, thus reducing the level of the lake through increased evaporation (see Figure 5). The project also includes a return canal which would bring highly concentrated brine back into the lake. It is estimated that the project should increase evaporation by 820,000 acre-feet annually, and under normal climatic conditions, reduce the level of the lake by 12.8 in during the first year of operation and 6.6 in in each additional year (U.S. Army Corps of Engineers, 1986). The cost of constructing the project was $60 million, and it became operational in the late spring of 1987.

However, it is dangerous to consider the west desert pumping project the solution to the lake-level rise problem. As a structural control measure, the project has important limitations. For example, it operates in a limited range of lake-level elevations, between 4,208 and 4,215 ft. In addition, if input into the lake is above 200 percent of normal, the lake will continue to rise (input was above 200 percent of normal in 1983, 1984, 1985, and 1986). If the level of the lake exceeds 4,215 ft, the pumping station will be flooded and the lake will begin to flow into a series of basins to the west of the lake (including the basin used for the west desert pumping project). While west desert pumping will help to alleviate the

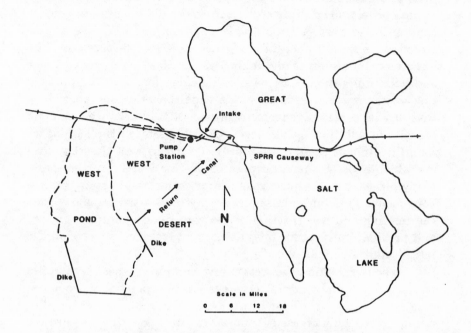

Figure 5. West desert pumping project.

problem, the fact remains that if the precipitation pattern of the recent past persists, the lake will continue to rise.

Yet in endorsing the project many resource managers and policymakers see the west desert pumping project as a long-term solution. However, in essence the west desert pumping project is only another incremental adjustment designed to mitigate the potential crisis that would result if the lake were to rise another foot or two. The danger is that the project might create a false sense of security, and thus increase societal vulnerability to lake-level variability in the future (Morrisette, 1987). The west desert pumping project is an external control designed to reduce fluctuations in the level of the lake. Thus it increases societal dependence on controlling, rather than adapting to lake-level variability. With the west desert pumping project as the sole solution to the lake-level rise problem, the potential consequences of failure are high.

Since the implementation of the west desert pumping project in the spring 1987, drier conditions have prevailed over northern Utah and the lake-level rise problem has moderated considerably. The lake fell to 4,209.45 ft in November 1987 (approximately one foot of the 1987 decline has been attributed to the west desert pumping project), and as of July 1988 the lake had declined to 4,208.70 ft. With drier conditions and with the lake apparently receding, the west desert pumping project has received much criticism. The principal concern is not over the effectiveness of the project, rather it centers on whether the project was necessary in the first place. To many, the crisis conditions and record high lake level of 1986 are only a distant memory and the lake no longer represents a problem. However, the lake has clearly demonstrated its capability to rise rapidly. A return to wetter conditions could once again create a crisis situation.

The Option of Flexible Land-Use Planning

The state also investigated one flexible non-structural approach to the lake-level rise problem. This approach involves designating a floodplain or what the state is calling a Beneficial Development Area (BDA) around the lake (Utah Division of Comprehensive Emergency Management, 1985). The BDA would include all shoreland below an elevation of 4,217 ft. Development within the BDA would be restricted to avoid additional flood losses if the lake continues to rise or should rise again in the future. The BDA concept is perhaps the best approach to long-term planning, and it is clearly the most flexible long-term strategy yet suggested for mitigating the lake-level rise problem. While the BDA concept would do little to help existing developments in the lake's floodplain which have already been impacted, it could help curtail potential future losses. Furthermore, the BDA is the only alternative which recognizes the lake's potential for fluctuation based on a lake-level record longer than just the historic record of the past 140 years. The 4,217 foot level has been identified as a relatively stable high stand for the lake, reached at least two times during the past five hundred years (Mckenzie and Eberli, 1985).

The BDA, however, has only been implemented in a scaled down fashion. While the BDA has received general support at the local level, local governments have been reluctant to accept the full provisions of the plan, and are reluctant to surrender their independence and authority over land use. In addition, some local governments are reluctant to do anything that would restrict development on valuable industrial lands adjacent to the lake. In addition, the BDA has received little support from the governor's office or the state legislature, both of which would rather support a structural solution to the problem than restrict development around the lake. Thus the BDA as a unified approach to lakeshore development has not been implemented.

Summary

In responding to the lake-level rise problem, resource managers and policymakers have opted to adapt to the immediate impacts of rising lake levels. Given past decisions and the fact that they assumed that the level of the lake was relatively stable, decisionmakers have had no other choice (see Table 1 for a chronology of important decisions concerning the Great Salt Lake and lake-level variability). The set of alternatives that the state most seriously considered were the more traditional structural solutions that resource management institutions normally turn to in times of crisis. The west desert pumping project as such a traditional structural measure represents only the latest and most substantial step in a set of incremental adjustments to the problem of rising lake levels. The long-term view of the problem has nearly been lost because of the pressing need to find an immediate solution, and the belief that normal conditions will prevail in the long-run. Innovative ideas designed to prevent future impacts, such as the BDA concept, have not received much attention because they are based on a long-term perspective and do little to mitigate the immediate problem of preventing damages to existing facilities. Decisions concerning the lake are complicated and risky, and the future is uncertain. Thus policymakers have chosen to avoid the long-term implications of the problem by responding only to immediate crises.

LESSONS FROM THE GREAT SALT LAKE CASE

A key question raised by this chapter is whether society can correctly and readily identify the onset of and impacts from a greenhouse-induced climatic change, and respond accordingly; or whether the impacts of a greenhouse-induced climatic change will simply be perceived as an anomalous climate event requiring short-term adjustments but not long-term changes in action. This is a difficult question to answer; however, given the uncertainty that presently surrounds the greenhouse issue, particularly concerning regional impacts, it is likely, despite the high awareness of the greenhouse issue, that many regional and local impacts will not be immediately recognized. Thus society will likely adapt incrementally as needed to the impacts of a greenhouse-induced climatic change (at least initially). This is how society often responds to problems of climatic stress, and it is clearly how decisionmakers have chosen to respond to the rising level of the Great Salt Lake.

However, the Great Salt Lake case study offers perhaps a more subtle and compelling analogy to the greenhouse problem. In the Great Salt Lake case, resource managers chose to ignore the long-term implications of rising lake levels despite a growing awareness that lake-level variability, and particularly high lake levels, are problems that will likely plague future generations as well. Instead, decisionmakers have responded only to the immediate crisis. While the lake is likely to recede in the future, to assume that it will not rise again is wishful thinking. The historical record indicates otherwise. Yet decisionmakers have only been willing to do what is necessary to mitigate the immediate problem and maintain the operational status quo, and have opted not to take action to prevent future impacts.

If the long-term goal in adjusting to climatic stress is to prevent or avoid future adverse impacts, then the response of resource managers and policymakers to the rising level of the Great Salt Lake is not encouraging. In a case where cause and effect, impacts and adjustments, and costs and benefits are well understood, decisionmakers have opted not to take action to prevent or avoid future impacts.

Society is clearly reluctant to change traditional ways of doing things despite warnings about impending crises. White (1986), for example, has noted this with the long-term response of farmers to the Dust Bowl of the 1930s on the Great Plains. In a recent assessment of adjustments to the 1930s drought, White (1986, 92) notes that the ability or willingness of society to change management processes and practices in the face of catastrophe was completely misjudged and overestimated. Similarly, with the rising level of the Great Salt Lake, decisionmakers are reluctant even to change existing use patterns to avoid future impacts and catastrophe. With a greenhouse-induced climatic change, where the cost of adjustments will be high, the task of implementing preventive and compensating policies will be enormous and difficult, particularly given the resistance to institutional change.

CONCLUSION

This chapter has discussed the response of decisionmakers to the rising level of the Great Salt Lake as an analogue of societal response to a greenhouse-induced climatic change. Clearly there will be significant differences between the recent rapid rise in the level of the Great Salt Lake and the problems and impacts that will likely be associated with the regional aspects of a greenhouse-induced climatic change—differences concerning onset, magnitude, and long-term consequences and impacts. Nevertheless, this analysis is still useful in understanding how society responds to climatic stress. While a direct analogy does not exist, and is not offered, both issues deal with how society responds to flux and change in the environment. In this context, it is hoped that this analysis of the response of decisionmakers to the rising level of the Great Salt Lake provides insight into the difficulties of formulating and implementing policy for dealing with long-term regional problems of environmental fluctuation and change that will likely accompany a greenhouse-induced global climatic change.

REFERENCES

Arnow, T., 1984: *Water-level and water-quality changes in Great Salt Lake, Utah 1847-1983.* U.S. Geological Survey Circular 913. Washington, DC: Government Printing Office.

Brooks, H., 1986: The typology of surprises in technology, institutions, and development. In W.C. Clark and R.E. Munn (Eds.), *Sustainable Development of the Biosphere.* Cambridge, UK: Cambridge University Press, 325–48.

Burnham, O.W., 1980: The Great Salt Lake comprehensive plan. In J.W. Gwynn (Ed.), *Great Salt Lake: A Scientific, Historical and Economic Overview.* Utah Geological and Mineral Survey Bulletin 116. Salt Lake City, UT: Utah Geological and Mineral Survey, 47–51.

Christensen, J.G., and R.T. Searle, 1974: *The Great Salt Lake: A Special Report for the Utah Legislature.* Salt Lake City, UT: Great Salt Lake Policy Advisory Committee.

Federal Emergency Management Agency, 1986: *Post-Flood Recovery Progress Report.* FEMA–720–DR–Utah, 14 March 1986. Denver, CO: Federal Emergency Management Agency.

Glantz, M.H., 1979: A political view of CO_2. *Nature, 280,* 189–90.

Holling, C.S., 1986: The resilience of terrestrial ecosystems: Local surprise and global change. In W.C. Clark and R.E. Munn (Eds.), *Sustainable Development of the Biosphere.* Cambridge, UK: Cambridge University Press, 292–317.

Jones, C.T., C.G. Clyde, and J.P.Riley, 1976: *Management of the Great Salt Lake: A Research Plan and Strategy.* PRWG 195–1. Logan, UT: Utah Water Research Laboratory.

Karl, T.R., and P.J. Young, 1986: Recent heavy precipitation in the vicinity of the Great Salt Lake: Just how unusual? *Journal of Climate and Applied Meteorology, 25,* 353–63.

Kay, P.A., and H.F. Diaz, 1985: *Problems of and Prospects for Predicting Great Salt Lake Levels.* Papers from a conference held in Salt Lake City 26–28 March 1985. Salt Lake City, UT: University of Utah, Center for Public Affairs and Administration.

Kellogg, W.W., and Z.C. Zhao, 1988: Sensitivity of soil moisture to doubling of carbon dioxide in climate model experiments: Part 1, North America. *Journal of Climate, 1,* 348–66.

Lindblom, C.E., 1959: The science of "muddling through." *Public Administration Review, 19,* 79–88.

Mann, D., 1983: Research on political institutions and their response to the problem of increasing CO_2 in the atmosphere. In R.S. Chen, E. Boulding, and S.H. Schneider (Eds.), *Social Science Research and Climate Change.* Dordrecht, The Netherlands: D. Reidel Publishing, 116–45.

Mckenzie, J.A., and G.P. Eberli, 1985: Late Holocene lake-level fluctuations of the Great Salt Lake (Utah) as deduced from oxygen-isotope and carbonate contents of cored sediments. In P.A. Kay and H.F. Diaz (Eds.), *Problems of and Prospects for Predicting Great Salt Lake Levels.* Salt Lake City, UT: University of Utah, Center for Public Affairs and Administration, 25–39.

Morrisette, P.M., 1987: Perception, public policy, and societal adjustment to fluctuations in climate: A case study of the rising level of the Great Salt Lake. Ph.D. dissertation. Boulder, CO: University of Colorado.

Morrisette, P.M., 1988: The stability bias and adjustment to climatic variability: The case of the rising level of the Great Salt Lake. *Applied Geography, 8,* 171–89.

Schelling, T.C., 1983: Climate change: Implications for welfare and policy. In *Changing Climate*, Report of the Carbon Dioxide Assessment Committee, Board on Atmospheric Sciences and Climate, National Research Council. Washington, DC: National Academy Press, 449–82.

State Road Commission of Utah, 1958: *Great Salt Lake Diking Study.* Salt Lake City, UT: State Road Commission of Utah.

State of Utah Office of Legislative Research and General Counsel, 1986: Great Salt Lake flooding. In *Interim Study Committee Reference Bulletin Report for 1986 General Session.* Reference Bulletin No. 7. Salt Lake City, UT: Office of Legislative Research and General Counsel.

Stauffer, N.E., Jr., 1980: Computer modeling of the Great Salt Lake. In J.W. Gwynn (Ed.), *Great Salt Lake: A Scientific, Historical and Economic Overview.* 265–72. Utah Geological and Mineral Survey Bulletin 116. Salt Lake City, UT: Utah Geological and Mineral Survey.

U.S. Army Corps of Engineers, 1986: *Great Salt Lake Utah: Reconnaissance Report.* Sacramento District, South Pacific Region. Sacramento, CA: U.S. Army Corps of Engineers.

Utah Department of Development Services, 1977: *Great Salt Lake Resource Management Study.* Salt Lake City, UT: Department of Development Services.

Utah Division of Comprehensive Emergency Management, 1985: *Hazard Mitigation Plan: Utah 1985.* Salt Lake City, UT: Utah Department of Public Safety.

Utah Division of Great Salt Lake, 1976: *Great Salt Lake Comprehensive Plan.* Salt Lake City, UT: Utah Department of Natural Resources.

Utah Division of Water Resources, 1984: *Great Salt Lake Summary of Technical Investigations for Water Level Control Alternatives.* Salt Lake City, UT: Utah Department of Natural Resources.

Utah Division of Water Resources, 1986: Alternatives for controlling flooding around the Great Salt Lake. Memorandum, 25 April 1986. Salt Lake City, UT: Division of Water Resources.

Wasatch Front Regional Council, 1973: *Lake com report.* Salt Lake City, UT: Wasatch Front Regional Council.

White, G.F., 1986: The future of the Great Plains re-visited. *Great Plains Quarterly, 6,* 84-93.

9

Future Sea-Level Rise and Its Implications for Charleston, South Carolina

Margaret A. Davidson and T.W. Kana

INTRODUCTION

It is becoming increasingly accepted that there will likely be a doubling of atmospheric CO_2 in the next century, which would raise the earth's temperature by 1.5°–4.5°C (3°–8°F). A global warming of even a few degrees translates into an increase in sea level because increasing atmospheric temperatures would cause seawater to warm and expand. Mountain glaciers, which have retreated in the last century, could melt more rapidly. Glaciers in Antarctica and Greenland could melt along the fringes, and portions of them could disintegrate and slide into the oceans (Meier, 1984).

In 1983, a report from the National Research Council estimated that worldwide sea level would rise 70 cm ($2\frac{1}{3}$ ft) in the next century (NRC, 1983), while a report from the U.S. Environmental

Margaret A. Davidson is the Executive Director of the South Carolina Sea Grant Consortium. As a lawyer and a science manager, she is interested in the relationship between science and decision-making, particularly at the state and local level. She has organized and participated in several workshops and symposia concerned with sea-level rise.

T.W. Kana is president of Coastal Science and Engineering, Inc. After receiving a Ph.D. from Johns Hopkins University, he taught coastal geology at the University of South Carolina. Among his many coastal projects and publications are two U.S. EPA-sponsored projects on sea-level rise.

Protection Agency (EPA) stated that uncertainties regarding the factors that could influence sea-level rise were so numerous that a single estimate of sea-level rise was impossible and instead developed high, medium, and low scenarios (Hoffman et al., 1983). The EPA report estimated that sea level would rise between 38 and 212 cm by the year 2075, with the likely range falling between 91 and 136 cm (3 and 5 ft), compared with a global rise in the last 100 years of 10 to 15 cm (4 to 6 in). Sea-level rise along the Atlantic coast of the United States has been 15 to 20 cm per century higher than the worldwide average, due to a subsidence trend that is expected to continue into the future.

Although predicting the actual process of global warming requires an international effort, it is at the local level that some basic reactions to this phenomenon will take place. The will and the ability of individual communities to respond to the range of effects engendered by global change is highly variable. Communities cannot prevent global warming; they can, however, aggravate or mitigate its impacts. Communities can pursue various land development patterns, construct physical barriers, or otherwise make adaptations to sea-level rise.

To meet the challenge of a global warming, society will need credible as well as reliable information concerning the likely effects of sea-level rise. Unfortunately, local areas generally do not have access to adequate information nor are they capable of anticipating and planning for long-term events like the localized impacts of a global warming of the atmosphere. So while the international and national communities debate and negotiate global responses to global warming, the effort at the state or local level is more fragmented.

Places like Charleston, South Carolina—the focal point of this chapter—are among those low-lying communities that stand to lose the most if strategic planning is ignored, or delayed too long. Fortunately, the rate of global warming and its pending impacts provide some lead time for planning. The question is what scenario is required to drive the planning process.

OVERVIEW OF THE CHARLESTON, SOUTH CAROLINA, REGION

The Charleston region encompasses a variety of topographic features and settlement patterns that would be affected by sea-level rise.

The historic and tourist areas—dating from 1670 and including downtown Charleston—are located on a peninsula at the confluence of the Ashley and Cooper Rivers, which combine to form Charleston Harbor (which, in turn, forms the Atlantic Ocean, according to local legend) (Figure 1). Much of the peninsular city is built over former marsh areas that were filled in the nineteenth and early twentieth centuries; more than one-third of the structures on the peninsula are on ground less than five feet above mean sea level (Basilchuk, 1986). The tip of the peninsula which contains the city's historic houses and most expensive residential real estate is protected by a seawall, built in the mid-1800s, and known as The Battery. During the present century, The Battery has been topped by waves associated with hurricanes.

The metropolitan Charleston area, with a population of approximately 500,000, spreads across several rivers and some distance inland; it also includes residential and resort development on a number of barrier islands. Elevation for much of this area averages approximately three meters (10 ft) above mean sea level (Michel et al., 1983), and the region's topography is characterized by a number of tidal rivers and creeks, expansive marsh resources, and large areas of freshwater wetlands (hence the name "Lowcountry," by which much of coastal South Carolina is known).

Over the past 20 years, rapid suburban development has been concentrated in the high ground and waterfront/marshfront locations within close (roadway) proximity to the peninsula, and development trends project both expansion and in-filling of the metropolitan area, mostly at low to moderate densities. Population and employment in the area are expected to grow by 16–28 percent within the next 12 years.

Waterfront development is varied. The Cooper River side of the peninsula, extending inland to the City of North Charleston, is primarily devoted to port activities and an extensive naval base, the third largest of the U.S. fleet. The Ashley River side of the

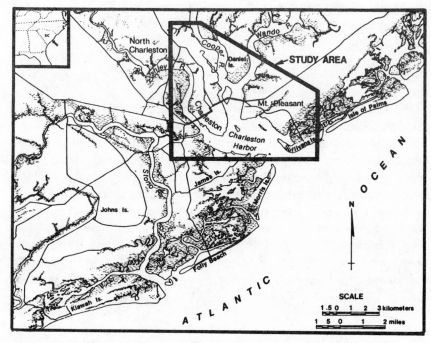

Figure 1.

peninsula includes a major arterial, a college, and residential development, while the side of the river opposite includes a state park, historic gardens, and substantial residential development. Additional waterfront development along the region's tidal rivers and creeks is predominately residential or recreational, or still undeveloped, but also includes commercial fishing berths, marinas, and shipyard operations.

In addition to low topographic relief, the area has a tidal range of 1.8–2.4 m (6–8 ft), with seasonal highs. As a result, the region is prone to flooding during periods of heavy rainfall—particularly when rainfall events coincide with high tides or strong easterly winds. There have also been several recent and notable winter storms with marked flooding and beach erosion.

Historical trend analysis has shown that the inner estuarine marshes in the area have been accreting since 1939 (Michel et al., 1983). The stability of the barrier island beaches varies—some are highly erosional, while others are stable or accretional. Human activities have influenced those trends. Computer modeling indicates that most of the region's barrier islands would be completely inundated by a moderately severe hurricane (winds exceeding 100 mph) hitting at high tide.

Drainage systems are inadequate in many areas. For instance, the peninsula city's gravity-based drainage system dates from the 1800s. A 1984 plan proposed the expenditure of $130 million to upgrade the system and install a series of pumps, but to date adequate funding has not been forthcoming.

As for wastewater treatment, there are four major treatment systems and a number of small ones in the region. The city of Charleston's system, which serves the city and a substantial suburban area, is based on an island in the Ashley River at 4.9 m (16 ft) above mean sea level. Due to an aging collection system, there is significant infiltration of floodwaters into the city's sewer system.

Through intensive local efforts and federal investment over the past decade, most urban and suburban portions of the region are currently served by sewer lines or will be in the next few years.

Water service is provided to most of the region by the city of Charleston (Commissioners of Public Works) via a distribution system originating inland. The Back River Reservoir, located approximately 30 miles up the Cooper River and the site of the region's major new industries, provides a secondary but increasingly important source of raw water for the city's system.

The recent completion of the U.S. Army Corps of Engineers' Cooper River rediversion project, transferring flows out of the Cooper River in order to reduce siltation in the harbor, has created the need to monitor and manage salinity in the Cooper River and Back River Reservoir, to ensure that the tidal salt wedge moving up the Cooper River is balanced by sufficient freshwater flow to protect the reservoir.

POSSIBLE IMPACTS OF SEA-LEVEL RISE

Global sea-level rise produces direct impacts along coastal areas as well as some less obvious, indirect impacts. These include flooding, erosion, saltwater intrusion, and habitat alteration. But the role played by sea-level changes is often difficult to distinguish from other processes which overshadow the ocean's general rise over short time scales. For example, Charleston experiences daily tide changes of more than 1.5 m (5 ft), certainly a fluctuation that is hundreds of times greater than the yearly rise in sea level for the area. Storm tides occasionally raise water levels more than 3.0 m (10 ft) above mean sea level.

The relevance of sea-level rise is that each of these phenomena is superimposed on the global, mean water level. Should sea level rise at a faster rate, new land areas and habitats will become exposed to tidal and storm surge processes. Much of the planning that must take place in the next few decades will necessarily begin with the baseline conditions existing today. For Charleston, water levels effectively control development, habitat type and various infrastructure or administrative functions. These range from the "normal" limit of tidal inundation, which delineates critical and protected wetland habitat or marshes, to the predicted 100-year tidal flood elevation, which establishes minimum elevations for habitable structures (Figure 2). Sea-level rise changes the baseline; it raises the height of normal and storm tides. Importantly, it exposes new areas to tidal inundation or it increases the frequency and duration of flooding in presently vulnerable areas.

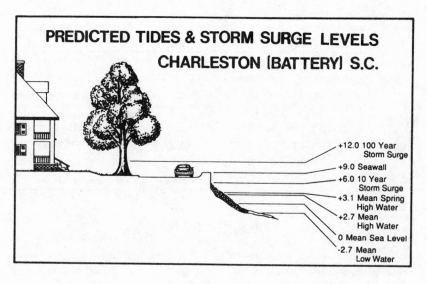

Figure 2.

Increased Flooding

Generally, increased property damage from periodic flooding is possible in low-lying areas. Up to one-half of the area could be frequently flooded if there are no anticipatory responses to sea-level rise (Kana et al., 1984). Expanded stormwater runoff management (mainly retrofitting existing developed areas) will be necessary in some residential areas.

The potential for storm flooding and periodic tidal inundation of certain key transportation arteries (including the causeways to the barrier islands) will be increased (Kana et al., 1984). The scope of this problem includes periodic inundation, resulting traffic inconveniences, enhanced degeneration of streets and the interruption of emergency service delivery, and episodic but potentially catastrophic reduction in evacuation capability from barrier islands and low-lying areas.

Waterfront properties will experience increased flooding as a result of normal tidal inundation, as well as increased risk of catastrophic flooding from episodic storms (hurricanes and northeasters).

Data indicate that the present division between transition tide levels and "high" ground is approximately 2.1 m (7 ft) above mean sea level for Charleston (Kana and Baca, 1986). Properties at elevations of 2.1–3.0 m (7–10 ft) will be the first to experience the effects of rising sea level.

Destabilization of Waterfront Property and Activities

Higher water levels will produce higher wave action. As a result, improvements may be needed in coastal structures ranging from minimal repair and upkeep to complete reconstruction, depending on exposure. It is generally accepted that the technology is available to provide structural solutions to sea-level rise. The question is what property value is sufficient to justify the very large cost of structural protection.

If a structural response is not economically or environmentally feasible, property abandonment may be necessary. The determination of which avenue to pursue will be dictated primarily by the frequency and severity of inundation. Sparsely developed areas susceptible to frequent inundation are more likely to be abandoned because of the economics of protection.

Activities will also become destabilized with a rising sea level. For example, the need for maintenance dredging of ship channels may be lessened by a moderately rising sea level. However, given that some ships currently have only a two-foot clearance under the Cooper River bridges at low tide, the utility of port and naval facilities may be diminished if a rising sea level prevents passage of some vessels to existing port facilities.

Increased Risk from Hurricanes

Hurricanes generally produce the highest water levels in coastal areas. Damages from hurricanes occur because of the combination of high waves, strong winds, and high water levels. The vulnerability of the Charleston region to hurricane damage is likely to be increased under any scenario of sea-level rise. Because much of the area is already subject to flooding and hurricane surges, higher sea level is likely to increase surge and flood heights, placing more areas at risk and exposing structures which are now higher than "baseline" surge heights to the possibility of wave-related damage.

Perhaps the greatest impact of sea-level rise in the context of hurricanes would be the reduction in evacuation time as a result of causeway flooding. The causeways to both Folly Beach and Sullivans Island have low spots at 1.8 m (6 ft) above mean sea level, and general elevations of 2.1–2.4 m (7–8 ft). Higher mean sea level would reduce the available time in advance of hurricane arrival that these causeways would be passable. This potential effect may be mitigated through highway construction that elevates the causeways. The final impact may be increased cost.

The higher and the more rapidly that sea level rises, the greater the potential for increased hurricane damage, and the prospects that such episodic risk may generate changes in land use and activity patterns. The range of these changes would be largely dependent upon the investment that public and private interests are willing to make to protect existing uses and activities. The scope of these investments is significant and includes moving electric generators at the area's numerous hospitals (most of which are located on the ground floor) and flood-proofing both active and inactive hazardous waste management facilities on the Charleston peninsula. Increased atmospheric warming will also increase the frequency of hurricanes in the area (Titus and Barth, 1984).

Increased Beach Erosion

A rise in sea level would cause shorelines to retreat (i.e., move inland) in two ways. First, they would retreat through simple inundation of the land presently at the margins of tide and wave action. Additionally, they would retreat through accelerated erosion of dunes which occurs in response to higher wave action associated with higher water levels. Scientists have shown that a rise in sea level of one foot could cause some beaches in the Charleston area to erode one to several hundred feet (Michel et al., 1983). The rate will depend upon the magnitude of sea-level rise as well as other factors not related to sea-level rise. The increased height of tidal crests associated with full- and new-moon tides is likely to increase erosional impacts on barrier island beaches. Islands which are currently erosional, such as Folly Beach, may see increased rates of erosion. Islands that are accreting, such as Sullivans Island, will see that rate reduced, and possibly reversed, as sea level rises. Stable islands such as the Isle of Palms or Kiawah Island may begin to experience erosion.

In addition, the episodic phasing of high tides and north-easters is likely to cause greater erosion and island flooding than presently occurs. Choices will have to be made between abandoning some property or replenishing eroded beaches.

Marsh Destruction

Wetlands in the Charleston area have been able to keep pace with the recent historical rise in sea level. Sea level during this century has risen an average of 3.6 mm/yr (Hicks and Crosby, 1974). This is equal to the approximate sedimentation rate in South Carolina marshes (Sharma et al., 1987). Thus, salt marshes have been accumulating sediment at a rate equal to sea-level rise, while migrating slowly inland over terrestrial soils. There is no evidence that the total area of salt marsh has expanded, so the rate of inland migration is probably balanced by erosion on the seaward boundary.

The success with which coastal wetlands adjust to rising sea level in the future will depend upon the ability of wetland sedimentary processes to keep pace with an accelerating rate of sea-level rise and the extent to which human activities prevent new marshes from forming as inland areas are flooded. Most of the marshes in

Charleston would be transformed to subtidal estuarine habitat if sea level were to rise 1.5 m (5 ft) and walls were built to protect existing development (Figure 3). In fact, even a conservative estimate of sea-level rise could permanently inundate coastal wetlands, because local salt marshes vegetated with *Spartina alterniflora* extend upward only to about 70 cm (2.3 ft) above mean sea level and down as low perhaps as mean sea level (J.T. Morris, unpublished data). Increased development along an estuary's fringe enhances the probability that bulkheads or dikes will be built to protect investments, which in turn increases the likelihood that estuarine marshes will be inundated and prevented from migrating inland (Titus, 1986).

The other threat to coastal wetlands is that they may not be able to rise as quickly as sea level may in the future. It is difficult to imagine that coastal wetlands could trap sediment at a rate three to five times greater than at present, which is what will be required if they are to keep up with rising sea level. A three- to five-foot rise in the next century would almost certainly upset these ecosystems in a fashion similar to that occurring in Louisiana, which every year loses over 100 square kilometers to the sea. There, freshwater wetlands are dying as a consequence of saltwater intrusion, which results in greater subsidence, erosion, and an accelerating rate of destruction (see Meo, this volume). South Carolina's rivers have extensive tidal and nontidal freshwater wetlands that would be threatened by saltwater intrusion associated with rising sea level.

Much of the sediment entering into the Charleston area is derived from suspended sediment originating primarily from the Cooper River which, since 1940, has carried the diverted flow of the Santee River (U.S. Army Corps of Engineers, unpublished general design memorandum). However, the Army Corps of Engineers has recently rediverted much of the Cooper River flow back into the Santee River. This will reduce sediment input, possibly reducing the rate of marsh accretion in the future, which would accelerate the loss of coastal wetlands (Kana and Baca, 1986). Lying between the sea and the land, tidal wetlands will experience the direct effects of changing sea levels, tidal inundation, and storm surges.

We can project several ecological as well as economic consequences of these losses. For example, the destruction of riverine wetlands would mean the loss of important wildlife habitats and would destabilize sediments. Mobilization of sediments will have

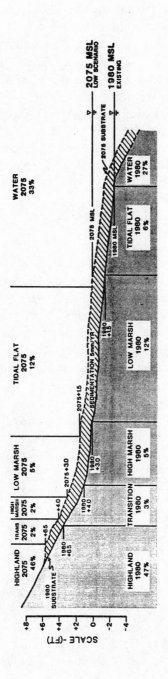

Figure 3. Conceptual model of the shift in wetlands zonation along a shoreline profile if sea-level rise exceeds sedimentation rates (using U.S. EPA scenarios). In general, the response will be a landward shift and altered areal distribution of each habitat because of variable slopes at each elevation interval.

a negative impact on navigable waterways and could release trace metals and other substances trapped in sediments. Finally, the productivity of several important fisheries, like shrimp, that depend upon wetlands will decrease.

It is possible, however, that man-induced alterations to estuarine dynamics, such as depositing dredge spoil material (from harbor maintenance dredging) along the edge of fortification in order to raise bottom heights sufficiently to support marsh communities, could be available to manage some of the potential impacts on habitat and vegetative communities. Creation of an artificially high sedimentation rate through the use of dredge spoil has promise, but changes in sediment chemistry brought about by this method may not support marsh vegetation.

FRAMEWORK FOR RESPONSE

As the magnitude of the effects of sea-level rise increase, the likelihood of national and state regulatory agency involvement also increases. These levels of government possess both greater fiscal resources and the necessary political jurisdiction to deal with complex environmental issues, *if* they choose to use them.

Local governments, however, have the tools available to set their own policy directions regarding the effects of sea-level rise. These include zoning ordinances, building codes, and local property tax structures. Moreover, local jurisdictions must be aware that in the face of pervasive sea-level rise impacts, state and federal agencies may be unable to address many local impacts, or to address them in the ways most responsive to local interests. This is especially true in South Carolina, where there is little or no centralized planning.

To a large extent, fiscal constraints dictate the substance of the responses. Local governments will have to take a pragmatic approach in deciding their range of commitments in the allocation of scarce resources. Private sector investment will play a large part in the decisionmaking process as it will dictate to a large extent the development patterns that occur once the impacts of sea-level rise begin to appear.

Political constraints will also play an important role. A multiplicity of local jurisdictions complicates a coordinated planning effort. In the Charleston region, there are more than 44 local government units and special purpose districts.

The vulnerable tenure in office of local decisionmakers adds to this complexity. Typically, those who make governmental decisions and financial commitments at the local level are in office for no longer than ten years. It seems difficult enough to gain consensus on policy decisions concerning events in the next decade, let alone the next century. The sophistication of information provided by the scientific community thus will play an important role in determining how and when local communities will prepare for and respond to the impacts associated with sea-level rise.

FACTORS AFFECTING LOCAL RESPONSES TO SEA-LEVEL RISE

In the absence of perceived manifestations of sea-level rise, the scientific community must provide local policymakers with detailed and logical projections that are good enough to give assurance that the phenomenon and its anticipated impacts are real.

If coastal impacts are localized, as would occur under low to moderate projections of sea-level rise, local policymakers need information on exactly how physical impacts are likely to be manifested. For example, increased flooding potential would be addressed quite differently if the manifestation is periodic flooding, as opposed to a theoretical, episodic increase in the height of a hurricane storm surge.

To the extent that certainty cannot be provided, and translated to inexpert and untrained decisionmakers, societal responses to the impacts of sea-level rise are likely to be postponed.

In addition to reliable projections, decisionmakers need comparative information, that is, a baseline. Local policymakers need documentation to show (1) that the phenomenon is actually happening, and (2) that very localized effects are occurring. Moreover, projections must be refined based upon observed conditions, before policymakers can be expected to make major decisions that are often irreversible in the short term.

Projections of and responses to the impacts of sea-level rise involve a variety of different time frames, which must be identified and coordinated as a precondition to meaningful local response. These time frames include:

- Rate of sea-level rise and related change
- Useful life of structures and infrastructure
- Financial life of structures and infrastructure
- Political tenure of decisionmakers
- Technological life and technological changes

RECENT PERCEPTIONS AND RESPONSES

The Sea-level Rise Forum in Charleston

In the early 1980s, popular media coverage of a global warming was still sporadic. In South Carolina, there was minimal awareness and discussion of this issue outside the academic community. However, EPA initiated studies of the potential physical and economic impacts of sea-level rise on Charleston, South Carolina, and Galveston, Texas, and then convened a national conference in Washington, DC, in March 1983, where these studies were presented.

Subsequently, a symposium on sea-level rise was held in Charleston during February 1984. This symposium featured a diversity of speakers from federal and state resource agencies, local government, and local economic development and real estate interests. There was advance publicity, and the meeting was attended by more than 100 people, many of whom had little prior awareness of global warming and sea-level rise.

Shortly before the meeting, there was considerable pressure by local development interests upon the local sponsors to cancel the meeting. Their concerns were based on apprehensions that highlighting the EPA case study would have a profound negative impact on residential and commercial development in the area.

During the course of the meeting itself, local public and private interests demonstrated the strongest resistance to being concerned about impacts associated with sea-level rise. In contrast, the representative from the state's coastal management agency (South Carolina Coastal Council) demonstrated the greatest reflection about sea-level rise effects and potential responses. Nonetheless, the general consensus of most local officials at that meeting was that the possibility and range of impacts from sea-level rise was too vague and too remote to affect any near-term planning (unpublished transcript of symposium, South Carolina Sea Grant Consortium).

Associated Beach Erosion

This localized resistance began to change, however, over the next few years as the state's beaches were adversely impacted by short-term meteorological phenomena. The natural, balancing processes of erosion and accretion were accentuated during a series of severely erosional winter storms. The combination of accelerated erosion and imprudent development along parts of the South Carolina coast set the stage for major erosion damages during a storm on New Year's Day 1987 (Kana, 1988).

This storm caused considerable damage all along the eastern seaboard and received extensive coverage in national media, including *USA Today*, *Time*, and *Newsweek*. The highly erosive effects were the result of a rare planetary alignment coupled with a winter northeastern storm. Marked erosion occurred throughout the South Carolina coast: residential and commercial structures were threatened and, in a number of places, toppled.

Public interest in these natural processes and their societal and environmental impacts increased rapidly. Concomitantly, coastal residents began to notice the increased media coverage of global warming and sea-level rise. Local communities began to discuss more openly the impacts of both short- and long-term natural processes on the beaches that contribute substantially to the state's $3 billion tourism industry. A statewide Blue Ribbon Committee on Beachfront Management was established by the South Carolina Coastal Council to address the complexity of issues affecting the coast: sea-level rise was recognized in the Committee's final report as a significant factor affecting erosion.

Institutional Response: Beachfront Legislation

The report of this Blue Ribbon Committee then was crafted into proposed state legislation which emphasized a retreat policy accompanied by a restriction on new erosion control devices such as seawalls. The bill that finally passed South Carolina's General Assembly maintained a commitment to a retreat policy but differed from the original bill which began its journey through the legislature in January 1988.

The bill was subjected to compromise and extended debate as it made its way through the legislative process and finally to

Conference Committee. The struggle over the passage of the legislation pitted members of the environmental community against lobbyists representing the lending industry and coastal property owners. Much of the debate focused on the establishment of a minimum setback line and what reconstruction, if any, should be permitted in front of that line. The lending community felt that mortgages they held might be devalued by the passage of a bill that adopted a strong "retreat" policy.

Another point of debate centered on the requirement to renourish the beach whenever a damaged or destroyed "erosion control device" is repaired or replaced. The requirement is the yearly renourishment of the beach in front of the property by the property owner. The amount of sand added to the beach cannot be less "than one and one-half times the yearly volume lost due to erosion." This may prove to be a cumbersome and expensive requirement in the face of accelerated erosion rates associated with sea-level rise.

The legislation also prohibits the rebuilding of vertical seawalls if damaged by more than 50 percent. The legislation mandates that all vertical seawalls be replaced with an approved erosion control device 30 years after the effective date of the legislation.

Other provisions of the Beach Protection Act of 1988 require the South Carolina Coastal Council to create a "long-range and comprehensive beach management plan" for the South Carolina coast, provide guidance on the distribution of renourishment funds, and stipulates that within two years of the effective date of the legislation, local governments must also prepare local beachfront management plans in coordination with the South Carolina Coastal Council.

Thus, it seems likely that local interests in understanding and planning for sea-level rise will now increase as a result of impacts not necessarily caused by a CO_2/trace gases-induced global warming.

SUMMARY

It appears reasonable to conclude that a three- to five-foot rise in sea level could seriously impact Charleston, South Carolina. Even if it is too early for local government to take many (if any) specific actions in anticipation of future sea-level rise, it is not too early to prepare for the time when specific, proactive decisions will need to be made. As such, more detailed and less variable projections are needed (including very localized impact assessments) in order to give policymakers the information they need to make responsible decisions about sea-level rise. While the scientific and the climate-related impacts communities are refining their projects, the interested public should be kept informed—but not alarmed—about the likely time frames when actions will be needed, and about the magnitude of impending impacts.

Ultimately, the local responses to the impacts of sea-level rise will involve decisions about fundamental priorities and the allocation of funding to preserve elements of local land use, lifestyle, and activities. Just as sea level is projected to rise gradually, local responses will be made over time, in reliance upon the developing products provided by the scientific community, and in hope that the measures implemented in response to sea-level rise will be sufficient to protect the economically developed coast and its residents' ways of life.

Substantial environmental and economic resources can be saved if better predictions become available soon, easily justifying the cost (though substantial) of developing them (Titus et al., 1983). However, deferring policy planning until all remaining uncertainties are resolved is unwise.

The knowledge that has been accumulated in the last 25 years has provided a more solid foundation for expecting sea level to rise in the future. Nevertheless, most environmental policies assume that coastal ecosystems are static. Incorporating into our environmental research the notion that ecosystems as well as social systems are dynamic need not wait until the day when we can accurately predict the magnitude of the future environmental changes.

REFERENCES

Basilchuk, N., 1986: Charleston's troubling sea change. *News and Courier*, Sunday, 24 August.

Hicks, S.D., and J.E. Crosby, 1974: *Trends and Variability of Yearly Mean Sea Level, 1893–1972*. NOAA Technical Memorandum NOS-13, COM-74-11012. Rockville, MD: NOAA.

Hoffman, J.S., D. Keyes, and J.G. Titus, 1983: *Projecting Future Sea Level Rise: Methodology, Estimates to the Year 2100, and Research Needs*. EPA 230-09-007. Washington, DC: U.S. Environmental Protection Agency.

Kana, T.W., 1988: *Beach Erosion in South Carolina*. Charleston, SC: South Carolina Sea Grant Consortium.

Kana, T.W., and B.J. Baca, 1986: *Potential Impacts of Sea Level Rise on Wetlands around Charleston, South Carolina*. EPA 230-10-85-014. Washington, DC: U.S. Environmental Protection Agency.

Kana, T.W., J. Michel, M.O. Hayes, and J.R. Jensen, 1984: The physical impact of sea level rise in the area of Charleston, South Carolina. In M.C. Barth and J.G. Titus (Eds.), *Greenhouse Effect and Sea Level Rise*. New York, NY: Van Nostrand Reinhold Co., 105–50.

Meier, M.F., 1984: Contributions of small glaciers to sea level. *Science, 226* (4681), 1418–21.

Michel, J., T.W. Kana, M.O. Hayes, and J.P. Jensen, 1983: *Hypothetical Shoreline Changes Associated with Various Sea Level Rise Scenarios for the United States, Case Study: Charleston, South Carolina*. Prepared by ICF, Inc., under contract to U.S. Environmental Protection Agency. Columbia, SC: Research Planning Institute, Inc.

National Research Council, 1983: *Changing Climate*. Report of the Carbon Dioxide Assessment Committee, Board on Atmospheric Sciences and Climate. Washington, DC: National Academy Press.

Sharma, P., L.R. Gardner, W.S. Moore, and M.S. Bollinger, 1987: Sedimentation and bioturbation in a salt marsh as revealed by ^{210}Pb, ^{137}Cs, and ^{7}Be studies. *Limnol. Oceanography, 32*, 313–26.

Titus, J.G., 1986: Greenhouse effect, sea level rise, and coastal zone management. *Coastal Zone Management Journal, 14*, 3, 147–72.

Titus, J.G., and M.C. Barth, 1984: An overview of causes and effects of sea level rise. In M.C. Barth and J.G. Titus (Eds.), *Greenhouse Effect and Sea Level Rise*. New York, NY: Van Nostrand Reinhold Co., 1–56.

10

Institutional Response to Sea-Level Rise: The Case of Louisiana

Mark Meo

INTRODUCTION

In the last few years, the scientific literature has expanded rapidly on the potential for an increase in the ocean's mean sea-level as a result of global climate change. Most, though not all, of the research has focused on the physical aspects of sea-level rise. Where will it occur? How great will it be? During what time span? What will be the associated effects? While these studies are, of course, important to anticipating and guiding societal response to sea-level rise, in and of themselves scientific studies are not sufficient for strategic planning. Any response will be constrained by the existing body of laws and regulations, social customs and behaviors, and environmental programs. Federal, state, and local governmental entities will be involved and each cannot be expected to pursue a coordinated strategy for mitigating the adverse consequences of sea-level rise. Similarly, the impacted economic interests of property owners (primary and vacation), local businesses, as well as major industries dependent on the coastal zone (e.g., the offshore oil and gas industry with its requirement for shoreside staging areas), will play a role in determining public

Mark Meo is currently a Research Fellow in the Science and Public Policy Program and holds an appointment as Assistant Professor of Civil Engineering and Environmental Science at the University of Oklahoma. His past research interests have included the use of wetlands for pollution control and wastewater treatment, planning criteria for estuarine and coastal resources management, renewable energy resource planning and evaluation, and the role of scientific and technical information in fostering economic competitiveness.

policy. These interests can be anticipated to favor strategies that will at times be in conflict with those adopted by various levels of government. Finally, interest groups such as conservationists or local chambers of commerce are likely to be drawn into the debate about the appropriateness of any given response.

Understanding and analyzing the interplay of these elements generally requires an understanding of the institutions involved. The knowledge gained from an institutional analysis helps to delimit the policy debate to what is socially practicable as well as acceptable. Such an institutional analysis can suggest what the ramifications of both inter- and intra-group action and reaction might be.

Clearly, the main difficulty in conducting an institutional analysis of sea-level rise lies in forecasting events that may occur in the future. This makes assessment of anticipated institutional responses almost impossible. But, this difficulty is not insurmountable. Through the use of a detailed examination of case studies of analogous situations, analysts can make informed statements about the nonobservable problem. In this instance the nonobservable phenomena are the incremental and collective institutional responses to climatically induced sea-level rise.

This chapter uses as its case scenario study the high rate of relative sea-level rise currently encroaching upon coastal Louisiana that has arisen primarily from land subsidence and wetlands erosion. From what was perceived to be a scientific curiosity 50 years ago, the cumulative disintegration of Louisiana's coastal wetlands has in time generated a variety of responses from local and state governments with new initiatives requested of federal agencies. Responding to public calls to halt, or at least slow down, the advancing sea, each level of government has sought to effect remedies within its power or has advanced changes in existing institutions to do so. Given the unique and novel nature of relative sea-level rise and the challenge it has presented to public leaders, the recent pace of institutional change has accelerated as the long-term implications of the phenomenon to the state's future have become more clear. An understanding of the gestation and development of different institutional responses to the present crisis confronting Louisiana may help readers to begin to understand what kinds of knowledge, resources, and actions will be required for society to respond to climate-induced eustatic sea-level rise over the course of the next century.

SEA-LEVEL RISE

Scientists concerned with forecasting and evaluating the range of social, economic, and environmental impacts engendered by changes in global climate have issued a number of reports in recent years (Revelle, 1983; Robin, 1986; Titus, 1986a; U.S. Department of Energy, 1985). Of particular interest to ocean and coastal scientists is the likelihood that continued atmospheric accumulation of greenhouse gases (especially carbon dioxide) will lead to glacial melting, thermal expansion of ocean waters, and perhaps the slippage of the West Antarctic ice sheet into the southern oceans, with a subsequent rise in global eustatic sea level. Current estimates of projected sea-level rise indicate that by the year 2075, sea level could rise as little as 0.76 m (2.5 feet) or as much as 2.3 m (7.5 feet) (Hoffman et al., 1983). Regardless of whether a low or high scenario is adopted, the consensus prospect is that global mean sea level could rise appreciably. With any such rise, coastal areas will experience inundation, increased rates of shoreline erosion, salt water intrusion into freshwater aquifers, a greater susceptibility to storm damage, and other related phenomena that in turn will generate social, economic, and political problems (Hoffman et al., 1983; Barth and Titus, 1984; Titus, 1986b, Davidson and Kana, this volume).

To policy analysts interested in or responsible for coastal resources planning and management, an important consideration with regard to sea-level rise is to anticipate the magnitude, severity, and distribution of the environmental and socioeconomic impacts that will arise from shoreline erosion and coastal submergence. If remedial actions or policies designed to mitigate identified impacts are to be initiated, an analysis of the institutional framework within which alternative choices can be compared and evaluated needs to be undertaken concurrently with any scientific studies.

Although a comprehensive analysis of anticipated responses to sea-level rise has yet to be undertaken, some policy research has been conducted as to what the socioeconomic impacts might be. To attain both a clearer understanding and a quantified estimate of the probable future impacts generated by a rise in sea level, detailed site-specific interdisciplinary studies have been undertaken in two low-relief coastal locations: Charleston, South Carolina, and Galveston, Texas (Barth and Titus, 1984). Both case studies examined the entire range of expected environmental and socio-economic impacts likely to occur under different rates of projected sea-level rise. These studies included an examination of the costs and benefits stemming from the impacts generated by an incremental submergence of urbanized properties and hazardous waste disposal sites, an increase in saline intrusion and damage from more severe floods, population and housing relocation, as well as the costs of deploying structural mitigation measures such as dikes and dams.

Both studies were fairly exhaustive, based upon evaluations of different scenarios of sea-level rise. Planning horizons in excess of one hundred years were simulated with variable rates of sea-level increases between two and ten feet or more. At its conclusion, results from the South Carolina study were presented at a public forum held in Charleston. The purposes of the forum were to inform the public and to garner responses from public officials and citizens. The general public response to the possibility of future coastal flooding by the sea, however, was ambiguous. Public officials, while finding the scientific forecasts credible, were unable to indicate how they thought individuals, groups, or agencies ought to respond in the absence of an immediate crisis or the perception of one.

Without a documented history of how societal response to a similar change has taken place in the past, forecasters are unable to develop scenarios of institutional response to probable environmental threats in the future. Consequently, sea-level impact projections made by scientists in the absence of an institutional framework or social context, which would provide a clearer understanding of the strategic choices available to planners, are subject to a similar public reception.

INSTITUTIONAL CHANGE

Unlike other near-term impacts or extreme events engendered by climate change, sea-level rise will be irreversible, advance incrementally over a long (several decades) time horizon, and exert its influence simultaneously across a broad geographical range. These characteristics of the phenomenon suggest that adequate time will exist for adaptation and/or adjustment strategies to be planned and executed. They also suggest that institutional innovations will be induced or will evolve in such a manner that different strategies to mitigate physical and socioeconomic impacts can be adequately evaluated.

Changes in the institutions that mediate transactions within social systems occur in response to external demands exerted by both public and private sector interests, and are constrained chiefly by the distribution of social and economic benefits and costs that accrue to the status quo. Entrepreneurial agents promoting innovation in natural resource institutions often face separate tests that determine the overall likelihood that change will occur. Following Runge's (1985) typology, these include independent determinations of a new proposal's technical feasibility, its economic feasibility, and its political feasibility. These three criteria, taken together, can provide a useful policy analytic framework for assessing the degree to which promising innovations can successfully overcome existing impediments to change.

With respect to sea-level rise, the identification of physical impacts has generated a variety of possible technical responses. Indeed, an important and large segment of coastal research is devoted to studies of beach nourishment, breakwaters and jetties, dikes, relocation of coastal structures, and the use of subsurface barriers to retard saline intrusion (NRC, 1987). Less prevalent are studies of the economic implications of employing these technologies (Broadus et al., 1986; Gibbs, 1986), and even more rare are assessments of the political tractability of each. To focus only on one or two of the conditions underpinning institutional change may obscure an accurate portrayal of the critical factors required for successful innovation. Moreover, a partial analysis could also mislead decisionmakers by depicting future adjustment or adaptation measures as fluid and smooth societal transformations when in fact they may be quite difficult either to initiate or administer.

LAND LOSS IN COASTAL LOUISIANA

The broad expanse of Louisiana's coastal wetlands has arisen from the discharge of sediment- and nutrient-laden waters spilled by the Lower Mississippi River as it meandered over its low lying floodplain (Coastal Environments, Inc., 1982). In the course of the last 7,000 years or so, the coastal marshes have been built up through the growth, decline, and abandonment of 15 separate deltas that swept river-borne sediments out across the shallow waters of the Gulf of Mexico (Frazier, 1967). The present bird foot delta, which is constrained from further course changes by river and distributary levees, has passed its peak delta building stage and is undergoing a period of natural decline. Except for the young and rapidly accreting Atchafalaya River delta emerging west of the Mississippi's main channel, navigation and flood control structures built along the river in the last 50 years have precluded continued supply of nutrient-rich sediments to interdistributary marshes. Bereft of sediment and nutrients, these marshes have been found to be slowly and irreversibly subsiding (Gagliano et al., 1981). In addition, canals which have been excavated in the course of exploring for oil and gas deposits underlying the marsh have enabled saline waters to intrude into interior marsh habitats. Fresh water and intermediate salinity marsh plant communities have been relocated landward (Gagliano, 1973). Channel erosion along the unprotected canal walls has contributed significantly to an increase in wetlands loss (Deegan et al., 1984; Scaife et al., 1983).

As a result of the cumulative effect of these alterations, the Mississippi River deltaic plain has been "losing" land to natural subsidence and erosion at an accelerating rate over time. Annual losses which were estimated to be a little under seven square miles in 1913 had increased to almost 16 square miles by 1946. Between 1967 and 1980 annual land loss accelerated from 28 square miles to 40 (Turner, 1987). When all of coastal Louisiana is accounted for, coastal land is currently being lost to the sea in excess of 50 square miles each year. Every 16 minutes another 0.4 hectares (one acre) of marsh becomes open water (Davis, 1986). At this rate of loss (0.8 percent per year), an area the size of the state of Rhode Island would disappear within 21 years (Turner, 1987).

The present velocity by which land is subsiding has contributed to a rise in relative sea level at a rate equivalent to one meter per century (Nummedal, 1983). This is the fastest rate measured in the United States. If carbon dioxide and other trace gas accumulation in the atmosphere leads to a global warming through the greenhouse effect, eustatic sea level could rise as much as 1.22 m (four feet) by 2100 (Barth and Titus, 1984). Since the turn of the century, the state has lost 445,000 hectares (1.1 million

acres). Between 1955 and 1978, cumulative wetland loss has totaled about 227,000 hectares (560,000 acres). If the historic trends
are projected to the year 2040, coastal Louisiana could lose 400,000
additional hectares (one million acres) of wetlands. The effect of
a rise in sea level could cause the Gulf of Mexico to advance inland up to 53 kilometers (33 miles) (Hawxhurst, 1987). The land
area that could be inundated due to the projected sea level rise is
depicted in Figure 1.

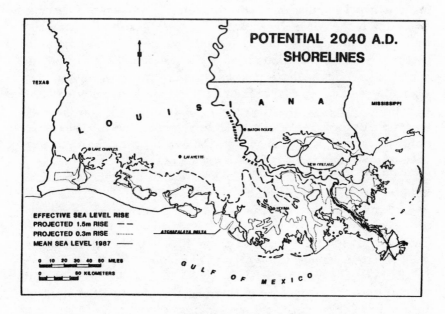

Figure 1.

However, there are several categories of negative consequences associated with coastal land loss that have already or could
soon affect Louisiana severely. Five of the more important ones are:
(1) the loss of land itself; (2) loss of economic activities associated
with the wildlife and habitats of coastal Louisiana; (3) losses of
state revenues derived from minerals on submerged lands; (4) salt
water intrusion into the water supply of metropolitan areas; and
(5) decreased protection from storm (hurricane) damage.

In considering land loss in Louisiana, the most salient point to keep in mind is that the land under discussion is in fact wetlands. This is important because wetlands tend to be one of the most biologically productive environments. As discussed below, they make economic contributions in many ways. Also important to remember is a second fact—south Louisiana's 6.5 million acres of coastal wetlands, which account for 40 percent of the nation's marsh ecosystems, produce approximately 28 percent of the national fisheries yield valued at $220 million annually, and provide habitat for about 66 percent of the Mississippi Flyway's wintering waterfowl population. An economically important fishery in Louisiana is the American oyster (*Crassostrea virginica* [Gmelin]). In recent years approximately 80,000 hectares (200,000 acres) of oyster beds have been harvested by more than 1,300 fee-paying lease holders. These oystermen produce on average 4 million kilograms (9 million pounds) of meat with an ex-vessel value of $3 to $4 million. The continued existence of this fishery is threatened by land loss because of the dual threat of saline intrusion; not only are natural predators such as the oyster drill (*Thais haemastoma canaliculata* [Gray]) able to invade previously protected oyster habitat, but vital low salinity seed oyster habitat is shrinking (Chatry et al., 1983; Davis, 1983).

Similar to oysters, brown and white shrimp (*Penaeid sp.*) are dependent on marshes and wetlands primarily for the nursery habitat and food resources that they provide anadramous postlarvae. With an annual value over $100 million, this is Louisiana's most important commercial fishery. Several of Louisiana's small coastal ports consistently rank among this nation's top ports in terms of the value of the shrimp landings (Craig et al., 1979).

The economic value of Gulf menhaden (*Brevoortia patronus*), whose survival is also estuarine dependent, is related to its high oil content. After capture, this species is processed principally into fishmeal and fish oil for use as animal feed. This fishery harvests between 270 million and 500 million kilograms (600 million and 1+ billion pounds) annually, which equates with a dollar value of approximately $10 million (Davis, 1983).

An aspect of the coastal Louisiana economy which is not widely known is its production of pelts from marsh dwellers such as the muskrat and the nutria (a fur bearer accidentally introduced from Argentina). This small area of the United States produces approximately 65 percent of the country's annual harvest of furs. In so doing, it provides important winter employment for several thousand trappers. The value of the yearly harvest can top $24 million (Davis, 1983).

Davis (1983) has indicated the importance of the area's sporting species to the economy of Louisiana. Sporting species include only the waterfowl, game, and fish which are valued by sportsmen. His calculations placed the species annual contribution to the Louisiana economy at between $175 and $200 million.

Another source of potential economic loss for Louisiana pertains to revenues of the state government derived from mineral resources (i.e., oil and gas) on state lands. Part of these lands lie under the northern Gulf of Mexico in the area between the shoreline and the outer edge of the state's three-mile territorial waters. Resources beyond the three-mile line fall under the jurisdiction of the federal government. With jurisdiction over fossil fuels comes the right to garner royalty and bonus payments. An unresolved issue is, if the coastline of Louisiana continues to recede, will the state lose revenues as resources currently under their jurisdiction technically become part of the purview of the U.S. Department of the Interior? If so, Louisiana could lose hundreds of millions of dollars in revenue (staff interviews, Louisiana Geological Survey, 1985).

Saltwater intrusion can also cause economic losses. As the coast moves landward, saltwater can intrude inland through the marshes and up the canals and bayous. Houma, in Terrebonne Parish, has already experienced elevated salinity levels in its water supplies during summer months (staff interviews, Terrebonne Parish Government, 1985). The city of New Orleans has expressed some concern that their water intake in the Mississippi River might also be threatened (staff interviews, New Orleans Planning Commission, 1985).

Finally, the wetlands and barrier beaches of the coast protect inland areas from the periodic ravages of storm and hurricane surges. Without the protection of these impediments to storm surges one could expect major increases in the costs associated with natural disasters either from increased damages or increased protection costs. For example, if Terrebonne Parish were to lose its barrier islands, "a drastic increase in the cost of providing hurricane protection for the 200,000 residents of the Terrebonne-Lafourche Metro Area can be expected" (Louisiana Department of Natural Resources, 1984).

THE RESPONSE TO COASTAL LAND LOSS AND SEA-LEVEL RISE

The public's emerging awareness of coastal land loss and its translation into political concern has been sparked primarily by three factors: fishermen's and trappers' visual observations of water encroachment upon commercially exploited wetlands; state politicians' belief that subsidence could lead to a landward relocation of the boundary separating Louisiana's territorial waters from those of the federal government; and finally, the skillful explanation by Dr. Sherwood Gagliano concerning the implications of incremental land loss to residents of coastal Louisiana in easily understandable terminology. Gagliano's characterization of the rates of land loss in terms of the years remaining for each coastal parish is credited with galvanizing local interests into action (staff interviews, LSU, Center for Wetlands Research, 1985; staff interviews, Plaquemines Parish Government, 1985).

Local Government Response

Public awareness has been converted into policies and projects at all levels of government. Those governments closest to the problem, that is, the coastal parishes, have been the most active. The two most threatened parishes are Plaquemines and Terrebonne. Each is projected to have less than a century of life remaining. Both parishes have initiated programs designed to dampen the major impacts associated with land loss. With Plaquemines Parish having less than 50 years of expected life, its coastal protection programs have received unanimous political support.

The parish, which straddles the Mississippi River just below New Orleans, has employed several freshwater siphons to retard saline intrusion. It is also actively promoting the use of strategic breaks in the levees along the river by the Army Corps of Engineers (ACE) to redirect river sediments into creating new wetlands. As part of their efforts to stabilize the coast, parish officials have attempted to replenish sediment along the Chandeleur Island barrier chain (an abandoned delta lobe remnant), but were denied permission by the National Park Service (staff interviews, Plaquemines Parish Government, 1985). To enable the citizenry sufficient time to conduct a strategic retreat from the encroaching waters, a backstop barrier dike running parallel to the Mississippi River is under consideration. Local government leaders appear to accept the fact that the parish will have to be ultimately abandoned when advancing waves begin to lap against this levee. In addition to structural and nonstructural options, Plaquemines Parish is also conducting a vigorous education campaign (staff interviews, Plaquemines Parish Government, 1985).

A second parish equally vulnerable to the Gulf's advancing waters is Terrebonne Parish which lies to the west of the Mississippi River. The region at present is experiencing a variety of subsidence related problems. In addition to natural subsidence occurring in the interdistributary basin of the abandoned La Fourche delta, wetlands are rapidly eroding due to the maze of oil and gas canals that permeate the marsh. Isles Dernieres and the barrier islands which mark the seaward terminus of the abandoned delta, have been ravaged by erosion and are moving landward rapidly (Nummedal, 1983). In 1984, it was estimated that Terrebonne Parish had lost as much as 42 percent of its barrier island acreage.

To address these interrelated problems the parish has devised a comprehensive response that has sparked a good amount of interest statewide. The parish plan is driven by four program goals: (1) develop a comprehensive data base; (2) educate the public; (3) preserve the wetlands; and (4) preserve the barrier islands. Achievement of the data base goal will include tasks such as an inventory of sand resources, a marsh valuation study, and an oyster contamination study. The broad education program includes slide presentations, handouts, use of billboards, the development of a foundation to support barrier island conservation, and a curriculum for junior high school students. The wetlands preservation goal involves fourteen separate management efforts which include private companies, state agencies (e.g., Louisiana Department of Wildlife and Fisheries, Department of Natural Resources), and federal agencies (e.g., ACE, U.S. Environmental Protection Agency).

Progress in barrier island beach nourishment and stabilization has been made. By 1983 the parish had completed two restoration plans. While the parish awaits completion by the ACE of its wetlands studies, the main focus for its activities has been barrier island restoration. Dismayed by what it perceived to be too slow a response by the state toward barrier island problems, the parish took unilateral action. By 1985 $1.1 million in local funds had been raised to help stabilize eastern Isles Dernieres with hydraulically pumped bay sediment (Edmonson, 1985).

However, the most provocative project proposed by the parish for protecting the wetlands consists of a closure dike, first suggested by the ACE as a possible structural remedy. The notion that a ring dike can be designed to protect the wetlands and at the same time maintain the ecological integrity of marsh habitat is not without controversy. Based on dikes and related engineering structures built in the Netherlands, the parish initiative, like many other innovative technologies, has not been adequately demonstrated in an environmental setting comparable to Louisiana's wetlands. The ring dike, once in place at a cost of $100 million, would resemble a Dutch polder and would provide for tidal exchange vital to the survival of estuarine-dependent fisheries. The state office of coastal zone management has undertaken a study to evaluate the technical feasibility of the proposal.

The State Response

Louisiana state government began to address the growing number of conflicts arising from the unmanaged exploitation of its coastal resources in 1971 with the establishment of an Advisory Commission on Coastal and Marine Resources. With the active involvement of local government officials, university scientists, and ACE personnel, the Commission released its report, "Louisiana Wetlands Prospectus," in 1973. At that point, the state began a five-year effort to develop a state coastal resource management program that was consistent with federal coastal zone management legislation. By 1978, after several lengthy sessions marked by political bargaining, Louisiana established a regulatory program to oversee the use of its coastal resources. Since then, the effectiveness of the state permit program has been harshly criticized for its lack of concern for protecting wetlands (Houck, 1983).

By 1981 several state legislators had become aware of the increasing magnitude of the land-loss problem. Fearful that the

state stood to lose oil and gas royalties if the state-federal boundary should be moved landward, Senator Samuel Nunez instructed
his legislative staff to prepare a report on the problem with a list
of specific recommendations. Pursuant to the preparation of this
report, legislation was enacted which provided for a Coastal Protection Trust Fund with an endowment of $35 million for project
development (Louisiana Revised Statutes, 1981). The responsibility for project management was accorded to the Louisiana Geological Survey in the Department of Natural Resources. Two years
after passage, Senator Nunez, whose districts include St. Bernard
and Plaquemines Parishes, held hearings in August of 1983 to review the progress made. Distressed both by the interest earned by
the trust fund and by the absence of any project under active development or any plan prepared for implementing the legislation,
the Geological Survey was given 90 days to produce a master plan.

The state's Coastal Protection Master Plan, developed in
1984, is separated into two distinct five-year phases. Program elements of phase I include the implementation of existing projects,
barrier and shoreline restoration projects, coastal vegetation program, and the development of a wetland protection program. It
has been estimated that barrier island stabilization will require
$131 million for the first five years. Program elements of phase II
include beach nourishment projects, dune vegetation and stabilization, and implementation of the wetlands protection program. This
latter element presents severe obstacles for the state program which
will require extensive intergovernmental cooperation to overcome.
This is because the issues to be dealt with in wetlands management are much more difficult than those in the shoreline protection
and barrier islands stabilization phase (staff interviews, Louisiana
Geological Survey, 1985). Figure 2 illustrates the location of the
different projects funded in the master plan.

As the state's response to land loss has gained political
momentum, interested legislators such as Representative Manuel
Fernandez, have coalesced into an informal group known as the
"coastal caucus." They have toured severely impacted areas of the
coast and have kept abreast of agency developments. Generally,
the caucus has served to broaden the base of political support in
south Louisiana, to maintain interest in the land loss issue, and as-
pires to create a national awareness of Louisiana's plight (Niebuhr,
1985). In the 1985 legislative session, for example, an additional

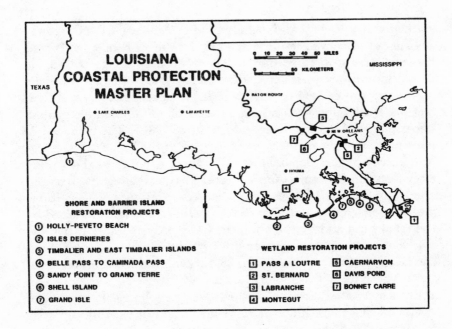

Figure 2.

$50 million was allocated to the Coastal Environment Protection Trust Fund from funds held in escrow pending the settlement of litigation between the United States and Louisiana over revenues from oil and gas deposits which underlie the current three-mile line in the Gulf of Mexico.

To assist the state in implementing its master plan, the Louisiana Senate in 1986 directed the Geological Survey to establish a Joint State-Federal Coastal Protection Task Force. This task force is comprised of all relevant agencies having jurisdiction over coastal resources (Hawxhurst, 1987). The following year, the governor established the Save Our Coastline Commission, the purpose of which is to leverage federal support through activities of Louisiana's congressional delegation.

The Federal Response

The federal presence in coastal Louisiana is constituted primarily by the efforts undertaken by the ACE to protect the Lower Mississippi River floodplain from periodic flooding, to maintain navigable waterways including inland ports, and to provide protection for the port city of New Orleans. The ACE has been involved in studies of the Louisiana coastal region since 1967 when both the U.S. Senate and House committees adopted identical resolutions empowering the ACE to "determine the advisability of improvements or modifications to existing improvements in the coastal area of Louisiana in the interest of hurricane protection, prevention of salt water intrusion, preservation of fish and wildlife, prevention of erosion, and related resource purposes." Since 1967 the ACE has sponsored a number of environmental studies of the coastal area. In addition to the ACE studies, research has been undertaken by the U.S. Fish and Wildlife Service, the Louisiana Department of Wildlife and Fisheries, the National Marine Fisheries Service, and the Center for Wetland Resources at Louisiana State University. Since 1980 the ACE has published several studies germane to understanding the land loss problem. These include studies of the impacts of canal dredging and feasibility reports on controlled and uncontrolled freshwater diversions from the Mississippi River into Barataria and Breton Sound Basins as well as for discharge into Lake Pontchartrain (U.S. ACE, 1984).

The controlled river diversion projects at Caenarvon and Davis Pond locations are especially notable for several reasons. They will: (1) help to retard saline intrusion and help restore valuable oyster habitat; (2) provide nutrients and some sediment for marsh creation which will partially offset subsidence; and (3) provide valuable engineering guidance for the state coastal protection program. The controlled diversions, when in operation, are expected to reduce estimated wetland losses by 16 percent in the next 50 years. The amount of wetlands saved would be almost 40,000 hectares (100,000 acres), equivalent to a 20 percent reduction in the statewide rate of wetland loss. Under current funding agreements the state will pay for the diversion structure's operating costs. Research undertaken by the Louisiana Department of Wildlife and Fisheries has confirmed the relationship between the 15 parts per thousand isohaline contour and oyster productivity (Chatry et al., 1983). The timing and rate of discharge of the river diversion, scheduled for completion by 1990 (U.S. ACE, 1987), is important to optimize oyster harvests (Chatry and Chew, 1985).

Although river diversions have been discussed for decades, they have failed to meet either economic feasibility tests (i.e., benefit-cost ratio greater than unity) or sociopolitical acceptance. The present ACE practice for estimating and measuring project benefits under the Water Resources Council's "Principles and Guidelines" does not avail itself of the inclusion of the nonmarket social benefits and values endemic to wetlands.

In the economic evaluation of the Davis Pond controlled diversion, for example, 90 percent of the complete set of economic benefits to be obtained from retarding saline intrusion and marsh creation is derived solely from oyster production; no measurable benefits are obtained from the ecological functions and nonmarket services provided by the restored wetlands. Moreover, the ACE planning horizon assumes overall environmental stability despite scientific consensus that sea-level rise will increase (U.S. ACE, 1984).

To further its goals, the state has started to formalize alternative conceptual and operational techniques for natural resource valuation and decisionmaking that include the measurement of nonmarket benefits. When complete, these studies should help to aid the revision of benefit-cost procedures used in project evaluation in the coastal protection program. Although it is constrained in practice by the limitations of its "Principles and Guidelines," the ACE has encouraged the use of new economic valuation techniques and has supported methods for ranking wetland functions and values. A revision of wetland evaluation techniques is an important aspect of the second phase of the state's master plan because *a priori* agreement should be reached on the comparative social benefits and costs that each potential course of action generates.

When completed, the controlled diversions will have to be managed carefully to supply the proper amount of water. Timed releases will have to be based upon temperature and flow regimes prevailing in each of the receiving basins that will optimize fishing productivity, marsh creation, and saline intrusion over the complete growing cycle. At present, there is some concern that the necessary scientific information for managing the controlled diversions is inadequate to the task. Moreover, the actual timing of freshwater releases will require closer collaboration between state resource management officials and the ACE to realize the full account of project benefits and to maintain local political support.

Another federal agency involved in the coastal protection program is the Federal Emergency Management Administration

(FEMA). Although it does not play a direct role in the coastal protection program, its separate activities affect coastal policies. Through its insurance programs, FEMA sets the rates for coastal properties within membership categories. Risk assessments are based upon hindcasts of hydrological and meteorological records and rates are determined to reflect the established risks.

Other Participants

In addition to the activities of key public officials such as Senator Nunez, other participants in the Louisiana response have come from academic institutions (perhaps better labeled as quasi-private consulting firms) and major land holders, primarily oil and gas exploration companies, in southern Louisiana. More recently, other federal agencies and citizen's groups have become involved.

The primary academic actors have included scientists from LSU's Center for Wetland Resources. The center, which includes a number of specialized units such as the Coastal Studies Institute, the LSU Sea Grant College program, and the Department of Marine Sciences, has pioneered research in coastal sedimentary environments (e.g., deltas) around the world and has played a key role in clarifying the ecological structure and function of coastal wetlands. In the course of their studies, scientists from the Center of Wetland Resources have documented Louisiana's coastal retreat through field studies and aerial photography, and have communicated research results to federal and state officials.

In a jointly published document in 1982, the center, along with the Gulf South Research Institute (GSRI) stated their reservations about the state's existing approach to land subsidence and coastal erosion. In essence, they argued that Louisiana was approaching the problem from a position of basic scientific and technical ignorance. Given the complexity of coastal processes in the state's wetlands, continuation of the program could only lead to failure. Correspondingly, the GSRI-LSU approach would see greater amounts of available funds dedicated to research on the region. Specifically, they proposed that it was necessary to better understand the processes at work in the region. Based on this improved understanding, it would then be possible to construct a model of wetlands loss. Then, using the model, it would be possible to predict future conditions and suggest appropriate management alternatives. The cost of this effort was projected at between $450,000 and $550,000 (GSRI and LSU, 1982).

Organized only in the last decade, the Louisiana University Marine Consortium, LUMCON, should also be considered among the important academic players. A symposium on coastal habitat changes in Louisiana was organized in 1982 by Dr. Donald Boesch of LUMCON in which the causes and consequences of land losses were explored (Boesch, 1982). Despite the lack of scientific consensus on different aspects of the land loss problem, the symposium afforded a timely opportunity for federal officials and state politicians to understand the causes of wetlands loss, and what measures to mitigate different impacts would be most appealing for further research and field demonstration.

Among the active private sector industries is Tenneco Oil Co., one of the largest land owners in southern Louisiana. Tenneco is a major corporation involved with a variety of natural resources. One of its major holdings is LaTerre properties (74,000 hectares or 183,000 acres) near the city of Houma. Primarily concerned with the exploitation of oil and gas resources, Tenneco-LaTerre also manages its extensive wetlands for wildlife resources which can be trapped or hunted. Tenneco's LaTerre project is basically aimed at mitigation banking. The company is setting aside 2,000 hectares (5,000 acres) which will be protected by weirs, bulkheads, and mud dams. These lands will be used to offset losses to wetlands resulting from oil canals that they dredge (Woodard, 1985).

Among individuals actively involved in the development of state and local government policies that address various aspects of coastal land loss a key one is Sherwood Gagliano. Dr. Gagliano's prior research for the ACE, other federal agencies, and the state over the last two decades has had a strong influence in shaping federal and state concerns and drafting of state coastal legislation. Both early on in his capacity as a university scientist and later as a private consultant, he has helped local and state government officials attain a clearer understanding of the implications of land loss for both local and regional economies as well as the wetlands in general. His consulting firm (Coastal Environments, Inc.) has helped numerous governmental agencies and private businesses solve coastal erosion and saline intrusion problems, and understand the nature of land subsidence.

Other important federal agencies include the Minerals Management Service and the Fish and Wildlife Service of the U.S. Department of Interior which operates the National Wetlands Research Center in Slidell, Louisiana. More recently, the U.S. Environmental Protection Agency has come to the assistance of the state. Through its collaborative efforts with the Louisiana Geological Survey, a Louisiana Wetland Protection Panel comprised of distinguished scientists was convened to organize a study to evaluate alternative strategies that would substantially reduce wetland loss in coastal Louisiana through the end of the next century. The panel's report (U.S. EPA, 1987) concludes that the coastal subsidence problem is in effect a national problem and requires a major commitment at the federal level to protect the state's coastal zone from relative sea-level rise.

Outside of the most threatened coastal parishes, public citizens groups have formed to review the adequacy of state and federal programs and issue recommendations of their own. In search of both a larger federal role and responsibility for protecting the coastal wetlands as well as a better funded and more coherent program for coastal protection and restoration, the Coalition to Restore Coastal Louisiana organized and issued its citizens' program in 1987 (Coalition to Restore Coastal Louisiana, 1987).

The coalition has argued that,

> the state lacks a strong focused political will to deal boldly with causes of land loss. Without such a state initiative, the federal government is not apt to do its part. There is no understanding at the national level of the magnitude of the biological and economic disaster occurring in coastal Louisiana.

The coalition placed the blame for Louisiana's plight primarily upon the oil and gas industry and the unwillingness of the state to regulate industry activities more carefully. The coalition also recognized the limited state institutional capability to deal with the multiple problems associated with land loss. To this end, they argued for increased sources of revenue to be raised from petroleum and gas pipeline and navigation use fees.

DISCUSSION

From the above assessment, it can be seen that a variety of strategies has been adopted by those taking part in the Louisiana response to subsidence and erosion. Several strategies have been adopted or are seriously being considered to counter the problem of land loss. Among these are beach nourishment, freshwater diversions, dikes and polders, mitigation measures, and additional demonstrations or studies. While most of these are quite different reactions to the problem, in most cases they are not mutually exclusive.

The diversity of programs, however, reflects a changing pattern in which institutions have responded. Despite the long geologic record of deltaic subsidence, public recognition of the environmental, economic, and political ramifications was mostly obscured or overlooked until only a few decades ago. In this respect, the Louisiana case shares attributes analogous to climate change. Just as climate change is likely to mobilize and focus public attention slowly and occasionally generate a redistribution of social benefits and costs through an extreme event, so also does relative sea-level rise.

A number of organizational factors likely to play a role in climate change appear common to each of the different responses examined. These include aspects of organizational capacity, technical expertise, financial resources, and experience. Louisiana's historical dependence on the ACE for protection against flooding and the provision of navigable waterways, for example, has served to thwart development of the state's institutional capacity to undertake engineering projects of any significance and to prolong state reliance upon federal expertise. As much as the state has benefited from the federal government's distributive public works programs so also did it suffer from those programs' hidden costs. As long as the state requested federal assistance in making structural improvements, the market-based values underscoring its decision-making criteria (e.g., benefit-cost analysis) continued to serve the wetlands poorly. Furthermore, without the responsibility for operating and maintaining key projects, an adequate knowledge base essential for managing the wetlands in an environmentally benign manner was slow to develop. Only in the past decade has state-supported research begun to assemble the information necessary for protecting, preserving, and restoring the barrier islands and wetlands.

Louisiana's economic dependency on the oil and gas industry has also harbored unforeseen costs. As recent research has demonstrated, the network of cuts and canals dredged in the process of exploring and transporting petroleum has exacerbated natural subsidence processes substantially. In spite of industry arguments that subsidence is a natural process, local government officials now recognize the long-term implications to the area of poorly managed short-term petroleum exploitation.

Despite limited state capacity to deal with cumulative land loss, local governments took an aggressive approach when they began to sense the importance of the issue. Plaquemines and Terrebonne Parishes, both notably lacking in technical expertise and experience, have sought outside assistance to help them develop near- and longer-term solutions. At the same time, faced with the direct assault of the sea, they have rejected suggestions for additional research as overly time consuming and cost-ineffective.

The state's master plan appears quite attuned to local governmental concerns. Efforts now underway in Phase I projects are directed toward barrier island stabilization and beach nourishment. These projects are technically feasible, economically attractive, and because of their location across the coastline, politically appealing to both state and local interests. Less certain on all three counts, however, are the longer-term wetlands management and restoration activities which might engender conflicts with private interests. For example, a statewide moratorium on canal excavation or canal backfilling, which would help alleviate erosion, are two especially contentious proposals currently being debated.

Most participants in Louisiana's battle with the sea generally accept that ultimately the sea will win. The range of projects currently under development, as well as the growing number of requests for additional efforts and increased federal involvement, will not alter the final outcome substantially. A key question for state planners to address is how public agreement can be attained on the implications of land loss for strategic defense of specific areas and subsequent retreat or resettlement landward to higher ground.

The fragmentation of authority which exists within the state is a critical obstacle to attaining a more coherent state strategy.

Public and private interests will have to reach agreement on a number of unresolved issues pertaining to the coast. These include: developing an acceptable way to value the social services provided by barrier islands and wetland habitats; developing a workable mechanism for resolving disputes among and between federal, state, and private interests (who own 80 to 85 percent of the wetlands); developing a method for calculating the impacts of future land loss and sea-level rise on coastal communities; and attaining sufficient knowledge of the risks associated with coastal wetland restoration.

CONCLUSION

When viewed in the context of an analogue of future possibilities, the Louisiana case study holds several important lessons for guiding the formulation of public policies directed at mitigating the environmental and socioeconomic impacts of adjusting to long-term, irreversible environmental change. The various initiatives launched in Louisiana mark but the opening foray in what will likely be a long and protracted effort to conserve valuable and ecologically vulnerable resources. Yet, each of the approaches examined has addressed different aspects of technical, economic, and political feasibility, each of which is an important condition for fostering institutional change.

In a technical sense, the limited extent of scientific understanding of coastal wetlands and their adjacent marine, riverine, and terrestrial environments hampered the ability of public agencies to predict accurately the consequences both of natural phenomena as well as alternative approaches for taking corrective action. This uncertainty, in turn, contributed to a reassessment by which the socioeconomic value of the coastal environment and the criteria associated with alternative public actions could be evaluated. Different aspects of economic feasibility are clearly distinguishable among federal, state, and regional programs. Although sharp distinctions between different criteria exist, some movement toward a practical reconciliation is in progress which will enable the state to attain a leadership role. In addition, both the direct and indirect consequences of wetland loss were addressed in a political context once different stakeholders could identify specific threats to their interests and develop strategies around which disparate groups could coalesce. The degree to which different political interests have coordinated with specific programs has helped to further develop coastal protection efforts in a number of useful ways.

Noteworthy in this regard is the role of scientific communication. The public's perception of coastal land loss acquired greater intensity and much sharper focus when the remaining life expectancy of the coastal marshes was calculated. This raised a slight dilemma, however. Recognition of the personal impacts attributable to cumulative wetland loss not only catalyzed greater public demand for collective action and the strengthening of state and local institutional capacity to mitigate damages, but also served to diminish the legitimate and pressing need for additional scientific research.

As public awareness rose, different political interests and state agencies perceived the land loss crisis as an opportunity to further their own interests. In some instances this goal has been attained through lethargy and inaction. In other cases, technically questionable initiatives have been proposed under the guise of workable solutions. Additional research will be extremely useful to sort out conflicting claims and complexities.

Designation of the Coastal Protection Trust Fund signaled the state's commitment to accept a greater role in coastal planning and management. Despite numerous political obstacles and lengthy delays in implementation, the trust fund has provided a useful resource by which a "first cut" at a systematic state strategy could be initiated and implemented while the state further developed its institutional capacity to address the myriad political and economic issues associated with long-term land loss.

Two general lessons for climate change impact assessment are to be gleaned from the Louisiana experience. First, with any acute and predictable environmental problem such as sea-level rise, the "response" will be the aggregate of many responses emanating from all levels. The agencies which exist have jurisdictions and regulatory mandates pertinent to one or more aspects of the problem. Further, the agencies are unwilling to ignore the problem while others act. Inaction could result in a loss of bureaucratic "turf" or flexibility. Thus, environmental problems of this sort will always generate and require an intergovernmental effort.

Second, the federalism or intergovernmental aspect, while necessary, can create problems. These problems are derived from the anomalies that exist in the planning horizons of the various agencies taking part in the disjointed response. It is acknowledged that the most active agencies have been the local parishes. In part this reflects their planning horizons. While most of the relevant actors can agree on what should be done, they cannot agree on when and at what pace it should occur. Understandably, the parishes want something done now. Federal agencies do not labor under the same urgency.

As the ocean continues to rise, Louisiana's coastal protection setbacks and successes are likely to receive increasing scrutiny from other geographically low-lying regions similarly threatened. Given the cumulative and incremental nature of sea-level rise and the range and variability associated with the distribution of its socioeconomic impacts, the likelihood of developing a comprehensive strategy for motivating a societal response would appear to be slim until and unless the technical, economic, and political feasibility of any mitigation strategy has been duly addressed.

ACKNOWLEDGMENTS

This paper originated while the author was a postdoctoral fellow at the Woods Hole Oceanographic Institution (WHOI). Support was provided by the U.S. Department of Commerce, Office of Sea Grant, the J. N. Pew, Jr. Charitable Trust, and the Marine Policy Center at WHOI. The contributions of Maynard Silva are gratefully acknowledged.

REFERENCES

Barth, M.C., and J.G. Titus (Eds.), 1984: *Greenhouse Effect and Sea Level Rise*. New York, NY: Van Nostrand Reinhold Co.

Boesch, D.F., (Ed.), 1982: *Proceedings of the Conference on Coastal Erosion and Wetland Modification in Louisiana: Causes, Consequences, and Options.* FWS/OBS-82/59. Washington, DC: U.S. Fish and Wildlife Service.

Broadus, J.M., J.D. Milliman, S.F. Edwards, D.G. Aubrey, and F. Gable, 1986: Rising sea level and damming of rivers: Possible effects in Egypt and Bangladesh. In J.G. Titus (Ed.), *Effects of Changes in Stratospheric Ozone and Global Climate*, Volume 4, *Sea Level Rise*. Washington, DC: U.S. Environmental Protection Agency, 165–89.

Chatry, M., R.J. Dugas, and K.A. Eastley, 1983: Optimum salinity regime for oyster production on Louisiana's state seed grounds. *Contributions in Marine Science, 26*, 81–94.

Chatry, M., and D. Chew, 1985: Freshwater diversion in coastal Louisiana: Recommendations for development of management criteria. In C.F. Bryan, P.J. Zwank, and R.H. Chabreck (Eds.), *Proceedings of the Fourth Coastal Marsh and Estuary Management Symposium*. Baton Rouge, LA: School of Forestry, Wildlife, and Fisheries, Louisiana State University, 71–84.

Coalition to Restore Coastal Louisiana, 1987: *Coastal Louisiana: Here Today and Gone Tomorrow*. New Orleans, LA: Coalition to Restore Coastal Louisiana.

Coastal Environments, Inc., 1982: *Louisiana's Eroding Coastline: Recommendations for Protection*. Prepared for the Coastal Management Section of the Louisiana Department of Natural Resources. Baton Rouge, LA: Coastal Environments, Inc.

Craig, N.J., R.E. Turner, and J.W. Day, Jr., 1979: Land loss in coastal Louisiana. *Environmental Management, 3*, 133–44.

Davis, D.W., 1983: Economic and cultural consequences of land loss in Louisiana. *Shore and Beach*, October, 30–39.

Davis, D.W., 1986. The retreating coast. *Journal of Soil and Water Conservation*, May–June, 146–51.

Deegan, L. A., H. M. Kennedy, and C. Neill, 1984: Natural factors and human modifications contributing to marsh loss in Louisiana's Mississippi River deltaic plain. *Environmental Management, 8*, 519–28.

Edmonson, J., 1985: Terrebonne Parish begins barrier island reconstruction. Press release. Houma, LA: Terrebonne Parish Council.

Frazier, D.E., 1967: Recent deltaic deposits of the Mississippi River, their development and chronology. Gulf Coast Association of Geological Societies. *Transactions, 17*, 287–315.

Gagliano, S.M., 1973: Canals, dredging, and land reclamation in the Louisiana Coastal Zone. Hydrologic and Geologic Studies, Report 14. Baton Rouge, LA: Center for Wetland Resources, Louisiana State University.

Gagliano, S.M., K.J. Meyer-Arendt,and K.M. Wicker, 1981: Land loss in the Mississippi River deltaic plain. Gulf Coast Association of Geological Societies. *Transactions, 31*, 295–300.

Gibbs, M.J., 1986: Planning for sea level rise under uncertainty: A case study of Charleston, South Carolina. In J.G. Titus (Ed.), *Effects of Changes in Stratospheric Ozone and Global Climate*, Volume 4, *Sea Level Rise*. Washington, DC: Environmental Protection Agency, 57–72.

GSRI (Gulf South Research Institute) and LSU (Louisiana State University), 1982: *Position Paper on Approach to Projecting Future Coastal Conditions*. Baton Rouge, LA: Center for Wetland Resources, LSU.

Hawxhurst, P., 1987: Louisiana's responses to irreversible environmental change: Strategies for mitigating impacts from coastal land loss. In M. Meo (Ed.), *Proceedings of the Symposium on Climate Change in the Southern United States: Future Impacts and Present Policy Issues*. Norman, OK: Science and Public Policy Program, University of Oklahoma, 173–85.

Hoffman, J. S., D. Keyes, and J. G. Titus, 1983: *Projecting Future Sea Level Rise: Methodology, Estimates to the Year 2100, and Research Needs*. EPA 230-09-007. Washington, DC: U.S. Environmental Protection Agency.

Houck, O.A., 1983: Land loss in coastal Louisiana: Causes, consequences, and remedies. *Tulane Law Review, 58*, 3–168.

Louisiana Department of Natural Resources, 1982: Coastal Protection Task Force Report to Governor David C. Treen and to the Joint House and Senate Committees on Natural Resources and to the Legislative Budget Committee. Baton Rouge, LA: State of Louisiana.

Louisiana Department of Natural Resources, 1984: Coastal Protection Task Report to Governor Edwin W. Edwards and to the House and Senate Committees on Natural Resources and to the Legislative Budget Committee. Baton Rouge, LA: State of Louisiana.

Louisiana Revised Statutes, 1981: Vol. 30, Chapter 5-A, November 23. Coastal Environment Protections Trust Fund. Baton Rouge, LA: State of Louisiana.

NRC (National Research Council), 1987: *Responding to Changes in Sea Level: Engineering Implications*. Committee on Engineering Implications of Changes in Relative Mean Sea Level. Washington, DC: National Academy Press.

Niebuhr, G., 1985: Louisiana's coastline threatened by erosion. *Boston Sunday Globe*, February 10, 24.

Nummedal, D., 1983: Future sea level changes along the Louisiana coast. *Shore and Beach, 51*, 10–15.

Revelle, R., 1983: Probable future changes in sea level resulting from increased atmospheric carbon dioxide. In *Changing Climate*. Report of the Carbon Dioxide Assessment Committee, Board on Atmospheric Sciences and Climate, National Research Council. Washington, DC: National Academy Press.

Robin, G. deQ., 1986: Changing the sea level: Projecting the rise in sea level caused by warming of the atmosphere. In B. Bolin, B. Döös, J. Jäger, and R.A. Warrick (Eds.), *The Greenhouse Effect, Climate Change and Ecosystems*. Scope 29. New York, NY: John Wiley and Sons, 323–59.

Runge, C.F., 1985: Institutional innovation and land resource management in the 1980s. In E.B. Liner (Ed.), *Intergovernmental Land Management Innovations*. Cambridge, MA: Lincoln Institute of Land Policy, 47–57.

Scaife, W.W., R.E. Turner, and R. Costanza, 1983: Coastal Louisiana recent land loss and canal impacts. *Environmental Management, 7*, 433–42.

Titus, J.G. (Ed.), 1986a: *Effects of Changes in Stratospheric Ozone and Global Climate Change*, Volume 4, *Sea Level Rise*. Washington, DC: U.S. Environmental Protection Agency.

Titus, J.G., 1986b: Greenhouse effect, sea level rise, and coastal zone management. *Coastal Zone Management Journal, 14*, 147-72.

Turner, R.E., 1987: *Relationship Between Canal and Level Density and Coastal Land Loss in Louisiana*. Biological Report 85(14). Washington, DC: U.S. Fish and Wildlife Service.

U.S. ACE (Army Corps of Engineers), 1984: *Louisiana Coastal Area, Freshwater Diversion to Barataria and Breton Sound Basins*. Feasibility Report. New Orleans, LA: U.S. Army Corps of Engineers.

U.S. ACE (Army Corps of Engineers), 1987: *Status Report on Coastal Activities, New Orleans District*. New Orleans, LA: U.S. Army Corps of Engineers.

U.S. Department of Energy, 1985: *Glaciers, Ice Sheets, and Sea Level: Effect of a CO_2 Induced Climatic Change*. DOE/ER/60235-1. Washington DC: U.S. Department of Energy.

U.S. EPA (Environmental Protection Agency), 1987: *Saving Louisiana's Coastal Wetlands: The Need for a Long-Term Plan of Action*. Report of the Louisiana Wetland Protection Panel. Washington, DC: U.S. Environmental Protection Agency.

Woodard, J., 1985: Testimony of Tenneco Oil Exploration and Production Before the House Committee on Merchant Marine and Fisheries, Subcommittee on Fisheries and Wildlife Conservation and the Environment—Update on Tenneco's Mitigation Banking Project. 20 June. Houma, LA: Tenneco LaTerre.

11

Climate Variability and the Mississippi River Navigation System

William Koellner

INTRODUCTION

The Mississippi River navigation system is entirely dependent on its ability to transport commodities efficiently. The system's capabilities are in part directly related to climate. Extreme meteorological events in the region generate stresses on the system that often reduce its efficiency. The United States is very dependent on the movement of different kinds of commodities, both for internal national use and for export. The Mississippi River system links the Gulf of Mexico to the Great Lakes, as well as to other waterways.

The major portion of all tonnage originating or terminating in the system is comprised of five commodity groups: grain, agricultural chemicals, coal, cement or stone, and petroleum products. If any of the sub-systems that these commodities move through are affected by adverse climatic conditions, the efficiency of the whole system may be reduced drastically and even rendered inoperative for some period of time. Constraints on the system usually become

William Koellner is a hydraulic engineer with the Rock Island District of the Army Corps of Engineers, Rock Island, Illinois, and is Chief of the Regulation Section, Hydraulics Branch, Engineering Division. He is responsible for the hydraulic operation of the navigation dams and several flood control reservoirs on the Illinois and Mississippi Rivers as well as managing a large data base of meteorological and hydraulic data for the Upper Mississippi River Basin.

activated under conditions of either too much or too little water. Climate variability affects navigational commerce—the supplier, the shipper, and the purchaser.

This chapter examines these climate-related constraints and their impacts on parts of the Mississippi River drainage system, as well as on the Great Lakes system as it interfaces with the inland waterway network for the transportation of commodities. Historical analogues can be used to identify potential climate-related problems that might arise as a result of a change in climate in the Mississippi River system's drainage basin.

NAVIGATION SYSTEM DESCRIPTION AND HISTORY OF THE MISSISSIPPI RIVER AND GREAT LAKES BASIN

The Mississippi River is 2,340 miles long and has a watershed area of 1,245,000 square miles. Forty percent of the area of the United States and 13 percent of the area of Canada drain into this river system which provides an outlet for various commodities. The River is divided into three distinct reaches: (1) the Lower Mississippi River from its mouth at New Orleans to the mouth of the Ohio River at Cairo, Illinois; (2) the Middle Mississippi River from Cairo to St. Louis, Missouri, at the mouth of the Missouri River; and (3) the Upper Mississippi River above St. Louis, Missouri. Figure 1 is a map of the inland waterway system including the Mississippi River system and the Great Lakes (National Waterways Study, 1981).

The Great Lakes system is comprised of the five lakes (Michigan, Superior, Huron, Erie, and Ontario) and the St. Lawrence River. The Great Lakes–St. Lawrence River Basin extends a distance of 2,000 miles from the western end of the Lake Superior Basin in Minnesota to the Gulf of St. Lawrence on the Atlantic Ocean. Thus, from east to west the basin spans nearly one-half of the North American continent. The five Great Lakes with their connecting rivers and Lake St. Clair have a water surface area of about 95,000 square miles above the head of the St. Lawrence River at the eastern end of Lake Ontario. The total area of the Great Lakes Basin, including both land and water, is approximately 295,000 square miles. Of this area, 174,000 square miles (60 percent) are in the United States. The Great Lakes–St. Lawrence

Figure 1. Inland waterway system including the Mississippi River
system and the Great Lakes.

River system is bordered by eight states (Minnesota, Wisconsin, Illinois, Indiana, Michigan, Ohio, Pennsylvania, and New York) and by two Canadian provinces (Ontario and Quebec). The international boundary between Canada and the United States passes through all of the Great Lakes with the exception of Lake Michigan, which is located entirely within the United States.

Brief History

As early as 1705, trappers from the western Great Lakes area as well as from the Ohio and Illinois regions used the Mississippi River to move cargo downstream to relatively distant market areas. Transportation in the system has evolved from the canoe to the flatboat, to the keelboat, to the steamboat, and, finally, to the modern towboat. Early development of the region's natural resources led to the establishment of military posts along the Mississippi River by the United States Government. Fur trading and the mining of lead from Galena, Illinois, resulted in the first users of the inland waterway system. More than 450 lead mines (Lewis, 1967) had a total annual production of up to 55 million pounds, and in 1847 the inland waterways transported lead that was valued at $1.7 million (Peterson, 1979). Galena prospered from the shipment of lead on the waterways until the 1850s. Just as fur and lead were the leading commodities from 1800 to 1850, passenger traffic then dominated the river until the Civil War, with the migration of adventurous easterners who wished to start a new life by "going west." By 1860, more than 2 million people inhabited the states adjacent to the Mississippi River (Hartsough, 1934). Agricultural commerce then began to boom, and millions of bushels of grain, primarily for domestic use, moved on the inland waterway system.

Several significant events since 1705, which aided in the development of the inland waterway system, are listed below:

1705   The first recorded cargo, consisting of 15,000 bear and deer hides, was floated down the Mississippi River.

1803   The Louisiana Purchase from France provided direct American interest in the waterway.

1819   The U.S. Congress approved funding for a survey of the Mississippi River and its tributaries.

1824 The U.S. Congress enacted the first Rivers and Harbors Act, the General Survey Act of 1824. This law provided the Corps of Engineers with the authority to clear important rivers of obstructions.

1836 The State of Illinois began construction of the Illinois and Michigan Canal between Chicago and LaSalle, Illinois.

1878 The U.S. Congress authorized construction of the 4.5-foot channel on the Upper Mississippi River to be accomplished by dredging, and by wing and closing dams.

1907 The U.S. Congress authorized construction of the 6-foot channel on the Upper Mississippi River, to be accomplished by additional wing dam construction and by new locks at Minneapolis (Minnesota) and the Rock Island Rapids (Illinois).

1927 The U.S. Congress authorized construction of a 9-foot channel on the federal section of the Illinois Waterway and on the Mississippi River between the mouth of the Ohio River and St. Louis. Surveys were also to be conducted between St. Louis and Minneapolis.

1927 Record flooding on the Mississippi River provided Congress with the reason to authorize a comprehensive flood control system in the Basin to complement the authorized navigation system.

1930 The U.S. Congress authorized construction of a 9-foot channel north of St. Louis using a stairway of locks and dams. This project was completed in 1939 with 28 locks and dams on the Mississippi River and 7 locks and dams on the Illinois River.

1953 Construction was initiated on the last locks and Dam 27 at St. Louis on the Mississippi River.

1960 Construction was initiated on the Thomas J. O'Brien Lock and Controlling Works on the Calumet River as a link between Lake Michigan and the Illinois Waterway.

Canalization

The purpose of canalization is to provide a dependable 9-foot navigation channel for the safe movement of barges and towboats with widths suitable for long-haul common carrier service. The 1930 U.S. Congressional Act authorized the formation of slackwater pools (i.e., pools formed above a navigation dam during periods of very low river flow) by means of the 28 dams on the Upper Mississippi River and 7 dams on the Illinois Waterway that would be used for the passage of tows and barges. This legislation provided for a "Stairway of Water" on both the Mississippi and Illinois Rivers. Figure 2 depicts the "Stairway of Water" for the Illinois River system. The last dam downstream in the system is at Alton, Illinois. The navigation channel in the Middle Mississippi and Lower Mississippi River reaches is maintained by means of dredging and by contraction of the low water channel by rock jetties called training dikes, which direct the movement of the current. This increases the velocity of water in the channel in order to keep the sediment moving and the channel open.

Figure 2.

Because the Mississippi River channel is relatively shallow, its floodplain is relatively wide. A large portion of the floodplain below Muscatine (Iowa) has been protected from flooding by means of flood control levees which protect highly productive agricultural and industrial properties. Low river banks and probable damage to productive bottom lands from inundative overflows and lateral seepage limit the system's dams to relatively low vertical lifts. Of the two high lift dams contained in the authorized 9-foot project, the one at Keokuk (Iowa) was constructed primarily for hydropower generation. Prior to adoption of the canalization project, it was in a reach of the Mississippi River called the Des Moines Rapids where the natural low water river slope was steep and the water depth was hazardous to navigation traffic during much of the year.

The second high lift dam, located at Minneapolis (Minnesota), is used for power production. The dam is located in a deep gorge where overflow damage is negligible. In order to minimize flow damages in the balance of the canalization system, low head dams were constructed with movable gates offering the least possible obstruction to flood flows, but providing the minimum 9-foot navigation depth. Except in the vicinity of Minneapolis, Rock Island, and Keokuk, the river flows across a bed of sand where bedrock is at too great a depth to bear the structural loading of a dam in its operating capacity. Thus, the majority of the dams are built on pile structures. Special precautions such as steel sheet piling, cutoffs, wide aprons of concrete in the dam section, together with stone protection below the dams and spillways, were taken to prevent undermining of the structure and endangering the stability of the dams during extremes in the flow regime.

A typical navigation dam site consists of the following: a lock section containing a main lock with dimensions of 110 ft by 600 ft and an auxiliary lock which was not usually completed; a section of submergible roller gates sufficient to pass ordinary winter flows, ice and drift; a section of tainter gates (some of which are submergible) to complete the main regulation section; a fixed overflow weir of concrete or protected earth embankment; non-overflow earth dikes to complete the closure.

At two dams, LaGrange and Peoria on the Illinois Waterway, Chanoine wickets with needles are used to form 80-mile pools. During flows above 15,000 to 18,000 cfs, the Chanoine wickets are lowered, and the river is in an open pass status, allowing towboats to pass over the submerged wickets without having to lock through the lock structure. This occurs, on average, about 60 percent of the time.

Two of the determining criteria for the use of wicket dams (compared to the use of movable gate dams) are the slope of the river and the amount of annual flow passing through each reach of the river. The lower Illinois River between LaSalle (Illinois) and the mouth of the Illinois River has a fall of about 0.1 feet per mile, while the Mississippi River has a fall of 0.5 feet per mile or (in some locations) greater. The dendritic river system of the Illinois Waterway has very flat contouring and, in general, tributary discharge levels change more slowly than they do for the Upper Mississippi River tributaries. Figure 3 illustrates a typical dam on the Mississippi River.

Further Legislative Action

Conflicts over river usage have occurred at least since the 1930s because of diverse, sometimes competing, uses by a variety of interest groups. Adding to this problem is the fact that societal and environmental uses of the river have changed over time. A 1974 environmental impact statement (EIS) for the 9-foot navigation project revealed that little information was available regarding human activities and interactions with the river resource (U.S. ACE, 1974). A Great River Study Team was authorized by the U.S. Congress' Water Resources Development Act of 1976 (PL 94-587). This act authorized the development of a river system management plan. The main purpose of the study was to resolve conflicts arising from multiple demands on river usage. This plan would provide for the widest degree of benefits to users without system impairment or degradation. The study focused on channel maintenance areas and associated biological, economic, and social impacts. A management plan was developed and implemented.

Additional requirements for the coordinated development (and enhancement) of the Upper Mississippi River were provided by Congress in the Upper Mississippi River Management Act of 1986 (Section 1103, PL 99-662). This act declared that it was the intent of Congress to recognize the system as a nationally significant ecosystem as well as a nationally significant commercial

U.S. Army Corps of Engineers
Rock Island District
L/D No. 21
Mississippi River
Rock Island, Ill.

Figure 3.

navigation system. The purpose of the legislation was to assure that the system was administered and regulated in such a way as to fulfill its several purposes.

System Assessment

The capacity of the inland waterway is defined as the maximum amount of commodities, in tonnage, that the system can sustain. It is usually the locks which limit the maximum tonnage throughout the system. Figure 4 illustrates the number of rail cars and trucks required to fill a single barge and to make up a total tow for transporting commodities on the Upper Mississippi River (Iowa DOT, 1978). Table 1 shows the number of fleeting areas and terminals by river pool location in the Upper Mississippi River system (U.S. ACE, 1987). A fleeting area is a location where barges are either parked for loading or unloading or to await being picked up by a towboat.

The locations of these fleeting areas and terminals are important for the economically centralized collection and stockpiling

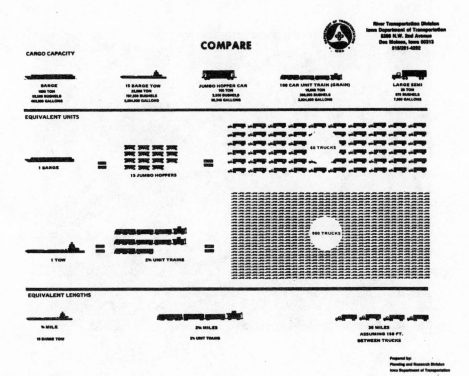

Figure 4. Comparison of total trucks or rail cars to fill a single barge or make up a tow.

Table 1

River	Number of Fleeting Areas	Number of Collection Terminals
Upper Mississippi	83	245
Illinois	21	37
St. Croix	0	4
Black	3	9
Minnesota River	2	11
Total Upper Mississippi River System	109	306

of each major commodity and for their efficient movement through the system. The limitations of a given lock are determined by any one of several characteristics: the physical lock dimensions, tow sizes and configurations, the time required for locking, the number of vessels with empty barges, the number of fleeting areas and collection terminals, recreational usage, and the seasonal nature of the transported commodity. Table 1 illustrates that the number of fleeting areas and collection terminals can concentrate the traffic in various pool locations close to the sources of the commodity, thereby creating traffic problems for certain locks. If a problem were to exist because of either high or low water, then a large amount of barge traffic as well as towboats could be stranded in an area. This would create an economic hardship on the shippers as well as on the terminal, which needs additional space to accept incoming commodities by truck or rail. There would not only be a direct effect on river collection sites, but also on the arterial transportation system and on the inland storage system, if commodities could not move at a normal pace.

Channel Maintenance

The dimensions of the navigation system's waterway channel were authorized by Congress under the various acts mentioned earlier. The dimensions of the navigation channel that were originally established by the 9-foot navigation project are continually subjected to the dynamic forces of the riverine system. Changing hydrological and meteorological conditions, towboat traffic, and variable channel dimensions, among other factors, result in the hydraulically unsteady-state condition which exists throughout much of the navigation system. However, the primary factor affecting the ability to maintain the 9-foot navigation system is the rainfall-runoff relationship of the Mississippi River Basin. Climatic conditions over the several thousand square miles of drainage area determine the system's eventual water yield. The areal extent of rainfall in the basin, combined with the erosive forces of surface water runoff, cause extensive movement of both suspended and bedload sediment. Sediment is deposited in channel locations, which can reduce the efficiency of the movement of towboat and barge traffic. Occasionally the channel becomes blocked because of large amounts of sediment deposited in the channel conveyance area.

During prolonged dry periods when river flow is very low, it is possible for an entire channel to be closed off because of sediment aggradation. Low water levels do not have sufficient carrying velocity to sustain the movement of suspended and bedload sediment. An extensive channel closure occurred from July 19 to July 29, 1987 (*Des Moines Register*, 29 July 1987). Such closures cause economic losses on the order of millions of dollars. Statements from barge line officials indicated that the channel closure cost the tow companies between $250,000 and $300,000 per day. The flow at that time was not at an historically low level, but at a flow level that occurs about once in 10 years. An article from the *Waterways Journal* (27 July 1987, 1) not only describes the same shoaling which blocked river traffic near Buffalo, Iowa, but also identifies several other areas in which the channel was about (or slightly less than) 9 feet in the Upper Mississippi River system. Figure 5 is a photograph of barges temporarily fleeted until the area could be dredged.

Figure 5.

The perpetuation of a dependable navigation channel requires the establishment and maintenance of a stable alignment which conforms to authorized project dimensions and in which a minimum amount of shoaling occurs. To maintain stable channel alignment, training works have been constructed to prevent bank erosion, limit meandering, and constrict the main channel, thereby concentrating flows and deepening the navigable portion of the river. While training works aid in the maintenance of project depths, rapid and fluctuating flow levels associated with variable climatic conditions still result in shoaling. Therefore, periodic maintenance dredging is required to maintain a navigable depth.

Towboat Operations

The constraints to navigation that determine the type and size of vessels that can operate on the inland waterway are to a certain extent a function of the river's dimensions (its width, depth, bend radius, channel alignment, etc.). Substantial economic savings can be realized by the use of large tows, in part because manpower requirements for tows are nearly constant over a wide range of tow sizes. In the Upper Mississippi River region 15 barges are easily maneuvered downstream to refleet into larger 25-barge tows below St. Louis, Missouri, at Lock 27, and finally into 40-barge tows below Cairo, Illinois. It is common for carriers to do everything possible to increase tow size in order to fully utilize the ability of the channel to accommodate tows. Adverse changes in hydrologic conditions often lead to the decision to reduce tow size or to push a smaller total load.

Channel dimensions, navigational constraints, and tow size all combine to determine the waterway on which a particular tow can operate. A tow of maximum size operated on a waterway with multiple constraints is associated with risks of increased travel time, higher total fuel consumption, delays in commodity movement, and reduced safety. Several waterways, including the Illinois Waterway and the Upper Mississippi River, tend to be limited to a maximum tow size of 15 barges. These waterways have relatively poor bends, controlling widths narrower than the ideal width for two-way traffic, and a number of narrow bridge spans. At the other extreme are rivers that are nearly ideal transportation arteries, such as the Mississippi River from St. Louis to the Gulf of Mexico.

U.S. Coast Guard accident statistics indicate that 86 percent of past waterway accidents have occurred on four of the major inland waterways. Two of those waterways are the Illinois and the Upper Mississippi Rivers (U.S. DOT, 1979). The Coast Guard also found that accidents were clustered over about 10 percent of the rivers' navigable length. The accident-prone segments of the river system tend to have the following characteristics: one or more bridges; a bridge and a lock; a bend or intersection of very narrow width. The high accident rates on these rivers can be ascribed to adverse and changing hydrologic conditions which are a direct result of climatic variability. It is very difficult to navigate a river with tricky currents and cross-currents which have been created by unsteady river conditions. Rapid changes in the flow regime can change river conditions in a relatively short period of time.

Commodities

Studies completed in 1979 and re-examined for the 1986–87 time period have shown that the total waterborne commerce of the United States has risen at a rate of about 3 percent annually (Institute of Water Resources, 1981; Kearney, Inc., 1974). This trend, however, does not accurately reflect the increase in foreign commerce trade, which has been between 5 and 6 percent. International trade comprises about half of the total waterborne commerce of the United States. The greatest increase of a single commodity is in crude petroleum. The level of imports of crude petroleum was at 44 million tons in 1953 and 90 million tons in 1970. That level increased dramatically in 1977 to 404 million tons (Institute of Water Resources, 1981). However, in 1979, market requirements dropped to 379 million tons of imported petroleum, and they have fluctuated since then because of a host of factors, including the increased domestic use of North Slope oil, as well as oil from other domestic sources, and the implementation of many domestic conservation programs. The major petroleum products include gasoline, residual fuel oil, distillate fuel oil, and jet engine fuel. Residual fuel oil has been the major component of petroleum product imports.

The major waterborne commerce commodities that are exported are grain, coal, and petroleum. The leading national export commodity is grain, the most important of which are corn, wheat, and soybeans. The export of these significant grains (nationally) has increased more than ten-fold since 1953. At that time, the export (nationally) was 10 million tons (U.S. ACE, 1986). In 1950, the Upper Mississippi River and the Illinois Waterway transported about eight million tons of commodities. By 1980, this figure had risen to 100 million tons with traffic equally divided between the Upper Mississippi River and the Illinois Waterway (U.S. ACE, 1986).

Four commodity groups (grain, coal, petroleum, and chemical products) constitute 75 percent of waterborne tonnage. Agricultural products, primarily grains, are the main commodities moving out of the midwest region of the United States served by waterborne commerce. The largest concentration of grain moves by the Mississippi River and its tributaries to the Gulf of Mexico for export. Only a small amount of waterborne grain moves to domestic markets. The inland water transportation system extends into the principal grain-growing areas in the Mississippi River Basin. Illinois (the number one state for grain sales), Iowa (second in grain sales), Minnesota, Missouri, and Indiana together provide the high volume of grain exports flowing down the Mississippi River. Grain moves from the areas of agricultural production and from the head of navigation on the Mississippi River downstream for export via the Mississippi and the Gulf of Mexico. Grain also moves from these states by rail and truck to ports such as Duluth-Superior on Lake Michigan for export shipment by way of the St. Lawrence Seaway. The major routes for waterborne grain shipments to New Orleans and other lower Mississippi River ports are the Illinois and Mississippi Rivers. These shipments account for about 45 percent of the grain exported. Corn, soybeans, and wheat comprise 40 percent of the traffic on the system (U.S. ACE, 1986).

Climate anomalies throughout the world have required a large increase in U.S. grain exports to supply basic foods to populations at risk to food shortages in areas that have suffered from climate-related food production problems. The vast majority of the commerce is moved by the inland waterway system. Many of the commodities are moved to a transfer or collection location where the inland waterway system collects both truck and rail tonnage and either moves or temporarily stores the commodities. The commodities are eventually moved downstream to the point of export where they are outloaded onto transcontinental freighters.

The Mississippi River is clearly a major artery for the transport of these and other important commodities.

CLIMATE VARIABILITY

A climate change could be either real or perceived. The scientific literature has pointed out the pitfalls of detecting trends using short historical records (see Katz, this volume). In addition, a climate scenario is not easily converted into a water resources scenario. For example, changes in temperature and precipitation regimes cause changes in cloud cover and radiation transfers that can alter existing patterns of regional agricultural and industrial development. Population distribution and movement can also change as climate regimes shift. As a result of a climatic shift, the types of crops that are grown in certain regions may change as heating and cooling degree-day totals change. Land use practices could also change, thereby causing an increase in the amount of runoff. Such regional climatic changes might cause changes in streamflow patterns that would surely result in social, environmental, and commercial changes in the inland waterway system.

Extreme conditions such as "too much" or "too little" river flow cause the greatest stress on the economic base of the Mississippi River system. A long streamflow record, the basis for analyses of the Mississippi River and Illinois Waterway systems, constitutes a tenuous set of information, in that it represents both deterministic and stochastic data. The variability in runoff, as shown in Figure 6, does not follow any easily forecasted trend. A smoothed ten-year running average generally shows some longer-term cycle of the watershed yield. These data, however, cannot necessarily be directly related to climatic trends of regional warming or cooling.

It is very difficult to estimate the potential direct effects of climate variability or change on the Mississippi River Basin. Such changes are difficult to measure, particularly when using unadjusted precipitation and temperature data. One difficulty is the existence of noise, ambiguity, or extreme outliers (very extreme, outlying, meteorological events) in the data set. Another difficulty is the complex delaying process of water retention because of groundwater zones, impoundments, evaporation, and other features of the hydrological cycle. Raw precipitation data from areas affected by water retention would have to be adjusted before the

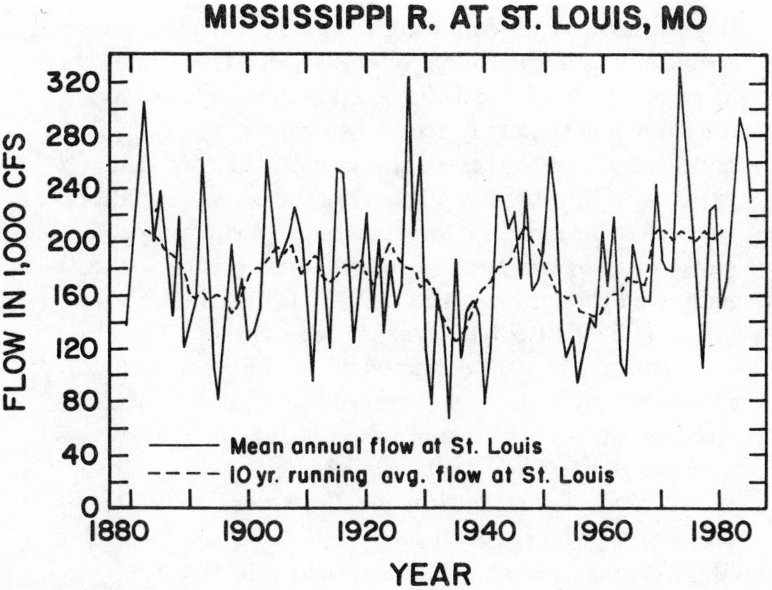

MISSISSIPPI R. AT ST. LOUIS, MO

Figure 6.

data could be applied to evaluate variations in streamflow. The effects of climate variability or change on streamflow can only be identified imprecisely without such adjustments. In fact, a small increase or decrease in mean annual precipitation or temperature may not be recognizable in streamflow data.

Societal and environmental impacts of climate change could be noted over an extended period of time as a result of associated changes in vegetative cover, shifts in residential and commercial land development, and changes in agricultural land-use practices. Exports of products and materials dependent on waterborne commerce, as identified in the Waterborne Commerce Statistics (U.S. ACE, 1986), can help to identify long-term shifts in the activities and responses of societies. For example, the stockpiling of grain at collection points or terminals has resulted from such shifts. More specifically, such stockpiling might occur for any one (or a combination) of the following reasons: waiting for a market price change; the cost of stockpiling is preferred to shipping by another more expensive mode of transport; the lack of availability of other transportation modes; tax purposes; government policies, etc. An article in the *Waterways Journal* (31 August 1987) stated that

Record farm yields following record farm yields, not to speak of the influence of roller coaster export markets in recent years, have U.S. officials bracing once again for problems that may require storage in ground piles, barges, rail cars and caves throughout the corn belt. (If exports hadn't increased this year the problem could be worse.) A major bin manufacturer expects "panic" because of the projected 7.2 billion-bushel corn harvest. And even though that's 12 percent less than last year's crop, The *Wall Street Journal* reports, it is far more than the government expected after idling some 70 million acres of land. And a record five billion bushels of surplus corn exists from previous harvests. Wheat production is expected to be up about 5 percent to 2.1 billion bushels and cotton up 33 percent to 12.9 million bales. The soybean harvest is expected to be flat at 2 billion bushels. The scenario sets the stage for additional barges to be taken out of operation and the possibility of improved rates for those remaining.

With the increased capability for high levels of grain production in the Mississippi River drainage area, additional stockpiling of grain could result if a climate change occurs. Climatic shifts may not necessarily produce new conflicts of interest, but they could exacerbate existing conflicts. Several federally funded programs have been resorted to in order to resolve conflicts among competing users of the river system. However, existing conflicts might be further aggravated by a climatic change that could cause existing waterways to experience more frequent extremes of streamflow (either higher or lower flows) for extended periods of time than have usually been experienced.

CLIMATE IMPACTS

Historical Experiences

The climate of the Upper Mississippi River Basin has been a major factor in the region's agricultural, commercial, and industrial development. From the very early period of navigation, shippers were subject to the effects of climate variability, and on occasion they experienced considerable difficulty navigating the length of the Upper Mississippi River. Early shipping development took place during a period recorded as the driest 20 years (1808–1827) of the reconstructed historical record (Waite and Partington, 1986). During this time, the use of the waterway rapidly increased, providing increased accessibility to buyers who were willing to pay high prices for commodities. The types of transportation also changed rapidly. After 1827, the climate was wetter, and navigation had few problems maneuvering through the system, except for submerged trees and snags which damaged (or sank) boats and barges.

During the Civil War, the historic low water period of 1863–64 occurred, and traffic was unable to proceed north of LaCrosse, Wisconsin. Cargo rates increased dramatically and, consequently, the railroad system began to carry increasing amounts of freight (Belcher, 1947). Low rainfall and subsequent low water caused merchants to pressure their congressmen to appropriate $400,000 for river improvements on the Upper Mississippi River system. The construction of a safer passage through the Des Moines Rapids was part of these improvements and was undertaken in order to lower shipping costs and to increase the quantities that could be shipped on each load.

The Des Moines Rapids Canal at Keokuk, Iowa, was completed and opened in 1877, but shippers continued to pressure Congress for further improvements. The low water in 1863–64 remained paramount in users' minds and resulted in the 4.5-foot channel being authorized by the U.S. Congress in 1878, with several small locks to be constructed at the steepest portion of the river. This congressional decision followed a general low water period that extended from 1856 to 1872. The rainfall recorded at stations in the Upper Mississippi River Basin for this period was below normal and yielded a period of low run-off (U.S. ACE, Rock Island and St. Louis Districts, historic stage records).

Rainfall increased in 1880 and remained considerably above normal until 1885, only to be followed, again, by a substantial decade-long low rainfall period from 1886–1896. Even with record low flows during this period, steamboat traffic peaked in 1892. At that time, 5,468 steamboats, 1,000 barges, and 2,000 log- and lumber-rafts were recorded in the Mississippi River system (Hartsough, 1934). In November 1906, a convention in St. Louis, Missouri, formed the Lakes-to-Gulf Association, which petitioned

President Roosevelt to improve the inland waterway. In March 1907 the 6-foot channel was authorized.

During the spring of 1927, the largest magnitude flood ever recorded was experienced on the Mississippi River. Later in 1927, Congress authorized a survey for increasing the depth to a 9-foot channel. Through the efforts of President Herbert Hoover, and with the support of several special interest groups, the project became a reality in July 1930. The project was undertaken at the right moment, as a prolonged midwest drought produced very low water conditions on both the Mississippi River and the Illinois Waterway from 1930 to 1937. The low water (and lack of precipitation) reinforced the need for the 9-foot navigation system to move commerce on the inland system. The majority of the locks and dams were constructed during this low water period.

In October 1956, the water depth at Alton lock in Illinois decreased to a point well below the 9-foot authorized navigation depth, because of a prolonged drought in the basins that are tributary to the Missouri and Upper Mississippi Rivers. As a result, navigation operations were severely restricted and about 200 barges, essential to interstate transportation of vital materials, were stalled at Alton, Illinois. The Metropolitan Sanitary District of Greater Chicago increased the flow from Lake Michigan by 1,900 cfs for 10 days beginning 23 October 1956. This was all that was possible if the diversion was to be held to the authorized average yearly limitation of release for 1956 of 1,500 cfs. In early December 1956, the State of Illinois filed a petition with the U.S. Supreme Court for an emergency 100-day increase in the diversion of Lake Michigan water to 10,000 cfs to counter the low water conditions in the Mississippi River. The Supreme Court granted modifications on 17 December 1956, to allow for an increase in the diversion of Lake Michigan water not to exceed an average of 8,500 cfs for 46 days. On 28 January 1957, the State of Illinois petitioned for an extension because of continuing drought. An extension was granted until 28 February 1957.

Severe low flow conditions on the Mississippi River during the winter of 1980–81 resulted in river stages insufficient for navigation. This produced significant losses and hardships not only for shipping interests but also for those who rely on the items transported along this major inland waterway system.

Several of the principal congressional decisions that have been made for improvement of the inland waterway system have occurred during periods of extreme meteorological or climatological events, demographic change, land use change, increased requirements for the movement of commodities, and stresses on the users of the system. During recent years, climate extremes in the Upper Mississippi River Basin have caused millions of dollars in lost production, in addition to other related economic costs and losses (Waite and Partington, 1986). The increasing frequency, intensity, or duration of extreme meteorological events experienced in recent years in the United States has also been reported to have occurred in Europe (Lamb, 1982). Regionally, Iowa records for the most recent years indicate increasing climatic stress (Waite and Partington, 1986).

The greatest climate stress to date occurred during 1959–1986. This period ranks as the wettest 28-year period in the climatic history of the Mississippi River Basin. The extreme wetness of this period is also apparent in the recorded streamflow, which was well above normal for the Upper Mississippi River Basin (United States Geological Survey Water Resources Data for the period 1879–1986). Examination of a mass flow curve (a technique for examining long-term trends in river flow (Butler, 1957)) indicates a change in total runoff, which increased from 1959 to 1986. This is a result of a much wetter climatic period in the Mississippi River Basin and, in turn, resulted in a greater number of flooding events.

During the period of design and operation of the navigation system, corn and soybean yields have increased (Table 2) as a result of moisture availability as well as the modernization of farming technologies and practices. Similarly, total crop production has set new harvest records, producing surpluses of each commodity.

The historical record of temperature for the Upper Mississippi Basin has also been examined. The data, however, do not reveal a continual warming trend. The average temperature for the last three decades has increased only about three-tenths of one degree. This is not conclusive regarding a trend toward a climatic warming or cooling. However, examination of snowfall data for the same period in the Upper Mississippi River Basin has shown an increased length of snow season during the last two decades. Snowfall has begun earlier in the fall, in October, and has lasted

Table 2

Year	Soybeans	Corn
	(bushels per acre yield)	
1930	16	40
1959	26	65
1984	36	110

Source: U.S. Department of Agriculture, 1984.

later into the spring, into April. In fact, some of the heaviest snow-falls have occurred in late March or early April. In addition, the last two decades have experienced greater variability in snowfall amounts. Previously, the snow season lasted from mid-November to late March.

Climate Change

A basic philosophical guideline used by all water-resource planners has been that the recent past is the best analogue for the near future. Engineers have long appreciated the fact that the climate is highly variable, and that sequences of observed river discharge are unlikely to be exactly repeated in future periods. Uncertainty regarding future flows has led engineers to use long accurate records and to design water projects conservatively when undefined trends exist in the climatic and hydrologic records. When a water resource system is finally built, it must be operated in such a way as to maximize the resilience of users to changing conditions. Although river discharge is closely related to climate, it is not a variable that climatologists usually consider when they discuss climate, climatic variability, and climatic change. However, even precipitation, which is a more popular indicator of climatic conditions than streamflow, has proved to be difficult to forecast.

Certain trends suggest that some change in the earth's climate is occurring. In addition to climate-related predictions that consistently appear in the media, social changes can be expected to occur. In identifying a climate change scenario that would affect the efficient use of the total Mississippi River and Great Lakes Navigation System, variations in streamflow which would result from a change in regional climate provide the most critical boundary conditions. Critical boundary conditions have been established by known physical criteria for safe navigation operation on the entire navigation system.

The boundary conditions that would affect the Mississippi River system can be defined as "too much" and "too little" water to allow for the continued operation of the system. The lack of water (i.e., the "too little" situation) would alter demographic and economic trends in the Midwest and in the areas that produce commodities. Inhabitants of this region have, in general, considered water to be an inexhaustible natural resource, and water-related projects have been designed (and used) accordingly. Unexpected prolonged regional shortages resulting from a climate anomaly would prompt ad hoc adjustments by society, as well as a depressing effect on the regional economy, as witnessed by responses to the prolonged reduced river flow in the summer of 1988.

Drought policies and contingency plans are not used very often because of the relatively high probability that such policies would be outdated by the time they would need to be enacted. Short-term droughts or periods of climatic stress such as occurred during 1976–77 and 1982–83 prompt societies to react. Such reactions, however, tend to be temporary and regional in nature, based on little advanced planning and with little, if any, preparation for the possibility of future similar encounters. Extended periods of low water have not yet been experienced in the Mississippi River Basin in combination with the present demands of society. Therefore, one can only speculate about some of the potential responses to such very prolonged dry conditions. The river's physical (morphological) response, however, can be identified more easily based on analogies of the responses of other river systems under prolonged conditions of "too little" water.

An excessive amount of water is experienced for short durations just about every year in the Mississippi River Basin. An extended period of high flows of more than a few months, however, has not been experienced regionally, but such a condition would also have an impact on the river system, particularly on its ability to efficiently move commodities. In the climate change scenarios of "too much" or "too little" river flow, several identifiable impacts on the river system would occur. However, a comprehensive listing of adverse impacts on the navigation system is not available, nor have all the areas that would be potentially affected by such adverse impacts yet been identified. Clearly, additional evaluations of many of the physical, chemical, biological, and economic impacts are necessary.

Possible Impacts of High Flows

In contrast to low flow levels, higher flow levels on the Illinois Waterway would primarily benefit the entire system, including the Mississippi River, except in the upper portion of the waterway called the canal. Higher flow levels there would increase the velocity in the system more than 5 feet per second, making navigation more difficult. In addition, the upper approach to the Marseilles Lock and Dam would experience higher velocities which could generate conditions that may prove unsafe for the movement of commodities by standard-sized towboats. Thus, maneuverability would be restricted in the upper portions of the channel reaches.

Impacts can also be identified in the Upper Mississippi River. For example, towboats with larger horsepower engines, such as shown in Figure 7, would traverse the system more easily during high flows. This would be true as long as the discharge levels were below the threshold levels that would require removal of the miter gate motors from the lock walls, thereby immobilizing the lock. The miter gate motors are used to open and close lock gates, allowing ingress and egress of boats and barges to the various pools. The controlling discharge level at which the motors are to be removed is 157,000 cfs. If this flow were reached, the lower miter gate lock motors would be removed at Lock 12 (at Bellevue, Iowa) and Lock 16 (at East Muscatine, Illinois) (U.S. ACE, 1982).

Higher discharge levels would increase channel velocities to such an extent that some towboats and full barge loads would have difficulty maneuvering the tighter turns in the system. The increase in velocities would also increase the potential risk of boat accidents because of vessel control, vessels passing in narrow areas, and the limited sizes of bridge openings and approaching entrances to the lock chambers. The chance of barges colliding with lock entrance guide walls would also increase as a result of the existing design of the lock structures and the entrance geometry.

A wetter climate would increase the erosive potential in the upland watershed, on the tributary streambanks, and in the main

Figure 7. Towboat with larger horsepower pushing 14 empty barges during higher than normal flow.

Figure 8. Dredged material on an island in the Mississippi River.

channel. The increase in the amount of suspended and bedload sediment carried by the river would enhance the potential for material to be deposited in the main channel, which would in turn slow down tow traffic because of reduced available depth or blockage. This situation would require dredging and the placement of the dredged material on an island (Figure 8).

The average side channel and backwater of a braided stream like the Mississippi River has a life expectancy of about 80 years (U.S. ACE, 1987). An increase in flows would cause greater sediment deposition in these backwater areas, reducing that life expectancy. Additional sedimentation would degrade aquatic habitats on all sections of the river. Thus, the marsh-type habitats that were created by the lock and dam system would be threatened by the accelerated rate of sedimentation. Backwater and side channel sediment deposits would adversely affect fish population as well. The Upper Mississippi River system has about 249,000 acres of bottomland forest (U.S. ACE, 1987). When flooded, these areas provide spawning habitat for fish. They also provide food and cover for many species of wildlife. Backwater sedimentation would provide new forest habitat by way of plant succession.

The total Upper Mississippi River commercial fish catch is about 13 million pounds with a value of $1.6 million (U.S. ACE, 1987). The bulk of the catch consists of carp, buffalo fish, catfish, and drum. The top five sport fish include bluegill, crappie, bass, sauger, and walleye. Total expenditures on the Mississippi system are about $55 million annually with about 11 million fisherman days. The loss of side channels because of sediment accretion would be a threat to the fishing industry. This is a major resource, the loss of which would affect its many users (U.S. ACE, 1987).

The Mississippi River is also a major flyway for waterfowl; the mallard is the most abundant duck on the flyway. The dollar value of this resource is estimated at about $9 million, a value derived from hunting fees. This source of revenue could be lost as a result of the loss of backwater because of sediment deposits (U.S. ACE, 1987).

Higher than normal flows for an extended period of time could cause erosion of streambanks and flooding of properties, scenic highways, and some municipal utilities that are situated adjacent to the river. A program for the protection of landowners would increase the need for federal and state involvement in protection from prolonged higher river levels (U.S. ACE, 1987).

Some of the policy responses to be considered during periods of high flow include the restriction of tow traffic movement on portions of the river for reason of safety, wave wash on levees, and erosion. Another policy response would be to take action in case of a national emergency such as the need to continue product movement to support the armed forces in the event of a national mobilization. In addition, the needs for emergency dredging and the suspension of environmental constraints levied by various states on dredge spoil placement would have to be addressed.

Possible Impacts of Low Flows and
Increased Diversions from Lake Michigan

During a prolonged low flow event in the Mississippi River Basin, maintaining navigation on the Illinois Waterway and the Mississippi River below Grafton, Illinois, is accomplished by releasing water from dams or the diversion of Lake Michigan water into the Illinois Waterway. The Great Lakes are a vast water resource. However, the amount of water that can be diverted directly from Lake Michigan is limited hydraulically by the level of the lake and the capacity of the controlling works at the lakefront points. Increasing the diversion of water from Lake Michigan at Chicago for emergency purposes would have a temporary effect of lowering the levels of the Great Lakes if done for a short period of time. As the levels of the Great Lakes naturally fall below their long-term average levels, this additional lowering could result in significant adverse environmental and societal impacts.

The river and canal system in the Chicago area is operated primarily for the purposes of flood control, navigation, and the maintenance of water quality. The downstream control structure at Lockport, Illinois, is regulated to maintain water levels at sufficient depths for navigation. One consideration that would affect the amount of increased diversion concerns the resultant increased flow velocities. As flow velocities in the canal system approach 5 feet per second, normal navigation practices in the Upper Illinois River would be restricted. This would limit the amount of increased diversion that could be allowed to pass through the canal system to a maximum total discharge measured at Lockport, Illinois, of approximately 10,000 cfs.

A concern of residents along the Illinois Waterway, downstream of Chicago, is the increased potential for additional flooding because of increased diversions. Additional flooding might occur as a result of increased diversion discharge still being in the waterway system when a rainstorm occurs over a portion of the lower Illinois Basin. To reduce the likelihood of such a situation, increased diversion rates could be allowed only when the Illinois Waterway flows were predicted to be less than 30,000 cfs at Meredosia, Illinois, regardless of the needs on the Mississippi River. This concern, however, would be of less importance in considering the climatic change scenario of low flow, because if low flow conditions were occurring on the Mississippi River, the Illinois River Basin would most likely also be experiencing low rainfall and, therefore, low flow conditions. However, continuous increases in diversion rates over extended periods of time greater than a few months might not be possible without causing environmental and societal damage on the Great Lakes and Illinois Waterway.

The most significant benefits to navigation of increased diversions would be realized immediately downstream from the confluence of the Illinois and Mississippi Rivers, where the flow contribution of the Illinois River to the Mississippi River is the greatest. The critical stages, with their corresponding flows, at which navigation becomes impaired for two major locations are listed in Table 3. These values are used to indicate when it would be necessary to divert Lake Michigan water to the Mississippi River. If water were not diverted during low flow periods, commercial boat operators would have to push loads carrying fewer commodities, dredging of the channel could be required, and in some cases navigation might be delayed.

The impacts of increased diversions on the physical, biological, and cultural characteristics of the Illinois River would be expected to be both beneficial and detrimental. Those impacts would be highly dependent on antecedent river conditions and would vary widely depending on the season, duration, and magnitude of the diversion increase. Surface water levels would be increased throughout the Illinois and Mississippi River systems to provide the authorized minimum 9-foot channel. Flow velocities would also increase slightly in the canal section of the Illinois River. Surface acreage

Table 3

Mississippi River
Low Flow Stages and Discharges
Requiring Lake Michigan Diversion

Location	Gage Height (ft)	Discharge (cfs)
St. Louis, Mo.	2.0	78,000
Cairo, Ill.	12.0	190,000

Source: Institute of Water Resources, 1981

of bottomland lakes would be expanded, thereby submerging adjacent mudflats. The Peoria slackwater pool would be most affected, with up to 100 percent of the mudflats inundated, while the lower pools would be somewhat protected by natural and artificial levees.

Increased diversion of relatively clean Lake Michigan water would improve the quality of water in the Illinois River, in general, with changes most pronounced at the upstream reaches. The primary parameters that would be changed with such a diversion are dissolved oxygen, turbidity, toxic material levels, temperature, and mineral content of the water.

The increase in river stages because of increased diversions would result in the reflooding of backwater areas, with suspended sediments carried into and deposited in these areas. Water level change would affect populations of ground-dwelling fur-bearing species. Reflooding of the backwater areas would destroy ongoing succession and areal expansion of certain terrestrial communities.

With increased diversion rates to support low Mississippi River stages, stages in the lower Illinois River would increase until the Illinois River could go to open pass navigation (lowering the Chanoine wickets to the bottom of the river) at the Peoria and LaGrange Dams. Cost savings from increased open pass navigation would be dependent on the season of the increased diversion because less navigation takes place during the winter months.

Diversion impacts on power generation would increase at the Marseilles power plant, until the flow reached a level at which the head would be lowered, reducing the efficiency of the hydropower turbines. However, at Lockport, power production is not limited by the loss of head, and increased flows would lead to increased power production.

Direct adverse effects on bottomland agriculture could result from increased diversions for both leveed and unleveed lands. The impacts attributable to increased diversions would be greatest in low water years. Detrimental effects to leveed lands, assuming that no additional overland flooding were to occur, would primarily be financial, because of the increased pumping costs that would be required as a result of increased water seepage from higher river levels. Unleveed lands would be adversely affected by both increased groundwater levels and overland flooding, particularly in low water years, because of increased diversions. River stage increases that would result from a diversion might delay or prevent crop planting, if the discharge increase occurred before planting. In some cases it might lead to the planting of a crop of lower value.

The diversion of water from Lake Michigan is not a simple process of opening the gates and releasing water into the Illinois Waterway system. Because of the political complexities associated with the diversion of water from the Great Lakes, and the lowering of those lakes, the federal courts have placed constraints on the magnitude of diversion releases. These legal constraints would have to be overcome before a change in the present operating plan could be accomplished.

The Mississippi River lock and dam system from St. Louis, Missouri, to St. Paul, Minnesota, on the Mississippi River, and from the mouth to Chicago, Illinois, on the Illinois Waterway was designed to maintain a minimum 9-foot channel in each pool. This would be accomplished by the closure of moveable gates on each dam. Each gate could be placed at a zero setting and, theoretically, no water would be released in that gate bay. Gates can be closed, releasing only the water flowing into the pool from the upstream dam and the local tributaries, while the authorized pool level is maintained. Figure 9 shows Dam 15 at Rock Island, Illinois, during low flow conditions.

On the Mississippi River where the pools are more braided, the backwater areas are not continuously replenished with moving high water under low flow conditions. When this occurs during

Figure 9. Mississippi River dam with gates nearly closed during lower water period.

a warm period, the algal population increases and fish kills occur. During the low flow period in the summer of 1987, two large fish kills were reported (at Cordova, Illinois, and at Keithsburg, Illinois).

Recreational boaters, who normally are not aware of river controlling works, collide with the submerged wing dams during low flow periods, which causes damage to their craft and occasionally injures boaters. With continued siltation of the backwaters, what is presently an excellent sport and commercial fishery would be destroyed as a result of the "too little" climate scenario.

Additional channel closures could be expected, which would slow the movement of commodities. If low river flows persisted for a long period of time, the supporting watershed area would also become increasingly dry, resulting in lower crop production throughout the Mississippi Basin.

The water quality problem generated during low flow periods would potentially be more serious. Presently, both communities and industries treat effluent before it enters the river system. However, the level of treatment is commensurate with the mixing level of the river, and if river flows are low for extended periods, the proportion of pollutants entering the system would increase. Overall health requirements would need to be evaluated, since this water is utilized in many aspects of daily human life.

Some of the policy questions that have to be examined during low flow events would center on competing interests, such as navigation, hydropower, irrigation, water supply, water rights, water quality, recreation, and fish and wildlife. Lack of formal commitment does not preclude the responsibility to define and establish minimum objectives that may be achieved without jeopardizing authorized project objectives. The minimum goals during low flow periods should be based on analyses which take into consideration all system and project impacts in order to achieve the best use of the limited resource.

CONCLUSIONS AND REFLECTIONS

In this chapter the climate change scenario has been examined with regard to potential effects on the Mississippi River navigation system. The information presented does not conclude that a climate shift has taken place or that a such a trend has already started. In this respect, the indicators of a climate shift that would be used by an operating agency like the Corps of Engineers would fall into the categories of real-time data and archived data. To determine whether a potential for a flood or a drought exists, the following data are examined: temperature, precipitation, groundwater levels, streamflow, and such indices as the Palmer Hydrologic Index. Typically, short-term information may yield indications of a trend, and analogues are searched for in the historical records. In many instances the situation proves to have been only a short-term trend. Governments and people have adjusted to these short-term trends and have made decisions accordingly. If an extended wet or dry period occurs, then some alarm is raised, usually by the media and the public. Eventually government agencies respond.

Funding to monitor continuously all available meteorological and hydrological parameters is not the responsibility of any one agency. Most agencies are only funded to obtain the necessary information to operate their own projects within their own jurisdictions. Most often, for both agencies and individuals, the response is one of crisis management, which, by the way, is not necessarily an inappropriate response. In many cases, the historical development of the inland waterway system has occurred during crises. With regard to a potential crisis some time in the relatively distant future, important questions must be answered: who has the responsibility to decide if a climate change is actually occurring? What agencies or groups have information to support that conclusion? How long will it take for a consensus to be reached about the change and its possible impacts? Will the nation's decisionmakers simply assume that it is just another short-term trend?

The high- and low-flow scenarios highlight for planners and policymakers some of the areas with which they should be concerned. Long-range planning for potential crisis situations that might arise decades in the future has not been the usual format within government (or, for that matter, any) organizations. In addition, the amount of time that passes between similar past events and evaluations of "what would we do if it were to happen again" has varied. Many observers take the attitude that "I don't believe that will ever happen again." Apparently people tend to wait for extreme meteorological events to occur before taking action.

Conflicts in usage of a system will always exist, whether associated with climate variability and climate change, or with social variability or social change, or with normal conditions. The projection of future water demands and multiple uses for water should be investigated more comprehensively in a regional framework study.

Changes in such disparate activities as urbanization, channelization of streams, flood control, upland conservation, and agricultural land-use practices will only heighten the importance of regional problems related to climate change and to climate variability. Identification, analysis, legislation, and implementation of an action plan can result in prudent national usage of the inland waterway system.

EPILOGUE

In 1987 decisionmakers as well as the public believed that the situation could not get much worse for the Mississippi River system, having witnessed numerous groundings of vessels and very low river flows. However, extended dry conditions plus long runs of record-breaking daily high temperatures continued throughout the spring and summer of 1988. Under these extreme meteorological conditions, river levels continued to decline in the Upper Mississippi Basin. Rainfall was also greatly reduced throughout the entire Mississippi watershed. These low flows caused numerous groundings of vessels and delays of traffic and movement of commodities in the inland waterways system. Several low-water records were set, both in stage and in flow. As of August 1988, new low-flow records were being set. The agricultural heartland, which produces a great deal of the world's crops, has suffered from the dry conditions and the prolonged extreme heat. As a result, agricultural commodities produced in the region will be greatly reduced in 1988.

The Upper Mississippi above St. Louis and the Illinois waterway both were able to sustain navigation traffic due to an efficient lock and dam system. Below St. Louis, however, considerable difficulties were encountered because of channel closures and shallow water conditions; towboats were bottlenecked as a direct result of extremely low-flow conditions. Dredging operations were carried out 24 hours a day during much of the summer in an attempt to maintain the movement of river traffic.

Mississippi River users as well as some government agencies continued to search for sources of additional water during the 1988 drought with such areas as the headwaters of the Mississippi River and the Great Lakes being considered. Chronic political conflicts over ownership and use of Great Lakes water surfaced once again between the United States and Canada as well as between and within U.S. states. It is most likely that no water will be released from the Great Lakes to support navigation on the lower Mississippi.

REFERENCES

Belcher, W.W., 1947: *The Economic Rivalry between St. Louis and Chicago 1850–1880.* New York, NY: Columbia University Press.

Butler, S.S., 1957: *Engineering Hydrology.* Englewood Cliffs, NJ: Prentice-Hall.

Des Moines Register, 1987: Mississippi is cleared, shipping resumed. 29 July.

Hartsough, M.L., 1934: *From Canoe to Steel Barge on the Upper Mississippi*. Minneapolis, MN: University of Minnesota Press.

Institute of Water Resources, 1981: *National Waterways Study: A Framework for Decision Making*. Final Report NWS(T)-81-2, January. Fort Belvoir, VA: U.S. Army Engineer Water Resources Support Center, Institute of Water Resources.

Iowa DOT (Department of Transportation), 1978: *River Transportation in Iowa*. May. Ames, IA: Iowa Department of Transportation.

Kearney, A.T., Inc., 1974: *Domestic Waterborne Shipping Market Analysis*. February. Chicago, IL: A.T. Kearney, Inc.

Lamb, H.H., 1982: *Climate, History and the Modern World*. New York, NY: Methuen and Co.

Lewis, H., 1967: *The Valley of the Mississippi*. St. Paul, MN: Minnesota Historical Society.

Peterson, W.J., 1979: *Steamboating on the Mississippi*. Cranbury, NJ: A.S. Barnes and Co.

U.S. ACE (Army Corps of Engineers): *Mississippi River Historical Stage Records*. At selected stations, 1875 to 1986. Files located in the Rock Island District Office, Rock Island, IL.

U.S. ACE (Army Corps of Engineers), 1974: *Final Environmental Impact Statement, Upper Mississippi River 9-foot Navigation Channel Operations and Maintenance*. Rock Island, IL: Rock Island District.

U.S. ACE (Army Corps of Engineers), 1982: *Master Reservoir Regulation Manual, Mississippi River*. Rock Island, IL: Rock Island District.

U.S. ACE (Army Corps of Engineers), 1986: *Waterborne Commerce of the United States through 1986*. Washington, DC: U.S. ACE.

U.S. ACE (Army Corps of Engineers), 1987: *Draft Environmental Impact Statement, Second Lock at Locks and Dam 26, Mississippi River, Alton, Illinois, and Missouri*. St. Louis, MO: St. Louis District.

U.S. Department of Agriculture, 1984: *Statistical Bulletin.* Statistical Reporting Service, Crop Reporting Service. Washington, DC: U.S. Department of Agriculture.

U.S. Geological Survey, *Water Resources Data*, one volume produced annually for each state. Reston, VA: U.S. Geological Survey.

U.S. DOT (Department of Transportation), 1979: *Human and Physical Factors Affecting Collisions, Rammings, and Groundings on Western Rivers and Gulf Intercoastal Waterways.* Report CG–D–80–78. Washington, DC: U.S. Department of Transportation.

Waite, P.J., and M.M. Partington, 1986: *Iowa Precipitation Variations, Past, Present, and Future.* Des Moines, IA: Iowa Department of Agriculture.

Waterways Journal, 1987: Record farm yields. 31 August, 4.

Waterways Journal, 1987: River bottom's up, traffic is down. 27 July, 4–5.

12

Climate Variability and the Colorado River Compact: Implications for Responding to Climate Change

Barbara G. Brown

INTRODUCTION

The Colorado River system, one of the most important U.S. water systems due to its location in a semiarid region of the country, is controlled by a complex system of laws, treaties, compacts, and agreements. The history of the formation of these regulations is dominated by fierce competition for control of the water, determination not to "waste" a single drop of water by allowing it to flow to the sea (Fradkin, 1981), and changing perceptions regarding the amount of water that is available. Perhaps the last of these factors—changing perceptions of water availability—has had the greatest influence on the regulation of the river and has led to a large extent to the current problems of managing the Colorado River Basin. Early misperceptions that the river's flow would be adequate to meet all future conceivable needs has resulted in overallocation of the river flow and severely reduced water quality.

The size of the Colorado River is a poor indicator of its importance as a U.S. river system. In particular, the Colorado River

Barbara G. Brown is an Associate Scientist with the Environmental and Societal Impacts Group at the National Center for Atmospheric Research. She holds M.S. degrees in environmental sciences (University of Virginia) and statistics (Oregon State University). Her research interests are in applications of statistics in environmental sciences; climate impacts; and the value and use of meteorological information.

has only the sixth largest flow among U.S. rivers; the total volume of water carried by the river is only about one-tenth the volume of the Columbia River (Skogerboe, 1982). However, the semiarid climate of the Colorado River Basin forces farmers and municipalities to depend on the river to nourish local agriculture and to feed city water supply systems. The river has thus gained great importance to individuals as well as to local, state, and national governments.

The Colorado River flows 1,400 miles from the high mountains of Colorado and Wyoming to the canyons of Utah, Arizona, and New Mexico, until finally, as a trickle, it flows into Baja California in Mexico. The total basin covers about 243,000 square miles (Figure 1), an area that is approximately equivalent to the size of the Columbia River basin (Skogerboe, 1982). The headwaters in the upper basin states (Colorado, Wyoming, Utah, and New Mexico) contribute approximately 83 percent of the annual flow of the Colorado River (Skogerboe, 1982), an annual flow that averages about 13.5 million acre-feet (maf) (Stockton and Jacoby, 1976). Annual precipitation in this region is relatively low, with averages ranging between 10 and 16 inches.

While filled with stories of fierce competition for the scarce regional water resources, the history of regulation of the Colorado River is also a story of unique agreements for sharing those resources. This history is eloquently and thoroughly described by Hundley (1975). The primary agreement for sharing the water of the Colorado River is the Colorado River Compact of 1922 (henceforth, the Compact) which divides the runoff between the Upper and Lower Basin states, where the Upper Basin includes Colorado, Wyoming, Utah, New Mexico, and part of Arizona, and the Lower Basin consists of Arizona, California, and Nevada. The Colorado River Compact was the initial use of the compact clause of the U.S. Constitution, which allows states to make treaties among themselves with Congressional approval, to divide the waters of an interstate stream. Thus, it represented a unique application of federal law toward settling disputes among the states regarding their water rights.

However, the Colorado River Compact based its division of the runoff on a rigid strategy, dependent on a specific average flow of water through the river system. That is, the division of water is on an absolute rather than proportional basis. Furthermore,

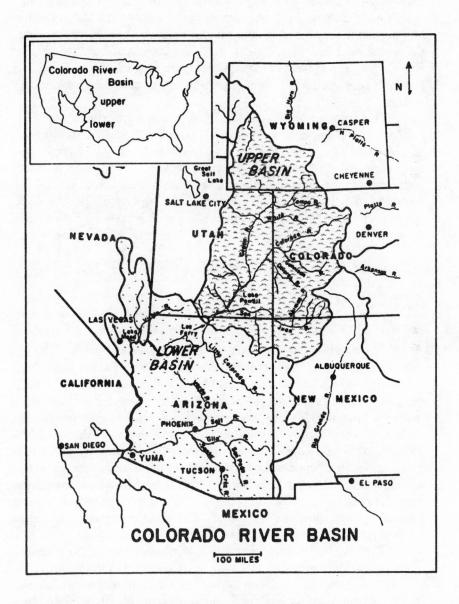

Figure 1. Map of the Colorado River Basin showing the division between the Upper and Lower Basins.

the Compact contains no provisions for possible re-negotiation, regardless of any changes in environmental, climatic, or demographic conditions. Hence, the provisions of the Compact are not sensitive to long-term variations in climate, to climate change, or to alterations in water-use priorities. Later components of the "Law of the River" (i.e., the system of regulations governing the use of Colorado River water) similarly reflect this rigid approach to division of the water.

The story of the Colorado River—and particularly the Colorado River Compact—represents a potentially useful analogy for the impacts of climate change on society in that it illustrates the long-term effects of imposing a rigid regulation strategy and the consequences of ignoring climate variability and the potential for climate change. This chapter explores the history of the Colorado River Compact and its implications regarding responses to future climate change, and it considers how long-term information about runoff has changed the perspective of policymakers regarding the outlook for the Colorado River and uses of Colorado River water.

THE COLORADO RIVER COMPACT

Brief History

The need for the Colorado River Compact arose out of the desire of the states in the Lower Basin—particularly California, which was developing and growing quickly—to construct dams and canals to make use of the water in the river, to divert water to California without entering Mexico, to generate electricity, and to prevent floods. Coincidentally, the Upper Basin states feared that such development would establish prior rights to the water by California (i.e., through prior appropriation) and preclude future uses and development of water resources by the Upper Basin states (Howe and Murphy, 1981; Hundley, 1975). Thus, states in both parts of the basin sought some type of agreement regarding individual states' rights to Colorado River water.

Rather than resorting to the federal courts to settle their differences, as had been the solution to such disputes in the past, the Colorado River Basin states agreed to attempt to use a clause in the U.S. Constitution which allows states to establish agreements—compacts—between themselves with final Congressional approval. In August 1921 Congress approved the compact process for use by the Colorado River states, and the states began formal meetings in January 1922 to formulate a compact. Representatives of the states met several times throughout the year and held several public hearings before they finally reached agreement—after much disagreement—in late November.

Following agreement by the commissioners themselves, the Compact had to be ratified by each of the seven states and finally approved by Congress. Ratification turned out to be an extremely difficult process, with individuals and groups within each state expressing suspicion or some disagreement regarding potential impacts of the Compact provisions. In particular, the state of Arizona refused to ratify the Compact because that state felt it would not be allocated a fair proportion of the water. Specifically, Arizonans desired control of all water arising from Colorado River tributaries in Arizona.

In addition, California required passage of the Boulder Canyon Project Act—which authorized the construction of Boulder Canyon Dam (now known as Hoover Dam)—before the state would ratify the Compact. Following passage of this Act in December 1928, California and Utah (which had earlier rescinded its ratification of the Compact) ratified the Compact and it took effect in June 1929 under a six-state agreement. Finally in 1944, Arizona also ratified the Compact due to concerns about the supply of water in the river and the desire to use power from Hoover Dam (Hundley, 1975, 297–9).

The primary enforceable condition of the Compact requires the Upper Basin states to deliver at least 7.5 maf of water to the Lower Basin per ten-year period. The point of delivery is Lee's Ferry, Arizona, the dividing point between the Upper and Lower Basins, located just below the present site of Glen Canyon Dam near the border between Arizona and Utah (see Figure 1). The intent of the Compact, however, was to divide the water of the Colorado River equally between the Upper and Lower Basins (Howe and Murphy, 1981). At the time of the Compact negotiations it was generally believed that the average flow of the Colorado River at Lee's Ferry was much greater than 15 maf per year, so that the Upper Basin could easily afford to pass an average of 7.5 maf per year downstream to the Lower Basin. In fact, the Reclamation Service at the time estimated the virgin (i.e., undeveloped) flow at Lee's Ferry to be at least 16.4 maf per year (Hundley, 1975, 192).

In addition, some water from tributaries and precipitation would be added to the flow in the Lower Basin.

The Compact purposely did not specify that any obligations existed to either Mexico or the American Indians (although Mexico had been using Colorado River water for many decades, and the Indians had already been placed on dry reservations and would require the water to survive). However, it did state that such obligations to Mexico, if they were determined to exist in the future, would be met first using surplus water—which was definitely expected to exist—and that if the surplus water was inadequate, they would be met equally by the Upper and Lower Basins (Holburt, 1982a). The Compact also allowed the Lower Basin states an additional 1 maf per year to be taken from the tributaries and authorized that any surplus water could be allocated between the basins in 40 years (i.e., in 1963).

The Compact Negotiations

The players in the negotiations for the Colorado River Compact included commissioners from each of the seven states and their engineering and legal advisors. A federal representative was chairman of the proceedings, and a representative from the U.S. Reclamation Service provided scientific information regarding the flow of the river and the future water requirements of each of the basin states.

The two primary actors among the states' delegates were Delph Carpenter from Colorado and Winfield S. Norviel from Arizona. Carpenter was a famous water-rights lawyer and an early advocate of the compact approach to settling the Colorado River water rights issue. Norviel was Arizona's state water commissioner. One other delegate, Stephen Davis, Jr., from New Mexico, was also a lawyer, an expert in water law. The remaining four delegates were state engineers and thus had practical interests in and knowledge of the Colorado River. Each delegate had his own goal(s) in mind for the outcome of the negotiations. Some were interested in securing rights for hydroelectric power, others in industrial development, and still others in reclaiming drylands for agricultural use through massive water storage projects. Each fought fiercely to achieve his goals, many of which were in conflict with one another (Hundley, 1975, 139–43).

Herbert Hoover, who represented the federal government in the negotiations, was selected by the other representatives to be chairman of the Commission. Hoover was from California, and was trained as an engineer rather than as a lawyer—two factors that made him an unpopular choice as federal representative among some delegates to the negotiations (Hundley, 1975, 131). Arthur Powell Davis, the nephew of John Wesley Powell (who first explored the length of the Colorado River) was the Director of the Reclamation Service and, as such, he provided the delegates with estimates of runoff and water requirements of each state. Davis had long advocated the construction of a series of high dams to control the flow of the Colorado River, and he firmly believed that the flow was adequate to meet all present and future water needs within the basin.

The original intent of the Compact negotiators was to allocate water to each state. Hence, the negotiations began with an attempt to quantify the water needs of each state (based on irrigable acreage), in order to determine the amount of water each should be allocated. However, it soon became clear that it would not be possible to establish a fair allocation among the individual states that would be agreeable to all delegates. In particular, each state (except California) disagreed with the Reclamation Service's estimates of irrigable acreage, and naturally all of the states' estimates were higher than those of the Reclamation Service (Hundley, 1975, 147). Further investigation into precise estimates of irrigable acreage would have required too much time; hence, the direction of the negotiations eventually veered toward an even allocation between the Upper and Lower Basins (Howe and Murphy, 1981). It is to the credit of the negotiators that they recognized that they lacked sufficient data to make specific allocations to each state.

Several different proposals were suggested for division of the water between the two basins before the final compromise was reached in November 1922. As early as March, a proposal to divide the runoff on a 35–65 percent basis between the Upper and Lower Basins was suggested by two engineers of the U.S. Geological Survey. Their division was based on consumptive uses at the time by the individual states in the basin, and it would have required the Upper Basin to release approximately 10.7 maf to the Lower Basin each year (Olson, 1926, 24).

However, the two-basin approach to dividing the water of the Colorado River was not given serious consideration until August 1922 when Delph Carpenter initially proposed that the water be divided between the basins at Lee's Ferry, Arizona. Carpenter proposed that each basin be allocated 8.7 maf annually, based on the widely held assumption that the total runoff of the river was 17.4 maf per year, according to measurements at Yuma, Arizona. The Lower Basin would obtain some of their share of the water from tributaries, so Carpenter calculated that the Upper Basin would be obligated under this plan to release only about 6.3 maf per year to the Lower Basin.

Hoover counter-proposed that the Upper Basin be required to release to the Lower Basin half of the flow at Lee's Ferry, which totaled (according to Reclamation Service estimates) 16.4 maf. Thus, under Hoover's plan the Upper Basin would be obligated to release 8.2 maf per year to the Lower Basin, and the Lower Basin states would also have rights to their tributary waters. Hoover also suggested that the deliveries be determined on a 10-year basis to even out year-to-year variations in flow, so the Upper Basin would be required to deliver 82 maf in any ten-year period.

The Upper Basin naturally objected to this proposal, since it required deliveries about 2 maf per year greater than Carpenter's proposal. In addition, the minimum flow record indicated that the Upper Basin would have difficulty meeting this obligation in low-flow periods while still meeting Upper Basin needs. Thus, the Upper Basin states proposed another compromise, which would require the Upper Basin to deliver 65 maf per ten-year period to the Lower Basin. This suggestion was unacceptable to the Lower Basin.

Finally, Hoover offered a proposal stipulating that the Upper Basin be required to deliver 75 maf to the Lower Basin in each ten-year period. The intent of the proposal was to divide the flow at Lee's Ferry equally by allocating an average of 7.5 maf per year to each basin. Hoover made this proposal in his role as chairman of the proceedings, desiring to mediate between the basins (Hundley, 1975, 195). Hoover also assumed that the flow was much greater than 15 maf per year, but he suggested that division of the surplus waters be postponed until later.

The Hoover proposal was accepted by all of the delegates. However, the question of allocation of the water from the Lower

Basin tributaries had not been resolved, and this issue became a snag in finalizing the negotiations. At last, however, it was agreed to allocate the Lower Basin an additional 1 maf per year, to come from the tributaries. Other issues, such as obligations to the Indians and to Mexico, were "easily" resolved by essentially leaving them for the future, and the Compact was signed by all of the delegates on 24 November 1922.

Thus, it appears that the compact negotiations were dominated by the special interests of each of the states, and, more generally, the two basins. Agreement regarding the amount of water to be allocated to each basin was reached essentially through a bartering process, only loosely based on scientific evidence regarding flows and water requirements. As was discovered later, more attention probably should have gone toward obtaining reliable estimates of river flow. However, the determination of the Upper Basin states to protect their rights to future uses of the water, and the Lower Basin's determination to quickly develop the river, clouded the issue of water supply. It was not long, however, before the available water supply became a major issue.

LATER ASPECTS OF THE LAW OF THE RIVER

Later aspects of the Law of the River include various laws, treaties, compacts, and court decisions, such as the Boulder Canyon Project Act, the Upper Colorado River Basin Compact, and the *Arizona v. California* Supreme Court decision. Action in all of these cases was influenced to some degree by concern over the flow of the river as well as the desire to develop the water resources in the Colorado River Basin.

Passage of the Boulder Canyon Project Act in 1928 was required by California as a precursor to that state's ratification of the Colorado River Compact. This bill authorized construction of the first high dam on the Colorado River, now known as Hoover Dam, in Black Canyon of Nevada. In passing the legislation, Congress added a suggested allocation of water among the states of the Lower Basin. This allocation would give 4.4 maf per year to California, 2.8 maf to Arizona, and 0.3 maf to Nevada.

Mexican claims to water of the Colorado River were recognized in the Mexican Water Treaty of 1944. This treaty was pushed forward by the U.S. government both as a way of establishing good relations with Mexico and as a way of preventing Mexico from taking an even greater amount of water from the river (see Hundley, 1966, for a thorough description of the negotiations between the U.S. and Mexico regarding rights to the Colorado River). However, the treaty was seen as very objectionable by representatives of some of the basin states who felt either that Mexico had no rights at all to Colorado River water, or that the amount allocated to Mexico was too large. In addition, California feared that all of the water allocated by the treaty to Mexico would come from its share of the Lower Basin allocation, since no agreement had yet been made with Arizona to divide the Lower Basin share. Fears that the treaty would allocate water currently being used by Arizona, as well as new information indicating that the water supply was lower than originally believed, inspired the Arizona legislature to finally ratify the Colorado River Compact.

The Upper Colorado River Basin Compact, which divided Colorado River water among the Upper Basin states, was agreed upon by those states in 1948. Due to increasing uncertainty regarding the flow of the river, the states in the Upper Basin agreed to divide their water on a proportional rather than absolute basis, except for Arizona, which has only a small area in the Upper Basin and was allocated 0.05 maf per year (Howe and Murphy, 1981). In particular, the Upper Basin Compact allocated 51.75 percent of the Upper Basin flow to Colorado, 23 percent to Utah, 14 percent to Wyoming, and 11.25 percent to New Mexico. The Upper Basin Compact also authorized establishment of the Upper Colorado River Basin Commission to monitor and divide the yearly flows. On the basis of this compact, Congress in 1956 enacted legislation (the Colorado River Storage Project Act) to develop storage projects, including Glen Canyon Dam, in the Upper Basin (Holburt, 1982a).

The Lower Basin water was finally allocated among the Lower Basin states by the U.S. Supreme Court decision in the case *Arizona v. California* in 1963. In this decision, the Supreme Court allocated the water among California, Arizona, and Nevada according to the values suggested in the Boulder Canyon Dam legislation. In addition, the *Arizona v. California* decision recognized the rights of American Indians to water required for irrigable acreage on the reservations—rights which even now have not been finally determined (Reno, 1981; Smith, 1982).

Two other pieces of federal legislation complete the list of major components of the Law of the River. These two laws are the Colorado River Basin Project Act of 1968, which authorized construction of the Central Arizona Project, and the Colorado River Basin Salinity Control Act of 1974. The Salinity Control Act requires limits on the salinity of water entering Mexico and authorizes construction of a desalinization plant at Yuma (Holburt, 1982b). This law became necessary because Colorado River water entering Mexico had become increasingly saline as more of it was used to irrigate alkaline soils, and was consequently causing damage to Mexican soils and crops. The Central Arizona Project, which is not yet completed, will consist of a network of canals to bring Colorado River water to irrigable lands in Arizona. However, when the project is completed, it will require a large amount of water such that Arizona will finally be taking its full share of the Colorado River. This fact, in turn, will mean that California can no longer take more than its official allocation (California has been taking up to 1 maf more water per year than its annual allotment of 4.4 maf (Boslough, 1981)).

Thus, through the Law of the River, rights to a large amount of Colorado River water have been allocated since the time of the Compact negotiations. However, even these rights (e.g., the rights of the Native American Indians) have not yet been completely quantified. The ultimate implication of the regulations governing the use of the Colorado River is that the river flow has been overallocated. This fact appears to be true even if one believes that the original Reclamation Service estimates of the flow were correct.

COLORADO RIVER WATER SUPPLY

The Reclamation Service's estimates of the flows at Lee's Ferry that were used in the Compact negotiations were actually based on flows at Laguna Dam, near Yuma, Arizona (see Figure 1). The average flow estimates were computed using measurements from the river for the period 1899 to 1920. The Reclamation Service believed that evaporation losses between Lee's Ferry and Laguna Dam were essentially compensated for by the tributary waters entering the mainstream between these two points (Hundley, 1975, 193). The dialogue at the Compact negotiations concerning these facts indicates that this topic was given fairly rough and arbitrary treatment (Olson, 1926, 312). Little questioning of the Reclamation Service's flow estimates appears to have taken place.

The Reclamation Service contended that the water flow was sufficient to meet all possible future requirements within the basin. Apparently this attitude also reflected the views of most of the delegates to the Compact negotiations. Hundley commented on this attitude with respect to a meeting of the League of the Southwest which took place in 1920:

> The ease with which the delegates accepted the Recla-
> mation Service's rosy forecasts of water supply showed
> merely their strong desire to believe that the supply was
> sufficient. Interstate cooperation—and development of
> the basin—would obviously be more easily attained if
> the government experts were correct or, at least, if ev-
> eryone believed they were correct (Hundley, 1975, 92).

Obviously, however, there would have been little need for an agreement dividing the water if the supply actually were adequate to meet all possible future needs in the basin. Hundley continues:

> ... others in the upper basin regretted their earlier en-
> dorsement of a high dam on the lower river. None of
> them gave the major reason for the turnabout, but it
> was implied in every thing they said: the water sup-
> ply might be inadequate. If they could not predict
> with precision their own future water needs... then how
> could they be sure that the Reclamation Service could
> do so? (Hundley, 1975, 97)

Skepticism about the adequacy of the water supply apparently involved both uncertainty regarding the river flow as well as disbelief in estimates of the extent of potentially irrigable acreage (the primary measure used to estimate future water requirements).

Some skepticism was also voiced during the hearings and meetings of the Compact negotiators. For example, Norviel questioned, "If there is water enough for all, then why all this division and this restriction upon the amount of water flow?" (Olson, 1926, 24). However, such skepticism was never formally acknowledged by the commission. In fact, at one point Carpenter proposed that since everyone agreed that the supply was adequate, the Upper

Basin should be allowed unrestricted development. Hundley comments:

> Carpenter might have added that if there was sufficient water, the upper basin need not fear development below. He skirted the obvious, however, preferring to emphasize the innocuousness of his proposal (Hundley, 1975, 150).

The question of the adequacy of the supply became a major issue during the ratification hearings in the individual states, during which several individuals also expressed disagreement with the Reclamation Service's flow estimates. Some people believed that the estimates were too low, while others believed they were too high. For example, Nellis Corthell, a Wyoming attorney, believed that the average flow at Lee's Ferry was only 15.8 maf and he was concerned about the high variability of the river flow over long time periods. The Compact delegate from Wyoming, Frank Emerson—who clearly supported ratification of the Compact—countered Corthell's argument with more recent estimates of river flow from the Reclamation Service. In support of Emerson, Arthur Powell Davis estimated the total flow of the river to be 20.5 maf (Hundley, 1975, 220–1). In the end, of course, Wyoming ratified the Compact.

However, all of the arguments regarding the river flow that took place during the ratification hearings were not based on any more scientific data than had been used by the Reclamation Service to make the original estimates. The first scientific data that indicated the flow might actually be less than originally estimated came in 1928 in a report to Congress from a group of engineers and geologists who were studying the feasibility of constructing Boulder Dam. The report contained estimates of average annual reconstructed flow at the canyon of about 15 maf, 1 maf less than the flow estimated by the Reclamation Service in 1922. The differences between the estimates were attributed to the unusually high flows that occurred during the period of measurement by the Reclamation Service (i.e., 1899–1920) as well as to the use of poor gauging equipment at Yuma. The report also warned that "periods of high and low flow occur in cycles of very uncertain magnitude and duration" (Hundley, 1975, 275).

Later estimates of flow by the Bureau of Reclamation (formerly the Reclamation Service) in 1945 indicated that the average flow was even less than estimated in 1928. The Bureau's data now indicated that during a dry period the water supply below Hoover Dam would not be adequate to meet all Lower Basin needs (Hundley, 1975, 297). This evaluation was partly a result of the very dry period that occurred between 1931 and 1940.

Runoff measured between 1922 and 1956, which was used in the Supreme Court decision of the case *Arizona v. California* in 1963, indicated a further reduction in the average flow of the Colorado River at Lee's Ferry over that estimated by the Reclamation Service in 1922. In fact, the estimate used in this case indicated that the average annual flow was only about 14 maf, significantly less than the 16.4 maf originally estimated by the Reclamation Service.

However, the most recent estimates of virgin flow at Lee's Ferry suggest that the long-term annual flow is only about 13.5 maf. This estimate is based on the tree-ring studies of Stockton and Jacoby (1976), using a tree-ring record extending back to about 1560. The measured flows since 1922 also indicate that the average flow is smaller than was originally estimated. In fact, it appears that the Reclamation Service's estimates of flows were based on an unusually wet period—wetter than any period since and apparently wetter than any other period in the 450-year tree ring record (except for the most recent two or three years).

Figure 2 contains the time series of virgin flow based on tree-ring analyses presented by Stockton and Jacoby (1976). It is quite apparent from this diagram that the flow in the period between 1900 and 1922 was unusually high. Soon after the Compact was drafted, the yearly flows dropped to much lower levels. In addition, the flows have varied greatly over the period of the tree-ring record. Recent measured (reconstructed) flows at Lee's Ferry are shown in Figure 3. This diagram suggests that three periods with different flow characteristics can be identified within this century. First, the period on which the Compact was based as well as some years following (i.e., 1899 to about 1929) was characterized by unusually high flows. This period was followed by much drier years between 1930 and about 1982. Finally, the most recent period (1983–1986) appears to be a return to higher flows. In fact, some flows in this most recent period have been higher than those experienced in the

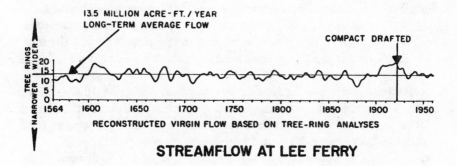

Figure 2. Streamflow at Lee's Ferry estimated from tree-ring analyses. From Stockton and Jacoby (1976).

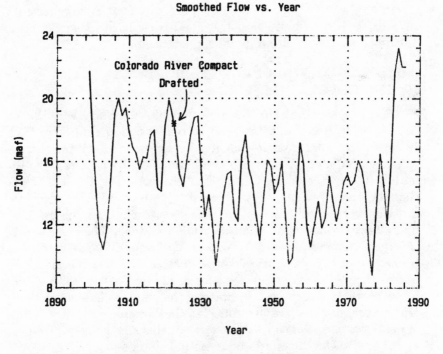

Figure 3. Reconstructed streamflow at Lee's Ferry. The data have been smoothed using a 3-point smoothing algorithm called hanning, with weights of $\frac{1}{4}$, $\frac{1}{2}$, $\frac{1}{4}$.

first part of the century. Thus, the flow of the Colorado River has exhibited considerable variation even in the most recent years.

The analysis by Stockton and Jacoby also indicates that periods with very low flows have occurred in the past. In fact, the lowest flows estimated using the tree-ring analysis were about 4.7 maf in a one-year period, or 9.7 maf per year over a ten-year period. Relatively low flows have also occurred in the post-1922 period. For example, the lowest one-year flow in this period was 6.6 maf in 1934, and the lowest ten-year flow, which occurred between 1931 and 1940, averaged 12.5 maf per year.

Thus, it appears that the average flow of the Colorado River was over-estimated to a large degree by the Reclamation Service in 1922. In fact, based on the tree-ring record, it appears that the Compact negotiators allocated at least 1.5 maf per year that will, on average, not exist in the river. This error could possibly have been avoided if the Reclamation Service had examined a longer streamflow record (apparently, gauge records extending back to 1878 existed for Yuma) or if they had considered other climatic data such as precipitation in the Upper Basin.

IMPLICATIONS FOR CURRENT PROBLEMS OF THE COLORADO RIVER

In essence, the error in measurement of the average flow of the Colorado River, and the gradual realization of that error, can be considered analogous to a change in climate in which the change was not predicted or in which predictions of the change were ignored. In the case of the Compact, the negotiators did not realize that the measurements they were using were based on an extraordinarily wet period. Hence, decisions were made on the basis of bad information about the future state of nature. The existence of this measurement-induced equivalent of a climate change gradually became apparent to policymakers as better data became available and as the region experienced a very dry period in the 1930s. That is, perceptions about the true state of the climate changed as more and better information was collected and greater variation in the climate was experienced. The effects of this change in perception are analogous to the effects of an actual change in climate.

Response to this new information about the climate (i.e., the measurement-induced equivalent of a climate change) has been limited by the terms of the Compact. According to the Compact, the

Upper Basin is obligated to provide 75 maf of water to the Lower Basin in each ten-year period, regardless of the amount of water actually available, and in spite of the occurrence of any climate change. This requirement, under the condition of average annual water availability of only 13.5 maf, added to the Upper Basin obligation to Mexico of 0.75 maf per year, means that the Upper Basin is only able to consume an average of 5.25 maf per year rather than the 7.5 maf that was intended by the Compact negotiators. The Compact contains no provisions for taking into account the possibility of better measurements of flow, or the possibility of a climate change that might affect the water supply. In fact, the Compact negotiators considered no other possibilities than that a surplus amount of water would be available.

The rigid nature of the Compact led to further rigid approaches to regulation and division of Colorado River water. For example, the Lower Basin water allocation was divided among the Lower Basin states (by the Boulder Canyon Project Act and the *Arizona v. California* Supreme Court decision of 1963) using an absolute rather than relative allocation system. That is, each state was allocated a specific rather than proportional amount of water. In addition, Mexico was allocated an absolute amount of water each year by the Treaty of 1944. On the other hand, the Upper Basin states, recognizing the occurrence of the measurement-induced equivalent of a climate change, in 1948 allocated water among themselves using a proportional approach which also required establishment of a permanent commission to monitor water usage, supplies, and requirements.

However, it is only now that the impacts of the measurement-induced climate change are being felt to a large extent. In particular, it is only now that the water of the Colorado River is being almost completely used (Holburt, 1982a). In addition, the Colorado River is facing numerous problems related both to water quantity and water quality. For example, the completion of the Central Arizona Project will add a further strain to the resources of the river (Boslough, 1981; Pearce, 1987) and will force other states (e.g., California) to limit their usage to their original allocations (Holburt, 1982a). American Indian rights to the water have not yet been determined, but they may be quite extensive (Reno, 1981; Smith, 1982); in fact, it has been suggested that the water rights for the Navajo reservation alone may be as large as the total Colorado River flow (Getches and Meyers, 1986; Reno, 1981; Smith, 1982). Extreme salinity problems have further reduced the usable supply of water (Law and Hornsby, 1982; Wescoat, 1986). Municipalities in the arid southwest, such as Phoenix and Tucson, as well as cities in southern California, are requiring more and more water to support their rapidly growing populations (El-Ashry and Gibbons, 1986). As a result of the demands of the cities, groundwater has been mined—and is being mined—at an alarming rate (Boslough, 1981; El-Ashry and Gibbons, 1986; Pearce, 1987). Strong competition exists between the various uses of the water (e.g., between agricultural and municipal uses, or between hydro-electric generation and wilderness/recreation uses) which will not be easily resolved (e.g., Nash, 1986). And, greater demands may still be placed on the resources of the river if it becomes economically feasible in the future to develop energy resources such as oil shale and uranium within the basin (Skogerboe and Radosevich, 1982).

Oddly, relatively little action has been taken to solve these problems. For example, the many possible water conservation tactics that could be pursued both to decrease salt uptake and reduce consumptive use of water (such as making greater use of drip irrigation systems) have not been put into widespread use. In part, this fact reflects the extremely low prices that farmers generally pay for water from Colorado River storage projects as well as limitations on their ability to sell any water that is conserved (Pearce, 1987; Wiegner, 1979). That is, farmers do not have much incentive to conserve water. In addition, due to its low cost and to restrictions on water trading, the water is frequently used to grow low-valued crops such as alfalfa or hay on marginal lands, rather than being used for higher-valued crops or other more economically valuable purposes in other locations.

In addition to the many problems facing the basin as a result of the measurement-induced climate change, concern has recently developed regarding the impacts of a possible real climate change in the region due to global increases in atmospheric carbon dioxide and other trace gases. For example, Dracup (1977) considered the possible impacts of a global climate warming on the management of water supplies in the basin. Such a climate change is likely to result in increased average temperatures in the Colorado River Basin as a whole. However, the effects of the climate change on precipitation and soil moisture in the region are as yet unclear, due to differences in the results obtained from the various global atmospheric circulation models that are being used to study this problem. In particular, some models suggest that soil moisture in the region would be decreased in both summer and winter, whereas others indicate it would decrease only in winter (Kellogg and Zhao, 1988). In any case, increases in the average temperature of the region would most likely result in significant reductions in the average flow of the Colorado River. For example, the results of a study by Revelle and Waggoner (1983) suggest that a 2°C increase in temperature alone (i.e., without a simultaneous decrease in precipitation) would result in about a 29 percent decrease in Colorado River runoff. If precipitation were to decrease by 10 percent as well, runoff would be decreased by 40 percent from the current average flow. Meko and Stockton (1984) indicate that such a reduction in runoff is not unprecedented in the past, even in this century.

According to the results of a study by Kneese and Bonem (1986), a long-term drought such as might occur as a result of a change in climate would have drastic impacts on the water uses of the basin. During such a drought, the Upper Basin would have great difficulty meeting its Compact obligations, hydroelectric production would be radically decreased, and farmers would have difficulty obtaining adequate water for irrigation. Furthermore, depletion of all or most of the water in the storage reservoirs during the drought period would have lasting effects on water usage for many years while the reservoirs are refilled (Clyde, 1986).

Gleick (1988) considers the impact that such a climate change would have on the agreement with Mexico. In particular, the 1944 treaty with Mexico makes no provisions for dealing with a change in climate, other than short-term fluctuations such as droughts, in which case Mexico agrees to decrease its water demands in correspondence with a similar response made by the United States. Hence, the rigidity of this document might allow Mexico to demand its full allocation in spite of a change in climate that would reduce the overall flow of the river. Thus, it appears that the anticipated climate change would increase the vulnerability of the basin states to shortages of water, beyond the vulnerability currently being experienced.

Several approaches have been suggested for solving the current problems of the Colorado River Basin—particularly the problems associated with the measurement-induced equivalent of a climate change, irrespective of future problems associated with a real change in climate. Among the suggested solutions are some possibilities that are unlikely to work in any event, and several others that are equally unlikely to be enacted. For example, one possibility is the transfer of water from another river basin, such as the Columbia River Basin, into the Colorado River Basin to supplement the Colorado River supply (Sheridan, 1981). However, this solution has effectively been removed from the realm of possibility by legislation prohibiting the Bureau of Reclamation from even investigating the importation of water from other basins to the Colorado River. Furthermore, the structural and other costs associated with such a plan would be prohibitively high. However, the Bureau of Reclamation is investigating the use of weather modification to increase the snowpack in the Colorado mountains as a means of increasing runoff into the Colorado River (Pearce, 1987; Sheridan, 1981).

More practical possibilities include providing incentives for conservation and charging farmers more realistic prices for water (Getches and Meyers, 1986; Pearce, 1987). However, currently such incentives do not exist since in most cases farmers could not sell the water they would save through conservation. In any case, it is not clear that conservation practices invoked by individual irrigators would significantly impact overall streamflow, due to the reduction in return flows that would accompany such practices. Recent legislation (the Reclamation Reform Act of 1982) requires the Bureau of Reclamation to begin charging more realistic operation and maintenance costs as well as water costs for storage project water used on what has been defined as "excess" amounts of land. However, the Bureau of Reclamation has apparently been reluctant to enforce the water-pricing provision of this legislation (El-Ashry and Gibbons, 1986).

Other possible solutions which would require further legal action and which would be complicated by many factors include the development of water markets and allowing water trading between the two basins (Gardner, 1986; Pearce, 1987). Such possibilities are attractive in that they would promote more efficient use of the water for applications that have the highest economic value. However, they might also result in an allocation of water that is inequitable based on other values (Ingram et al., 1986). Furthermore, vast legal difficulties would probably arise in the process of establishing and developing such markets and transfer options (Getches and Meyers, 1986).

Other suggestions involve the Compact itself. In particular, some Colorado River experts and policymakers have suggested that the Compact should be renegotiated. In fact, it has been suggested that the Compact could be invalidated because it was based on a "mutual mistake of fact" (Clyde, 1986; Getches and Meyers, 1986). Naturally, it is policymakers in the Upper Basin who have been most interested in such a renegotiation.

However, it appears unlikely that agreement will be reached to undertake such a renegotiation, particularly since the problems of the Colorado River are so much more severe and the issues are so much more complicated today than they were in 1922. In addition, the constraints that have been put in place since 1922 would make it very difficult to actually alter the Compact without simultaneously changing many other aspects of the Law of the River. Reaching agreement on so many regulations and agreements, which would require approval by the seven states, the federal government, and Mexico, appears to be an impossible task. The difficulties that the seven states alone had in reaching the simple agreement represented by the Compact suggest that a renegotiation is not a practical possibility.

Another option related to the Compact is the establishment of a permanent Colorado River Commission that would include representatives from each of the seven basin states, the federal government, and possibly Mexico (Bloom, 1986; Weatherford and Brown, 1986). This commission would monitor and attempt to adjust for changes in Colorado River flow and problems associated with water quality and supply. The idea of having such a commission is not new. In fact, it was originally suggested by W. F. McClure, the California Commissioner at the Compact negotiations. However, the idea was vetoed by Delph Carpenter who did not want to establish "super interstate agencies" (Hundley, 1975, 181). Later the idea of a commission was also suggested by Olson (1926) in his dissertation on the Colorado River Compact. Recent suggestions of establishing a commission have not progressed very far, perhaps because of the difficulties that would be involved in changing the Compact and other laws. In addition, such a commission would essentially represent another layer of bureaucracy to which the states would be subservient with regard to their use of Colorado River water, a burden which may be unacceptable.

CONCLUSIONS

At least two lessons of importance to policymakers concerned about the possible need to prepare for climate change can be learned from the experience and impacts associated with the Colorado River Compact. First, the rigid strategy applied by the Compact negotiators to allocating the river's water supply (i.e., the allocation of absolute rather than proportional amounts of water) has led to further rigid structures in other regulations governing the river. The approach taken by the negotiators in some ways determined the approaches to be taken in future decisions. This set of rigid regulations has limited the kinds of actions that could be taken in response to the measurement-induced climate change, and they will further limit the actions that can be taken in response to a real climate change. For example, it appears that it would be very difficult at this point in time to establish a Colorado River Commission to manage the river and, hopefully, to solve some of its many problems. If such a commission had been established in 1922, it would probably be able to respond to these problems in a more flexible manner than is possible now. Furthermore, a commission would have been able to respond to the measurement-induced climate change as it gradually became evident that the average flow observations were in error.

Perhaps the most important lesson to be learned from this experience is that it is extremely hazardous to ignore the variability that is inherent in natural systems and the change that is a natural part of social systems. The Colorado River Compact negotiators based their decisions on a very small "slice" of the history of the Colorado River. More information was available at the time than was used by the negotiators. Unluckily, particularly for the Upper Basin states, this "slice" happened to be the wettest period in at least 450 years. Furthermore, the Compact negotiators considered only average flows, generally ignoring the flow variability even in the short data record available to them. As Katz (this volume) emphasizes, the variance of a climate process is probably of much greater practical importance than the average. Howe and Murphy (1981) also noted in a study of the impacts of climate information on the development and management of the Colorado River that "... it is difficult to specify any one number as the average flow to use for planning purposes. Rather, attention must be paid to the nature and range of climatological variability likely to be faced and to the flexibility of the system being planned" (Howe and Murphy, 1981, 39).

The Compact negotiators also appear to have treated rather casually the information that they did have available. They made agreements using a bartering approach—each desiring specific concessions from the others—rather than seriously considering the implications of the data available to them. They were so certain that surplus water would exist, and they were so desirous of meeting their individual goals, that they simply closed off consideration of the possibility that the flow estimates might be too high. As Hundley (1975, 92) notes, they truly wanted to believe that the flow was adequate.

However, even with the limited data available at the time, the negotiators could have established a mechanism for checking the reliability of the flow estimates in the future and adjusting the Compact based on new information. For example, they might have written into the Compact a provision for adjustment of the allocations based on new data twenty or even ten years after the Compact was signed—in fact, the error in the flow estimates was beginning to be recognized within ten years after the Compact was signed (Hundley, 1975, 275). Or, the negotiators might have included the establishment of a Colorado River Commission as part of the Compact provisions, as a way of adjusting the regulation of the river on the basis of new information or changes in the needs of the basin states. Such a provision would have had the effect of both making the allocation system less rigid and allowing consideration of the impacts of changes in the river flow. In addition, a permanent commission would be able to respond to changes in water needs as well as to other social and demographic changes, such as the large increase in the urban population of the southwestern United States that has occurred during this century. The alternative approach of allocating the flows on a proportional rather than absolute basis also would have been much more responsive to variations in river flow. In this way, for example, each basin might have been allocated half of the flow at Lee's Ferry, regardless of the total amount of flow. Of course, such a system probably would have required the establishment of some type of commission to monitor flows.

Thus, the analogy of the experience associated with the Colorado River Compact with the response to a future climate change suggests two basic conclusions. First, it is much more difficult to adapt to changes in climate (or in society, etc.) when a rigid regulation strategy has been established. Furthermore, the "winners" in such a situation (i.e., the Lower Basin states in the case of the Colorado River) are unlikely to be willing to alter the terms of the regulations even when a change in climate has been identified. Secondly, ignoring climate variability—or the possibility of climate change—can have very serious consequences for the future when unexpected variation or change occurs. These lessons are unfortunately being learned now by those responsible for making decisions regarding the current water quantity and quality problems of the Colorado River.

REFERENCES

Bloom, P.L., 1986: Law of the river: A critique of an extraordinary legal system. In G.D. Weatherford and F.L. Brown (Eds.), *New Courses for the Colorado River. Major Issues for the Next Century*. Albuquerque, NM: University of New Mexico Press, 139–54.

Boslough, J., 1981: Rationing a river. *Science, 81*, 26–37.

Clyde, E.W., 1986: Institutional response to prolonged drought. In G.D. Weatherford and F.L. Brown (Eds.), *New Courses for the Colorado River. Major Issues for the Next Century*. Albuquerque, NM: University of New Mexico Press, 109–38.

Dracup, J.A., 1977: Impact on the Colorado River Basin and southwest water supply. In *Climate, Climatic Change, and Water Supply*, National Research Council. Washington, DC: National Academy Press, 121–32.

El-Ashry, M.T., and D.C. Gibbons, 1986: *Troubled Waters: New Policies for Managing Water in the American West*. Study 6. Washington, DC: World Resources Institute.

Fradkin, P.L., 1981: *A River No More. The Colorado River and the West.* New York, NY: Alfred A. Knopf.

Gardner, B.D., 1986: The untried market approach to water allocation. In G.D. Weatherford and F.L. Brown (Eds.), *New Courses for the Colorado River. Major Issues for the Next Century.* Albuquerque, NM: University of New Mexico Press, 155–76.

Getches, D.H., and C.J. Meyers, 1986: The river of controversy: Persistent issues. In G.D. Weatherford and F.L. Brown (Eds.), *New Courses for the Colorado River. Major Issues for the Next Century.* Albuquerque, NM: University of New Mexico Press, 51–86.

Gleick, P.H., 1988: The effects of future climatic changes on international water resources: the Colorado River, the United States, and Mexico. *Policy Sciences, 21* (in press).

Holburt, M.B., 1982a: Colorado River water allocation. *Water Supply and Management, 6*, 63–73.

Holburt, M.B., 1982b: International problems on the Colorado River. *Water Supply and Management, 6*, 105–14.

Howe, C.W., and A.H. Murphy, 1981: The utilization and impacts of climate information on the development and operations of the Colorado River system. In *Managing Climatic Resources and Risks*, Panel on the Effective Use of Climate Information in Decision Making, National Research Council. Washington, DC: National Academy Press, 36–44.

Hundley, N., Jr., 1966: *Dividing the Waters. A Century of Controversy between the United States and Mexico.* Los Angeles, CA: University of California Press.

Hundley, N., Jr., 1975: *Water and the West. The Colorado River Compact and the Politics of Water in the American West.* Los Angeles, CA: University of California Press.

Ingram, H.M., L.A. Scaff, and L. Silko, 1986: Replacing confusion with equity: Alternatives for water policy in the Colorado River Basin. In G.D. Weatherford and F.L. Brown (Eds), *New Courses for the Colorado River. Major Issues for the Next Century,* Albuquerque, NM: University of New Mexico Press, 177–99.

304

Kellogg, W.W., and Z.-C. Zhao, 1988: Sensitivity of soil moisture to doubling of carbon dioxide in climate model experiments: Part I. North America. *Journal of Climate, 1,* 348–66.

Kneese, A.V., and G. Bonem, 1986: Hypothetical shocks to water allocation institutions in the Colorado Basin. In G.D. Weatherford and F.L. Brown (Eds.), *New Courses for the Colorado River. Major Issues for the Next Century.* Albuquerque, NM: University of New Mexico Press, 87–108.

Law, J.P., Jr., and A.G Hornsby, 1982: The Colorado River salinity problem. *Water Supply and Management, 6,* 87–103.

Meko, D.M., and C.W. Stockton, 1984: Secular variations in streamflow in the western United States. *Journal of Climate and Applied Meteorology, 23,* 889–97.

Nash, R., 1986: Wilderness values and the Colorado River. In G.D. Weatherford and F.L. Brown (Eds.), *New Courses for the Colorado River. Major Issues for the Next Century.* Albuquerque, NM: University of New Mexico Press, 201–14.

Olson, R.L., 1926: *The Colorado River Compact.* Ph.D. dissertation. Cambridge, MA: Harvard University.

Pearce. F., 1987: A phoenix drowning in the Arizona desert. *New Scientist,* 28 May, 52–6.

Reno, P., 1981: *Mother Earth, Father Sky, and Economic Development. Navajo Resources and their Use.* Albuquerque, NM: University of New Mexico Press.

Revelle, R.R., and P.E. Waggoner, 1983: Effects of a carbon dioxide-induced climatic change on water supplies in the western United States. In *Changing Climate,* Report of the Carbon Dioxide Assessment Committee, Board on Atmospheric Sciences and Climate, National Research Council. Washington, DC: National Academy Press, 419–32.

Sheridan, D., 1981: The underwatered West: Overdrawn at the well. *Environment, 23,* 2, 6–13.

Skogerboe, G.V., 1982: The physical environment of the Colorado River Basin. *Water Supply and Management, 6,* 11–27.

Skogerboe, G.V., and G.E. Radosevich, 1982: Future water development policies. *Water Supply and Management, 6,* 221–32.

Smith, S.K., 1982: American Indian water rights and the use of Colorado River water. *Water Supply and Management, 6*, 75–86.

Stockton, C.W., and G.C. Jacoby, 1976: *Long-Term Surface-Water Supply and Streamflow Trends in the Upper Colorado River Basin.* Lake Powell Research Project Bulletin 18. Washington, DC: Research Applied to National Needs, National Science Foundation.

Weatherford, G.D., and F.L. Brown, 1986: High water, carbon dioxide, and pig feathers. In G.D. Weatherford and F.L. Brown (Eds.), *New Courses for the Colorado River. Major Issues for the Next Century.* Albuquerque, NM: University of New Mexico Press, 225–31.

Wescoat, J.L., Jr., 1986: Impacts of federal salinity control on water rights allocation patterns in the Colorado River Basin. *Annals of the Association of American Geographers, 76*, 157–74.

Wiegner, K.K., 1979: The water crisis: It's almost here. *Forbes*, 20 August, 56–63.

13

Climate Change and California: Past, Present, and Future Vulnerabilities

Peter H. Gleick

INTRODUCTION

Despite great advances in understanding the role of climate in society, we remain vulnerable to existing climatic variability and, by implication, future climatic changes. Climatic vulnerabilities are apparent throughout the world—there are disastrous floods in Bangladesh, lethal droughts in sub-Saharan Africa, recurrent crop failures in parts of the USSR, the Indian subcontinent, and the Brazilian northeast, and shortages of water in the western United States. In the future, such vulnerabilities may be exacerbated by major human-induced climatic changes caused by growing atmospheric concentrations of carbon dioxide and other trace gases. Because of the difficulty of predicting the detailed characteristics of such changes, we must look to other methods for identifying vulnerabilities to climatic changes. One of these methods is to evaluate the consequences of past climatic variability, such as extremes of temperature or precipitation. Historical analogues provide insights into where future changes might be most

Peter H. Gleick is the Director of the Global Environment Program of the Pacific Institute for Studies in Development, Environment, and Security in Berkeley, California. He received his doctorate from the University of California, Berkeley, and in 1988 received a MacArthur Foundation Research and Writing Grant to look at the effects of global climatic changes on international rivers. Dr. Gleick has published widely in the areas of greenhouse warming, regional water resources, and international politics.

strongly felt, and they suggest appropriate responses for mitigating the worst climate-related effects.

By reviewing the impacts of past climatic extremes, we can identify sectors of the economy most sensitive to changes in hydrologic conditions. Lessons learned from these periods can then be applied to scenarios of future climatic changes in order to identify appropriate societal responses for reducing vulnerabilities to climate. Such responses might include changes in the physical structure of resource management systems (such as reservoirs), changes in the operation of these systems, and a range of socioeconomic actions (such as pricing and market mechanisms, institutional initiatives, and regulatory responses) (Frederick and Gleick, 1988).

For many reasons, California is one of the best regions in the United States to look at the vulnerability of water resources to climatic change: California's high agricultural productivity is largely dependent on irrigation; development and growth in the southern part of the state is dependent on water transfers from the north; the rich ecosystems of California rely on water availability and quality; and water supplies are already constrained by a large and growing demand. Given the rising demand for water by both the agricultural sector and the growing population in water-poor southern California, any climatic change that altered the timing, magnitude, or quality of freshwater resources would be cause for concern.

California has all the necessary characteristics of a good study region: extensive climatological, meteorological, and geophysical data are available; there are pressures on existing resources; and there have been previous periods of climatic extremes that provide insights into the behavior of the physical and institutional systems under stress. Conditions in California range from dry desert to alpine climates, from sea level to high altitude ecosystems, and from pristine areas to densely populated urban areas. Climatological data are regularly recorded at thousands of stations, and other long-term data are available on surface runoff, the operation of reservoir systems, irrigation and industrial water demands, agricultural productivity, soil characteristics, groundwater demand and recharge, and ecosystem types and vulnerabilities. These data sets extend over a period of 50 years for almost all of California and are even longer for certain locales.

PAST CLIMATIC ANOMALIES

Figures 1 to 3 show the long-term annual precipitation, temperature, and runoff for the most important watershed in California (the Sacramento Basin) plotted on the water year (October to September). These data can be used to identify periods of interesting climatic variability. For example, as Figure 3 shows, distinct periods of drought have occurred in the last 80 years. The most severe drought on record in the period 1906–86 occurred during the period from 1928 to 1934; the most severe two-year drought occurred during 1976–77. The fourth driest year in this period was 1976, and 1977 was the driest year on record. The present drought in 1987 and 1988 also offers important lessons for water management.

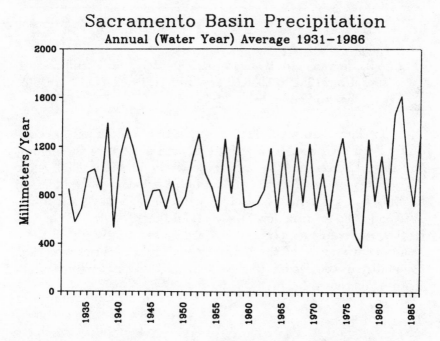

Figure 1. Sacramento Basin annual (water year) precipitation from 1931 to 1986. Note the drought in 1976–77 and the high precipitation in 1983–84. The water year extends from October to September.

Sacramento Basin Temperature
Annual (Water Year) Average 1931-1986

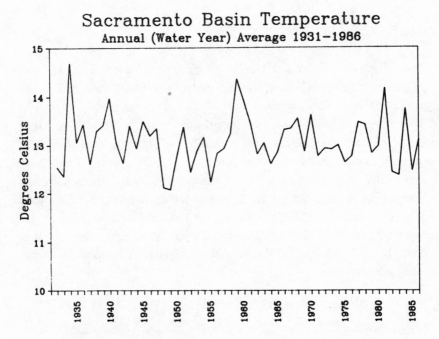

Figure 2. Sacramento Basin annual (water year) temperature from 1931 to 1986. The water year extends from October to September.

Similarly, unusually high precipitation and runoff have recently been recorded. The years 1983 and 1982 were the highest and third highest runoff years, respectively, in the last 80 years. The consequences of this combination of extreme years included flooding and soil damage. Both the extreme high-flow years in the 1980s and the extreme low-flow years in the 1970s led to changes in the way water supply systems are operated. Whether these changes have improved the resilience to climate of the system (i.e., the ability of the system to meet demands under extremes of climate) is discussed below.

Drought Effects

The most severe drought in California occurred from 1928 to 1934 and was long and widespread. The more recent 1976–77 drought, although shorter, combined two of the driest years on record. The focus of attention of this chapter will be on this

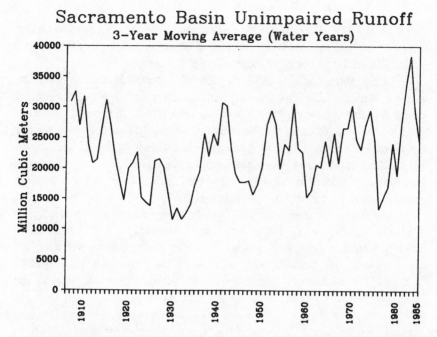

Figure 3. Sacramento Basin annual (water year) unimpaired
runoff from 1906 to 1986. The three-year moving average is plotted here in order to smooth some of the
annual variability. The water year extends from October to September. Unimpaired runoff is the runoff
that would have occurred in the absence of irrigation
withdrawals and reservoir operations.

more recent drought because of the lessons that can be learned
about the impacts on modern water management systems and on
the economic infrastructure of the region. Occasional reference to
earlier drought episodes will be made where interesting insights
can be obtained.

Normal (i.e., average) annual precipitation in California is
approximately 250 billion m^3 (200 million acre-feet). In the 1976
water year, precipitation decreased to 160 billion m^3 (130 million
acre-feet)—only 65 percent of the long-term average. This severe
shortage was compounded the following year when statewide precipitation totalled only 135 billion m^3 (100 million acre-feet), or
45 percent of average. Even more important, runoff in California's
rivers dropped to 47 percent of normal in 1976 and to 22 percent of

normal in 1977. Runoff in the Sacramento Basin dropped to under 30 percent of normal, while runoff in some smaller rivers dropped to less than 10 percent of normal.

The effects of the 1976–77 drought were felt throughout the state. Among the most important consequences were increased costs for energy to replace lost hydroelectric generation, penetration of salt water into the Sacramento–San Joaquin Delta, higher costs of water for Central Valley farmers, major livestock losses, and substantial forest damage from pests and fire. Among the more interesting institutional responses were innovative water transfers and changes in reservoir operating rules.

Because hydroelectricity is a major source of electricity in California, changes in water availability directly affect California's energy situation. During a normal water year, California's hydroelectric output provides approximately 20 percent of the state's total electrical energy production. In 1976, the first year of the drought, total hydroelectric output dropped to under 10 percent of total electrical production. In 1977, output dropped still further, to only 7 percent of California's total annual energy production. In 1976, some of the deficit was made up by importing surplus power from the Pacific Northwest, but electricity imports from the Northwest in 1977 dropped to one-third the 1976 level of imports because the Pacific Northwest was beginning to suffer similar drought effects. The reduction in hydroelectric production was made up largely by additional fossil-fuel use at an estimated direct cost of $500 million. No attempt to quantify the additional environmental costs of this fossil-fuel use have been made. The added economic cost was one of the largest direct impacts of the drought. Hardest hit were those utility districts most dependent on hydroelectricity. For the Sacramento Municipal Utility District, hydroelectric production in 1977 amounted to 186 million kWh, compared to 1,900 million kWh in a normal year (State of California, 1978).

Energy used in agricultural groundwater pumping also increased because of the drought. This increase took two forms: greater demand for groundwater because of decreased surface water supplies; and an increased cost of pumping as groundwater resources became depleted and pumping heights increased. In 1977, 4.2 billion m^3 (3.4 million acre-feet) more water than normal were pumped from groundwater reservoirs. In some groundwater reservoirs, levels dropped 6 meters or more during the two-year drought

(State of California, 1978). Increased pumping costs were estimated to be $25 million in 1977, and these costs continued into 1978, until excess groundwater recharge raised most groundwater to pre-drought levels.

Droughts in California also have a direct effect on water quality. The principal problem arises from the intrusion of salinity from the San Francisco Bay into the Sacramento–San Joaquin Delta. The reason for concern is that exports of water for southern California and for much of Central Valley agriculture come from the Delta. Any increase in salt content would mean a decrease in the quality of water available for export and would have impacts on important ecosystems in the Bay–Delta area.

Salt intrusion in the Delta is a natural function of tide and the volume of freshwater flow into the Delta from the Sacramento and San Joaquin Rivers. Salt intrusion was reported as early as 1775, when Juan Manuel de Ayala sailed on the *San Carlos* into San Francisco Bay and was unable to draw fresh water at the point reported by an earlier expedition (see Jackson and Paterson, 1977). The worst intrusion in the last hundred years occurred during the drought in the 1930s, when salt water extended many miles inland. At the other extreme, one account reports that freshwater fish were caught in the Pacific Ocean outside the Golden Gate after great flooding in 1862 (McGowna, 1961).

Salinity intrusion became a serious problem in the 1920s and 1930s when increased irrigation withdrawals and a series of dry years reduced freshwater flows into the Delta. At that time, proposals were made to build major salt barriers (e.g., dams, levees, lock systems, and so on) to keep salt water out of the Delta. Ultimately, these were all rejected for economic reasons. Finally, by the early 1980s, the worst salinity problems had, at least in theory, been eliminated by controlling the operating regime of upstream dams to permit the flushing of fresh water into the Delta during low-flow years. By operating upstream reservoirs in this manner, only 8,900 hectares were subjected to salinities exceeding 1,000 mg/liter in 1977, compared to over 81,000 hectares in 1931—two comparable low-flow years (State of California, 1978).

Unfortunately, at the same time that salinity intrusion into the Delta has been reduced through the operation of reservoirs, the vulnerability of the entire system has been increased in other ways. As a result, less severe salinity incursions can now have much greater societal effects than before. The two major problems are that (1) fresh water to feed the aqueducts to southern California is now withdrawn at pumping stations located where salinity problems still occur, and (2) in a severe drought situation, sufficient upstream release to control salinity cannot be maintained. The drought in 1976–77 led to such a problem.

In early 1977 water storage facilities could no longer be operated to meet salinity standards in the Delta because of insufficient water. As a result, the State Water Resources Control Board actually lowered the salinity standards to legally permit higher salt concentrations (State of California, 1978). By mid-1977, additional emergency regulations temporarily eliminated most water-quality standards. At the same time, temporary barriers were constructed to prevent salt penetration into certain parts of the Delta. The barriers maintained water quality at several critical locations.

In order to take pressure off the water-quality situation in the Bay and to provide sufficient water for northern California farmers and cities, the California Department of Water Resources permitted—even encouraged—water exchanges during the 1977 drought. Four southern California water agencies relinquished all or part of their State Water Project water to agricultural water users in the San Joaquin Valley and to San Francisco Bay urban users. The California Metropolitan Water District (MWD), for example, relinquished nearly 500 million m³ (400,000 acre-feet). The San Bernardino Valley Municipal Water District, the Coachella Valley County Water District, and the Desert Agency also gave up all or part of their entitlements. The MWD made up its demand by greatly increasing its reliance on water from the Colorado River. In 1977, nearly 90 percent of MWD's deliveries came from this source. Without these transfers, agricultural contractors in the San Joaquin Valley would have received only 40 percent of their normal allocation, instead of the 89 percent that they actually received during the drought.

There was an economic cost to these transfers and shortages. Farmers in the San Joaquin Valley who normally pay $7–$25 per acre-foot paid $40–$80 per acre-foot during the drought. Urban users who normally pay $40–$150 per acre-foot paid $50–$375 per acre-foot (State of California, 1978).

Floods

Flooding in California is a common phenomenon due to the nature of the region and the characteristics of the climate. Damaging rain-induced flooding occurred in parts of California in 1862, 1907, 1909, 1916, 1937–38, 1940, 1955, 1958, 1964, 1969, 1970, 1974, 1978, 1980, 1983, and 1986. Serious flash flooding occurred in 1939; a hurricane caused widespread flooding in 1976, and some of the most serious snowmelt flooding occurred in 1967 and 1969 in the southern San Joaquin Valley (State of California, 1980). In early 1986, heavy precipitation on top of the snowpack caused extremely severe flooding in the Sacramento Basin.

Flood control reservoirs are designed to operate at certain levels of reliability, usually computed as a frequency of failure. Reservoirs are typically built to handle floods that occur no more frequently than once in a hundred years or, for highly urbanized areas, once in two hundred years. The dilemma facing dam designers, however, is that there are rarely even 100 years of record from which to compute flood frequencies. Thus, expected flood magnitudes and frequencies are determined statistically from available streamflow records.

Not surprisingly, new events are constantly changing these statistical calculations. New extremes extend the historical record and improve the information used to operate existing facilities and to design new ones. In northern California, the Folsom Dam near Sacramento was originally designed to handle a design flood with a return frequency of approximately 500 years. Precipitation events during construction in the 1950s, however, produced runoff that would have exceeded this design flood. By the time the dam was completed, new calculations based on a longer record had reduced the return interval to around 100 years. Severe flooding in 1986 further reduced the design flood-return interval to well under 70 years (Riebsame, 1988).

In the case described above, changes in the expected frequency of the design flood came about simply because of the availability of better information about the actual variability of runoff in the basin, not because of an actual change in climate. In fact, long-term climatic variability may soon be changing with time as anthropogenically-induced climatic changes begin to appear, complicating the assessment of flood risks and probabilities.

FUTURE CLIMATIC IMPACTS IN CALIFORNIA

Information on future climatic changes in California comes from two sources: large-scale general circulation models (GCMs), and more detailed regional climatic impact assessments. In the first case, GCMs provide information on broad climatic changes that are likely to occur from increases in the atmospheric concentration of carbon dioxide and other trace gases. Thus, estimates of changes in temperature and precipitation patterns can be produced, together with other important climatic and hydrologic variables. Figures 4a,b,c show the changes in precipitation for northern California estimated by three GCMs, assuming an increase in the carbon dioxide content of the atmosphere (Manabe and Stouffer, 1980; Manabe et al., 1981; Hansen et al., 1983, 1984; Washington and Meehl, 1983, 1984). As these figures show, the precipitation estimates vary widely, although the temperature predictions are more consistent. The differences in precipitation result from the coarse resolution of the models and from the simplified hydrologic parameterizations used to estimate rainfall.

The hydrologic limitations of GCMs have stimulated research into developing methods for doing more detailed regional assessments (WMO, 1988). Some detailed hydroclimatological impact assessments have already been done in California and the western United States (Revelle and Waggoner, 1983; Williams, 1985; Gleick, 1986, 1987a; Flaschka et al., 1987), although their scopes are limited. This research does, however, provide insights into particular problems that might arise as the climate begins to change. In California, attention has been paid to evaluating possible changes in water availability and the impacts of rising sea level on the San Francisco Bay area. There is evidence to suggest that existing water supply systems will be vulnerable to future climatic changes (Gleick, 1986, 1987a; Williams, 1985).

Williams (1985) provides an overview of the impacts on the San Francisco Bay area of sea-level rise associated with increases in atmospheric concentrations of greenhouse gases. Among the consequences identified were the inundation of developed shoreline areas and the creation of an inland sea in the Sacramento–San Joaquin Delta. This would threaten freshwater supplies, reduce agricultural productivity, and accelerate ecosystem and wetland

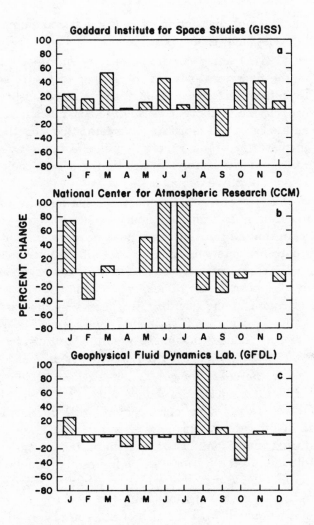

Figure 4. Percent changes over control precipitation estimated by three general circulation models for northern California following a doubling of atmospheric carbon dioxide. (a) Changes estimated by the Goddard Institute for Space Studies (GISS) model. (b) Changes estimated by the National Center for Atmospheric Research model. June and July precipitation increases are greater than 100 percent. (c) Changes estimated by the Geophysical Fluid Dynamics Laboratory model.

loss. Disruption and dislocation of industrial, commercial, and residential development is also possible.

Gleick (1987a) analyzed the sensitivity of runoff and soil moisture in the Sacramento Basin to a range of future climatic changes. Despite uncertainties about future changes in precipitation patterns referred to above, consistent and significant decreases in summer runoff and soil moisture were estimated, together with dramatic increases in winter runoff. These effects are attributed to temperature-induced changes in the timing of snowmelt runoff and increased evaporative demands.

Given this type of regional impact information, the standard planning assumption of climatic stability is slowly being replaced by the concern that all calculations of peak flood volumes, drought severity, recurrence intervals, and so on, will be made obsolete as the climatic effects of increased concentrations of greenhouse gases begin to appear. Indeed, looking at the long-term standard deviation of runoff in the Sacramento Basin suggests that there has been an increase in runoff variability (see Figure 5). Whether this increase is a "change" in climate or simply a return to normal variability after a period of relative quiescence is unclear.

Whatever the cause, however, increased variability has already led system managers to change operating rules. In California, such operating rules must contend with a contradiction inherent in many water resource systems: reservoir levels must be kept low during winter months to maintain protection against high flows and flooding, yet they must be as full as possible during the dry summer months to provide for hydroelectricity, irrigation, and water supply. Thus, the risk of filling the reservoir too soon in the winter is combined with the risk of not completely filling the reservoir in the spring.

In the late 1970s, following the severe drought, several changes were made in the way reservoir operators filled and emptied their reservoirs. These changes were supposed to ensure that the reservoirs were filled earlier and drained more slowly to increase their ability to handle multiyear droughts. This short-term response seemed appropriate, given the severity of the drought and the importance of water during dry periods. But in the early 1980s, two events forced water managers to reevaluate these rules

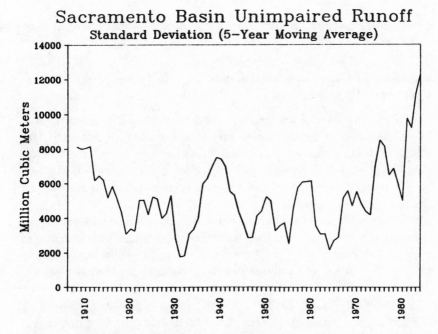

Figure 5. Five-year moving average of the standard deviation of Sacramento Basin unimpaired runoff 1906–86. Note the large increase in recent years in variability.

once again: (1) users complained that water allocations were being kept artificially low to maintain water in the event of a low-probability long-term drought; and (2) severe flooding occurred that was thought to result from filling flood-control space too early.

These complaints led once again to changes in the operating rules. Because of the rising demand for water, the decision was made to increase the risk of running out of water in a long-term, persistent drought in return for higher allocations of water in the short term. Thus, accepting that this higher drought risk will exist was a conscious decision. Unfortunately, the confusion evident in the above situation will be compounded by future climatic changes. In order to respond effectively to such changes, water managers must be prepared for climatic conditions different from those presented by past variability. In many cases, this will require nothing more than simply better planning for variability. In some cases, reviewing the ability of existing structures to handle new conditions may be necessary, and some physical changes may be needed.

Hypothetical Climatic Changes and Hydrologic Vulnerabilities

Existing hydrologic models used to evaluate reservoirs, water demand, and system reliability can be used to evaluate the effects of future changes in climate. At a simple level, internal assumptions about the mean and variance of water availability can be changed. This method can provide region-specific insights into hydrological vulnerabilities.

Figures 6 and 7a–d show some of the implications of future changes in the mean and variability of a hypothetical hydrologic variable—for example, runoff, soil moisture, or groundwater availability. This type of analysis has been used to look at the effects on food security of changes in water availability in different regions of the world (Gleick, 1987b). These figures show how changes in the hydrologic mean and variability can lead to changes in the frequency and severity of drought and flood events, defined as water availability less than or greater than some societally-defined limits. In the hypothetical base-case scenario of 50 "years" (Figure 6), there are three instances of "droughts" and three of "floods."

Case 1 shown in Figure 7a assumes increased mean and variability of water resources. This results in a dramatic increase in the frequency and severity of flooding events and an increase in the frequency of drought events. While an increase in mean water availability could be beneficial to water-short regions, extreme flooding events will have adverse societal effects and an increase in the frequency of droughts could be equally detrimental.

The second case (Figure 7b) increases the mean water availability while reducing the overall variability. This leads to a considerable reduction in both the frequency and severity of droughts compared to the base case and an increase in the intensity of the flooding events. Case 2 offers the greatest overall advantages for water-limited regions, but leads to increased risks in flood-prone regions.

If the variability increases, but the mean decreases, overall water availability may be reduced and the frequency of extreme events may increase, as shown in Case 3 (Figure 7c). This is the

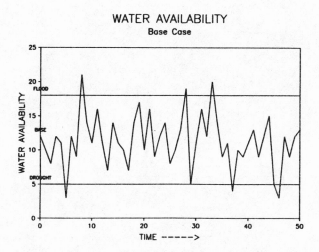

Figure 6. Hypothetical water availability—base case. In this
example, there are three "drought" events and three
"flood" events, defined as flows below or above some
arbitrarily defined level.

worst alternative for water-short regions, because it increases the
severity and frequency of droughts. Flood frequencies also increase,
although their severity may be reduced.

Finally, decreases in both the mean and variability of water
availability may offer relief to flood-prone areas at the cost of adversely affecting drought intensity. Figure 7d shows the increase
in drought severity brought on by such changes in hydrologic parameters.

These hypothetical cases serve to highlight the types of sensitivities that can result from simple changes in climatic conditions.
In California, given past experience, each example above raises serious questions about the resilience of the water resource system
and provides insights into appropriate responses. Within other hydrologic basins, water resource managers should begin to explore
just the same types of sensitivities in order to identify the points at
which existing reservoirs or water supply systems begin to fall below acceptable levels of reliability. Then, a wide range of physical,

322

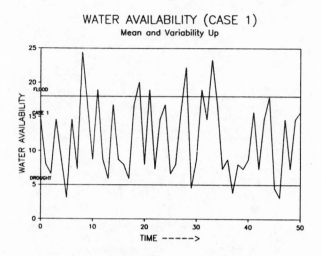

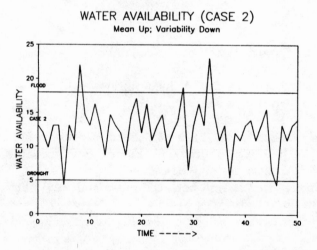

Figure 7. Changes in the mean and the variability of water availability lead to changes in the flood and drought frequency and severity. (a) Case 1. Increased mean and
variability leads to increases in the frequency and severity of flooding events and an increase in the frequency
of drought events. (b) Case 2. Increased mean and reduced variability leads to increase in flood severity and
a reduction in the frequency and severity of droughts.

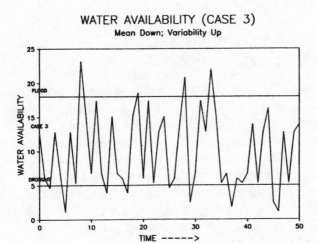

WATER AVAILABILITY (CASE 3)
Mean Down; Variability Up

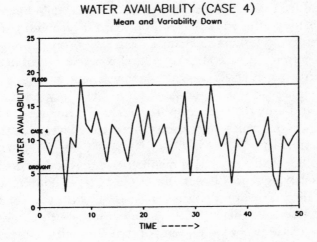

WATER AVAILABILITY (CASE 4)
Mean and Variability Down

Figure 7. (Continued) (c) Case 3. Increased variability and decreased mean leads to a reduction in overall water availability and an increase in the frequency of extreme events. The severity of flooding may be reduced. (d) Case 4. Decreased mean and variability of water availability leads to decreased flood severity and frequency and an increase in the severity of droughts.

economic, and institutional responses can be explored to determine how reliability can be maintained.

INSTITUTIONAL RESPONSES: SUMMARY AND DISCUSSION

There are dangers to drawing parallels between past variability and future climatic changes. For example, the value of climate analogues could be limited if future climatic changes have a root cause different from those that led to past climate variability, if the magnitude of the changes exceeds historical experience, or if new societal conditions change the ability of the system to respond. In each of these three cases, the validity of the analogue must be called into question. The greatest value to the analogue approach is that it can highlight the range of possible societal responses to different problems.

Water resource managers in California and elsewhere have learned from past climatic extremes. These events have led to the development of a wide range of options for responding to climatic variability and climatic change. For long-term climatic changes, these options include the design and construction of new physical structures such as dams and aqueducts, and a wide range of institutional responses. For short-term variability, such as extreme flood and drought events, options include changing existing reservoir operating rules, allocation agreements, levels of demand, and water transfers.

One of the greatest difficulties facing resource managers will be identifying a permanent climatic change, as opposed to climatic variability. The appropriate responses to these two conditions are often different (Gleick, 1988). Unless long-term climatic change can be distinguished from temporary short-term climatic variability, inappropriate management decisions could be made. Thus, a short-term drought could be mitigated by looking for surplus out-of-basin water. A long-term shortage, however, would require changes in demand, water pricing, and perhaps the water-supply infrastructure itself. The recent rule-curve changes in California, discussed earlier, are good examples of making decisions based on short-term climatic variability rather than long-term climatic change.

To pursue another example discussed earlier, during the 1976–77 drought southern California water agencies agreed to relinquish temporarily their water rights to northern California water *only* because they were able to replace that water with surplus water from the Colorado River. In the future, two complications will arise. First, a long-term climatic change that alters water availability in California may also reduce the overall water available in the Colorado River system. Second, growing demand in the Colorado River Basin will soon lead to the complete allocation of the available water, making out-of-basin transfers unlikely—especially in dry years. Thus, southern California users would be unable to rely on alternative water transfers to reduce their demand for northern California water, and northern California users might have to shoulder an even larger share of the drought burden.

In California, many of the problems that arose from past climatic variability had their roots in both structural limitations, such as dams or aqueducts improperly sized or located, and institutional constraints, such as inflexible operating rules and water markets. It is now understood that to try to eliminate completely damages from floods and droughts through physical means is economically and environmentally unrealistic. As a result, nonstructural management measures are receiving increasing attention. Addressing the problem of future shortages and floods will require innovative use of these measures in a manner that enhances the overall reliability and flexibility of our water management systems.

REFERENCES

Flaschka, I., C.W. Stockton, and W.R. Boggess, 1987: Climatic variation and surface water resources in the Great Basin region. *Water Resources Bulletin, 23*, 47–57.

Frederick, K.D., and P.H. Gleick, 1988: Water resources and climate change. In N. Rosenberg, P. Crossen, J. Darmstadter, and W. Easterling (Eds.), *Proceedings of the Conference on Controlling and Adapting to Greenhouse Warming.* 14–15 June 1988. Washington, DC: Resources for the Future.

Gleick, P.H., 1986: Methods for evaluating the regional hydrologic consequences of global climatic changes. *Journal of Hydrology, 88*, 97–116.

Gleick, P.H., 1987a: Regional hydrological consequences of increases in atmospheric CO_2 and other trace gases. *Climatic Change, 10*, 137–61.

Gleick, P.H., 1987b: Global climatic change, water resources, and food security. In M.S. Swaminathan (Ed.), *Proceedings of the International Symposium on Climate Variability and Food Security.* 6–9 February 1987. New Delhi, India: International Rice Research Institute.

Gleick, P.H., 1988: The effects of future climatic changes on international water resources: The Colorado River, the United States, and Mexico. *Policy Sciences, 21* (in press).

Hansen, J.E., D. Rind, G. Russell, P. Stone, I. Fung, R. Ruedy, and J. Lerner, 1984: Climatic sensitivity: Analysis of feedback mechanisms. In J.E. Hansen and T. Takahashi (Eds.), *Climate Processes and Climate Sensitivity.* American Geophysical Union Monograph 29, Maurice Ewing Volume 5. Washington, DC: American Geophysical Union, 130–63.

Hansen, J.E., G. Russell, D. Rind, P. Stone, A. Lacis, S. Lebedeff, R. Ruedy, and L. Travis, 1983: Efficient three-dimensional global models for climate studies: Models I and II. *Monthly Weather Review, 111*, 609–62.

Jackson, W.T., and A.M. Paterson, 1977: *The Sacramento–San Joaquin Delta: The Evolution and Implementation of Water Policy.* Contribution 163. Davis, CA: California Water Resource Center, University of California at Davis.

Manabe, S., and R.J. Stouffer, 1980: Sensitivity of a global climate model to an increase of CO_2 concentration in the atmosphere. *Journal of Geophysical Research, 85*, 5529–54.

Manabe, S., R.T. Wetherald, and R.J. Stouffer, 1981: Summer dryness due to an increase of atmospheric CO_2 concentration. *Climatic Change, 3*, 347–86.

McGowna, J.A., 1961: *History of the Sacramento Valley.* Volume 1. New York, NY: Lewis Historical Publishing Co.

Revelle, R.R., and P.E. Waggoner, 1983: Effects of a carbon dioxide-induced climatic change on water supplies in the western United States. In *Changing Climate,* Report of the Carbon Dioxide Assessment Committee, Board on Atmospheric Sciences and Climate, National Research Council. Washington, DC: National Academy Press, 419–32.

Riebsame, W.E., 1988: Adjusting water resource management to climate change. *Climatic Change, 12* (in press).

State of California, 1978: *The 1976–1977 California Drought: A Review.* Department of Water Resources Report. Sacramento, CA: The Resource Agency, State of California.

State of California, 1980: *California Flood Management: An Evaluation of Flood Damage Prevention Programs.* Department of Water Resources, Bulletin 199. Sacramento, CA: The Resource Agency, State of California.

Washington, W.M., and G.A. Meehl, 1983: General circulation model experiments on the climatic effects due to a doubling and quadrupling of carbon dioxide concentration. *Journal of Geophysical Research, 88*, 6600–10.

Washington, W.M., and G.A. Meehl, 1984: Seasonal cycle experiment on the climate sensitivity due to a doubling of CO_2 with an atmospheric general circulation model coupled to a simple mixed-layer ocean model. *Journal of Geophysical Research, 89*, D6, 9475–503.

Williams, P.B., 1985: *An Overview of the Impact of Accelerated Sea Level Rise on San Francisco Bay.* Project 256. San Francisco, CA: Phillip Williams and Associates.

WMO (World Meteorological Organization), 1988: *Water Resources and Climatic Change: Sensitivity of Water Resource Systems to Climate Change and Variability.* WCAP-2. Geneva, Switzerland: World Meteorological Organization.

14

Analyzing the Risk of Drought: The Occoquan Experience

Daniel P. Sheer

INTRODUCTION

The Fairfax County Water Authority (FCWA) is the sole water purveyor supplying some 650,000 people in Fairfax County, Virginia, a suburb of Washington, D.C. The primary source of water is a small dam and reservoir on the Occoquan Creek, a local tributary of the Potomac River (Figure 1). In 1977 the reservoir had been dropping steadily throughout the summer and by the end of August had receded to a record low level. Worried that the drought would continue, the FCWA and local governments instituted bans on nonessential water use, such as lawn watering and car washing. More drastic measures, such as the closing of schools and businesses, were actively being considered. It was under these circumstances that the author organized a technical effort to analyze the risk of extreme water shortages and to evaluate the policies necessary to reduce those risks to acceptable levels. (The team that undertook this effort was composed of Robert Hirsch, hydrologist

This chapter is reprinted from Journal *American Water Works Association, Vol. 5, No. 72, May 1980, by permission. Copyright ⓒ1980, American Water Works Association.*

Daniel P. Sheer is the founder and president of Water Resources Management, Inc. Dr. Sheer is a nationally recognized leader in the field of water resources management for his development and application of innovative analytical techniques applied to both large- and small-scale water resources systems. He was the director of the Cooperative Water Supply Operations of the Interstate Commission on the Potomac River Basin.

330

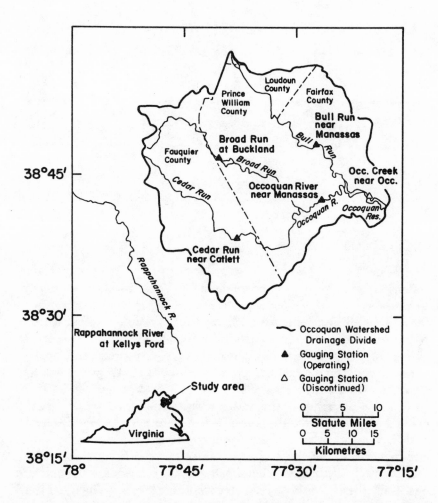

Figure 1. Map of the Occoquan Watershed showing locations of
streams, gauges, and reservoirs.

with the USGS, and John Schaake, Eric Anderson, and George
Smith of the National Weather Service.)

    This paper describes the development and implementation
of the risk analysis techniques, the subsequent interaction between
technical and political decisionmakers, generation of the required
information, and the application of these techniques and procedures to future drought problems.

BACKGROUND

In 1976, the year prior to the drought, the FCWA had commissioned a study to determine the safe yield of the Occoquan reservoir. This report pegged the safe yield at 247 ML/day (65 mgd) (Greeley and Hansen, 1976). By August 1977 the FCWA was averaging about 266 ML/day (70 mgd) withdrawal from the reservoir, with the expectation of reducing demand to 209 ML/day (55 mgd) by October as a result of normal seasonal reduction in water use. The average demand should have been below the safe yield, but during August the reservoir level fell precipitously.

Early in August the FCWA decided that voluntary conservation measures were prudent, and mandatory measures were instituted by September 14. That these measures were taken as early as they were is a tribute to the insight of the FCWA personnel. In late August, largely because of the publicity and controversy that accompanied the beginning of the restrictions, an independent analysis* of the safe yield was undertaken using readily available, computerized USGS streamflow data. It was believed that the results of that analysis would support the contention that the prevailing restrictions in water use were very conservative.

On the contrary, the analysis showed a historical safe yield of only 205 ML/day (54 mgd) over the 26 years of records available at that time. The discrepancy between the earlier report and the new analysis was due to the selection of the critical period used in the 1976 analysis. The difference of 42 ML/day (11 mgd) was critical.

Although it was significant, the historical safe yield analysis provided information only on the year that produced the most severe water supply problems (critical period) and on maximum yield from the start of the critical period to the end of the critical period. The fact that the current average withdrawal exceeded the historical safe yield implied that in at least one year in the historical record the reservoir would have gone dry if the FCWA had attempted to meet the current demand.

While this was important information, it did not relate directly to the problem at hand. No use was made of available data concerning the level of water in the reservoir or the relative severity of the existing drought as compared to the worst historical drought.

---

* By the author and R.M. Hirsch.

GENESIS OF THE TECHNIQUES

A simulation very closely related to the simulation used to determine historical yield was developed to incorporate information about the current situation. The new simulation was designed to answer the question of what would have happened if the demands had been as high as they were in 1977, the reservoir as low as it was in 1977, and the date were 1 September 1965 (one of the worst drought years), instead of 1 September 1977. The current storage was added to the inflows for September 1965, and the expected water use for September of the current year and an allowance for evaporation were subtracted. The result was the storage that would have remained on 1 October 1965. The calculation was then repeated for each subsequent month until the end of the 1965–66 drought. Table 1 shows the results of this simulation, indicating that the reservoir would have been dry during parts of the following January and February.

To determine in how many years in the historical record the reservoir would have gone dry, the simulation was repeated for each of the 26 years of record then available. In four of those years the reservoir would have been empty. Thus the risk of a dry reservoir was four in 26 years, or $4/26$—approximately 15 percent. The consequences of such a breakdown in the public water supply, particularly loss of fire protection and sanitary facilities, would have been disastrous.

This analysis was called the "position analysis" because it described the current position of the water supplier with respect to the risk of encountering a critical situation in the near future.

Almost as soon as the first analysis was complete it was apparent that additional information could be used to refine the estimate of risk. The historical streamflow record could be extended by using statistical techniques and records of long duration from nearby gauges. Serial correlation of streamflows, a measure of the persistence of previous droughts, could be taken into account by excluding from the analysis those historical years in which summer flows were high. Which years to exclude was the difficult question. The U.S. Geological Survey (USGS) agreed to develop techniques to answer this question using historical streamflow data.

Much of the persistence in droughts is caused by the lack of soil moisture and its impact on runoff from rainfall. For this reason,

Table 1

Position Analysis Simulation Using 1965–1966 Inflows

		Calculation of Reservoir Storage	
Position Analysis	Date	ML	mil gal
Actual reservoir storage	9/65	13900	3650
Inflow		+1900	+507
Withdrawal and evaporation*		−5700	−1500
Simulated reservoir storage	10/65	10100	2657
Inflow		+4500	+1192
Withdrawal and evaporation*		−5900	−1550
Simulated reservoir storage	11/65	8700	2299
Inflow		+1600	+433
Withdrawal and evaporation*		−5700	−1500
Simulated reservoir storage	12/65	4600	1232
Inflow		+1900	+493
Withdrawal and evaporation*		−5900	−1550
Simulated reservoir storage	1/66	600	175
Inflow		+2900	+753
Withdrawal and evaporation*		−5900	−1550
Simulated reservoir storage	2/66	0	0
Inflow		62000	16246
Withdrawal and evaporation*		−5900	−1550
Simulated reservoir storage	3/66	37200†	9800†

*At 190 ML/day (50/mgd)
†Maximum reservoir capacity

the assistance of the National Weather Service (NWS) hydrologic research laboratory was requested so that the NWS rainfall-runoff model and statistical analysis program (the NWS river forecast system, or NWSRFS) could be applied to simulate streamflows using historical meteorological data. The streamflows simulated were those that would have occurred, given the current soil moisture conditions. These simulated streamflows were then used in the position analysis instead of the historical streamflows, thereby providing a method of accounting for soil moisture conditions in the position analysis. A comparison of historical and simulated streamflows is given in Table 2.

In nearly all cases the simulated flows were lower than the historical flows. This was the result of the low soil moisture conditions used to initiate the model runs. It illustrates the dramatic impact that soil moisture can have on runoff, and thus the importance of considering soil moisture in establishing the risk of drought.

STARTING THE TECHNICAL-POLITICAL INTERACTION

The political climate eased the way for an independent evaluation of the seriousness of the drought. The question of how much additional development should occur in Fairfax County was and is a very hot political issue. The FCWA is charged with the responsibility of serving whatever growth does occur, but because its present facilities are running at their capacity, the FCWA was seeking approval (since granted) for the construction of a new water treatment plant on the Potomac River. Unfortunately, political groups opposed to further development viewed the control of provision of additional water and sewer service as the most effective way to control growth. In this context, warnings from the FCWA concerning the seriousness of the drought were seen as a ploy by the pro-growth forces to justify need for additional treatment capacity. With the 1976 safe yield report as a backup, the arguments of the anti-growth forces that the drought was a sham carried some weight.

The FCWA staff had been applying the best techniques at its disposal to attempt to predict the course of the drought. The U.S.

Table 2

Simulated Conditional Streamflow* Versus Historical Streamflow† for the Occoquan Reservoir in November

Year	Simulated Flow		Historical Flow	
	ML	mil gal	ML	mil gal
1949	12574	3309	80058	21068
1950	5890	1550	10682	2811
1951	13361	3587	39683	10443
1952	10617	2794	9633	2535
1953	55415	14583	87632	23061
1954	7323	1927	3029	797
1955	3876	1020	2922	769
1956	3203	843	9435	2483
1957	4636	1220	48613	12793
1958	3743	985	15743	4143
1959	2956	778	3386	891
1960	3728	981	13864	3648
1961	2736	720	4951	1303
1962	3397	894	9853	2593
1963	24426	6428	23100	6079
1964	38008	10002	31380	8258
1965	3245	854	7638	2010
1966	2223	585	1645	433
1967	2899	763	10591	2787
1968	3359	884	5491	1445
1969	4393	1156	18122	4769
1970	3276	862	8767	2307
1971	34717	9136	89334	23509
1972	5977	1573	55203	14527
1973	71254	18751	4366	1149
1974	3200	842	5586	1470
1975	3051	803	2660	700

*Soil moisture conditions as of 25 November 1977
†Reconstructed from Hirsch

Department of Agriculture Soil and Conservation Service rainfall-runoff curves were used with long-range weather forecasts to predict runoff (U.S. Department of Interior, Bureau of Reclamation, 1977). The curves plot runoff as a function of rainfall, given soil type and soil moisture conditions.

With this information, long-range weather forecasts were broken down into six-day periods. Starting with the first period, rain runoff was calculated from the curves, and the change in soil moisture conditions was estimated. The process was continued for ten periods (two months) to produce a predicted streamflow trace, and this trace was then used to predict the contents of the reservoir.

When the FCWA staff contacted the NWS to obtain long-range forecasts suitable for use in this manner, the NWS refused to provide them on the grounds that such forecasts could not be made accurately. The FCWA then contracted with a private weather forecasting firm for the forecasts. The FCWA staff was acutely aware of the dependence of the streamflow predictions on weather forecasts of doubtful accuracy, and the ability of the risk analysis technique to deal with this uncertainty was perceived to be advantageous.

The simplicity of the technique was also an advantage, as the results had to be explained to a nontechnical audience. FCWA staff formed an advisory group of local government staff members to guide the study effort and to keep responsible local officials aware of its progress. The first meeting of the group was held on 22 September 1977. At that meeting the study team described the kinds of information that could be made available and sought the group's aid in defining the objectives of the study effort and the critical conditions that formed the basis for the risk analysis.

OBJECTIVES OF THE ANALYSIS: DEFINING RISK

The primary objective of the study was to develop useful information concerning the future availability of water. As local governments are responsible for making and enforcing decisions concerning water use, it was appropriate that the local government representatives be consulted at the first meeting concerning what information would be most useful in the decisionmaking process.

Following the explanation of the techniques and the presentation of the preliminary results, the immediate question asked by the local officials was "How likely are we to run out of water?"

The study team responded to this question by saying that the chance of running out of water was highly dependent on policy decisions as to when and how much to further cut consumption. It was pointed out that as the available storage fell further, progressively more stringent water use restrictions would be imposed in order to preserve the remaining storage. This kind of operation would reduce the likelihood of exhausting the reservoir storage.

Prior to the meeting, the FCWA and the local governments had agreed on an operating policy to deal with the drought which called for the imposition of increasingly severe water use restrictions keyed to water levels in the reservoir (Whorton, 1977). Stage 1 of the restrictions called for voluntary restrictions on outside uses of water, including car washing and lawn irrigation. Stage 2A called for mandatory restrictions on the same uses. Stage 2B called for loosely defined restrictions on water use within the home, and stage 3 imposed the shutdown of schools and businesses. At the time of the first meeting stage 2A restrictions had just been imposed, and much discussion followed on what would be required in stage 2B and how such restrictions could be enforced. Stage 3 was thought of as a last resort, even though the risk of the reservoir's dropping low enough to trigger stage 3 restrictions was unknown.

In this context, the local officials decided that the risk of the reservoir's falling to the level requiring stage 3 restrictions was the risk to be controlled. The decision to be used to control this risk was the decision as to the severity of the stage 2B restrictions to be imposed. Developing information on the implications of implementing stage 3 restrictions was indeed possible; however, in the interest of developing the most pertinent information in the shortest time, all effort was directed to determining appropriate stage 2B restrictions.

TECHNIQUES USED BY THE USGS

The techniques used in the USGS analysis have been fully described by Hirsch (1978).* They are essentially a refinement

---

* This publication is available from the author upon request.

of the position analysis method. The initial work by the USGS focused on extending the streamflow record as far back into the past as possible. Longer flow records on nearby drainage basins were available. Linear regression of flows at gauges in the Occoquan basin with flows in these other basins formed the basis for the extension of the historical record. The record produced by this technique had less variability than the actual historical record, a fact that tended to reduce the severity of historical droughts as represented in the extended record. Nevertheless, the record did represent a best estimate of past streamflows. Using the regression techniques, a sequential streamflow record of 49 years was generated.

To account for the persistence of low streamflows, the autumn and winter flow series from the 49-year record were divided into two groups, those which followed dry Septembers (DS) and those which followed wet Septembers (WS). The dividing flow for determining whether a September was wet or dry was arbitrarily set at an average of 190 ML/day (50 mgd). Twenty-two years in the 49-year record met the DS condition, and these were used in the position analysis techniques to determine risk.

The choice of 190 ML/day (50 mgd) as the dividing flow was made as a compromise of two conflicting objectives. The first was to keep the number of years used in the analysis as high as possible to account for the inherent variability of streamflow. The second was to eliminate from the analysis those years dissimilar to the current year. A statistical procedure, the Mann-Whitney rank sum test (Seber, 1977), was employed to determine the similarity between the DS and WS flows. With the division set at 190-ML/day (50 mgd) average flow, the hypothesis that both groups came from the same sample population was rejected at the 1 percent level. While 190 ML/day (50 mgd) was not the largest dividing flow that still produced this level of confidence, it provided an appropriate balance between sample size and sample similarity. Because the DS group included all the worst drought years but excluded many high flow years, the risk estimates were approximately twice as great as the estimates generated using the entire historical record.

NATIONAL WEATHER SERVICE TECHNIQUES

During the course of the NWS study effort, existing NWS techniques were combined with the position analysis to produce risk estimates. Like the USGS effort, the major focus of the NWS effort was to produce a set of streamflow records that were more representative of what could happen in the future than the historical streamflow record. This record would then be used in a position analysis to determine risk. The NWS used the NWSRFS model (Curtis and Smith, 1976) along with historical meteorological data to produce the streamflow records. The NWSRFS is an integrated system of forecast models including the Sacramento soil moisture accounting model (Burnash et al., 1973), streamflow routing models, reservoir operations models, procedures for estimating average precipitation and temperatures over large areas, and statistical analysis techniques. The Sacramento soil moisture accounting model is the basis of the NWSRFS (Figure 2). This is a conceptual hydrologic model for the movement of water through the soil, through groundwater systems, and over the surface to the stream network. It accounts for evapotranspiration and maintains a detailed water balance in time increments of six hours.

Because the entire 1560-km^2 (600-sq mi) drainage area of the reservoir was treated as a single catchment, the flow routing routines available in the NWSRFS were not used. NWSRFS also contains a snow accumulation and ablation model (Anderson, 1973) (Figure 3) to account for the influence of snowmelt. Calibration of the model was performed using 26 years of historical streamflow and meteorological information. The object of the calibration was to adjust the model parameters to achieve the best match between flows predicted by the model using historical meteorological data and the observed streamflows for the same period.

The meteorological data used in the analysis are stored on magnetic tape by the NWS. The data represent six-hour averages for precipitation and temperature readings at NWS stations. Those readings for stations in and around the basin were converted to mean aerial temperature and precipitation for the basin by routines contained in the NWSRFS model.

Once the model was calibrated, the current values of the model's soil moisture parameters were estimated by running meteorological data for the preceding six-month period through the

340

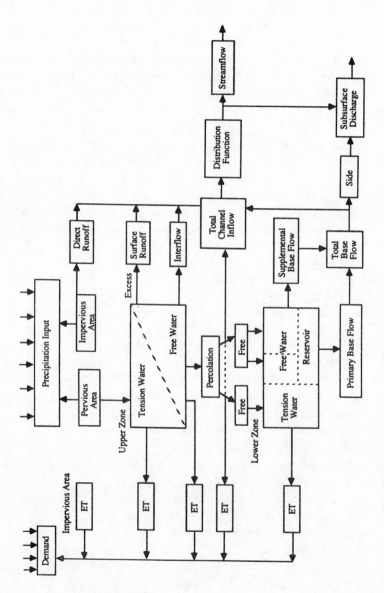

Figure 2. Sacramento soil moisture accounting model.

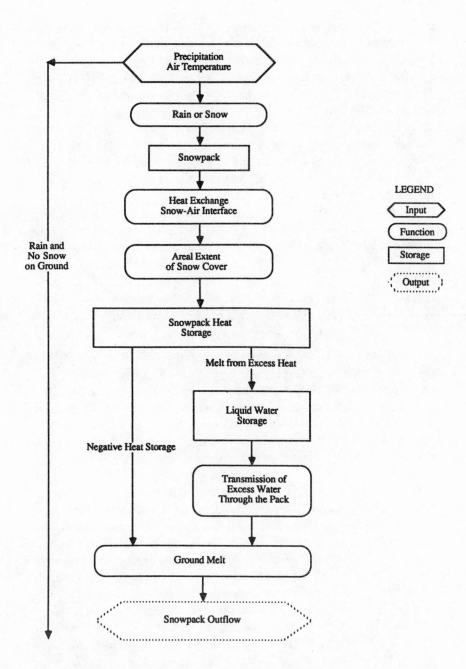

Figure 3. Snow accumulation and ablation model.

model. With the model parameters set to these estimates, meteorological data beginning at the current data and for one year in the historical record were run through the model. This produced the streamflow trace that would occur if the historical meteorological data repeated themselves in the immediate future. Running the simulation for every year in the historical record produced a set of 26 historical "conditional" streamflow records, each conditioned on current soil moisture conditions in the basin. It is important to note that although the streamflows generated were conditioned on soil moisture conditions, no actual soil moisture or groundwater level measurements were taken.

These streamflows were used in a position analysis to estimate the risk of encountering a critical condition. Statistical techniques used in the extended streamflow prediction program (Twedt et al., 1977) of the NWSRFS were used to further refine estimates of risk.

ATTEMPTS TO INCORPORATE LONG-RANGE WEATHER FORECASTS

NWS long-range weather forecasts are updated every fifteen days and cover the forthcoming month. They are intentionally imprecise, stating only that temperature and precipitation will be above normal, normal, or below normal for the month for large areas of the United States. The forecasts are based on upper air circulation patterns expected for the next 30 days and the weather that has been associated with those patterns in the past.

These forecasts cannot be directly incorporated into the NWSRFS because that system requires input on a six-hour basis in order to simulate runoff. Instead, the approach taken was to use the forecast to eliminate from the historical record those years that had upper air circulation patterns most unlike the forecast patterns. This is analogous to the techniques used to eliminate WS years from the USGS analysis. Because of the great uncertainties in the long-range forecasts, only the three most dissimilar years were eliminated from the historical record. Unfortunately, the upper air circulation patterns at the time of the analysis were not generally associated with either drought or wet periods, and none of the record years chosen as most dissimilar were particularly dry or wet. Therefore, the incorporation of long-range forecasts made little difference in the estimate of risk. Had the long-range forecast shown a definite trend, this probably would not have been the case.

No attempt was made to use the long-range forecast provided to the FCWA by the private forecasting service mentioned earlier. However, these forecasts were made for six-day periods, an interval much more suitable for direct use in the NWSRFS models. The forecasts are compared to the actual rainfall in Table 3. Although the monthly totals seem to agree fairly well, the regression line of the predictions on the actual rainfall is negative for the ten forecast periods. The correlation coefficient is not significant. This tends to support the NWS contention that such forecasts cannot be made accurately. Certainly, such forecasts should not be the basis for decisions crucial to the protection and well-being of large numbers of people.

Table 3

Private Weather Predictions Versus Actual Rainfall

Date	Predicted Rainfall		Actual Rainfall	
	cm	in	cm	in
September				
2–7	1	0.4	2.5	0.99
8–13	1	0.4	0.6	0.24
14–19	1	0.4	1.6	0.62
20–25	1.75	0.7	1	0.4
26–Oct. 1	1.25	0.5	0.4	0.16
Total	5.85–6.15	2.28–2.40	6.1	2.02
October				
2–7	3.6	1.4	0.17	0.07
8–13	4	1.56	1.85	0.72
14–19	3.8	1.48	2.77	1.08
20–25	2.25	0.9	0.37	0.15
26–Nov. 1	1	0.4	5.75	2.3
Total	14.65–15.05	5.7–5.9	10.7	4.17

CONTINUING THE TECHNICAL-POLITICAL INTERACTION

The second meeting of the advisory group was held on 7 October 1977. By this time, preliminary risk estimates were available from the USGS effort. Because withdrawals from the reservoir had been reduced by mandatory restrictions on outside water use, by public cooperation in conserving water, and by purchasing as much water as practicable from surrounding utilities through small existing interconnections, the risk of reaching the critical level in the reservoir had been reduced to about 10–15 percent despite the continuing drought. This risk level was still unacceptably high. The preliminary analysis indicated that the risk level could be reduced threefold by a reduction of only 30 ML/day (8 mgd) in reservoir withdrawals, or a 46-Lpcd (12-gpcd) reduction in water use.

The average water use was then about 330 Lpcd (87 gpcd), and no member of the advisory committee or the study team felt such reduction was unachievable.

Although the NWS results were not yet available, a decision had to be made as to whether or not the USGS information should be made public in its preliminary form. The probable results of delaying the implementation of the proposed water reduction program were analyzed. Delaying a 30-ML/day (8-mgd) reduction for two weeks would mean the use of an additional 425 ML (112 million gallons) of water. This represented only a three- to four-day supply, or at the most an additional week of time to wait for rain should the drought continue through the winter. The advisory group decided to wait the additional time to be certain that the further reductions were necessary. The study group supported this decision, knowing that the additional 425 ML (112 million gallons) of storage would not have a significant impact on risks.

The discussions at this second meeting also centered on what information could be made available to assist the local governments in deciding when it was safe to lift restrictions, should the level in the reservoir begin to rise. If the risk of reaching a critical level fell to less than 1 percent with normal consumption, it was decided that restrictions could be lifted. The task of finding the reservoir level at which risks fell to 1 percent as a function of time was then undertaken.

RESULTS

The agreement between the final NWS and USGS results was very good. This was encouraging, since the two techniques are quite different. The study group met and resolved the small differences that did occur.

The final results of the study effort were presented in a short report (Hirsch et al., 1977), and are given in part in Tables 4 and 5 and Figures 4 and 5. The report recommended immediate implementation of water conservation to achieve a 30-ML/day (8-mgd) reduction in demand.

IMPLEMENTATION

These results were presented to the advisory group at its third and final meeting on October 18. It was decided to implement the recommended 30-ML/day (8-mgd) demand reduction by voluntary measures rather than through enforcement. The members of the advisory group felt that, given adequate publicity, the public would understand the implications of the risk analysis and respond favorably to a request for a moderate reduction in water use. Much more severe reductions in water use were then being achieved in California, which was also in the midst of a drought.

Briefings were immediately arranged for the leaders of the affected jurisdictions to produce a unified call by all responsible officials for the appropriate reduction in demand. This unified front was deemed necessary if the voluntary reductions were to be effective, given the existing political situation.

A televised press conference was arranged, and the story was leaked to a *Washington Post* reporter the night before. Unfortunately, there was no chance to see if the publicity would work. On October 27 it began to rain—in torrents.

POSTMORTEM

By November 8, when restrictions were formally ended, the reservoir level had risen 5 m (17.5 ft) and was about 1.5 m (5 ft) above the level specified as safe in the target storage curve. The reservoir was filling so fast that the restrictions certainly would have been lifted in any case. However, in the two-month period between September 1 and October 31, the risk analysis techniques

Table 4

Probabilities of Occoquan Reservoir Falling Below Critical
Storage Levels at 150 ML/day (40 mgd) Withdrawal

Storage Level		Volume of Storage Remaining		Probability
m	ft	GL	bil gal	percent
28	93	6.5	1.7	13
27	88	4.2	1.1	10
25	83	2.7	0.7	8

Table 5

Probabilities of Occoquan Reservoir Falling Below Critical
Storage Levels at 122 ML/day (32 mgd) Withdrawal

Storage Level		Volume of Storage Remaining		Probability
m	ft	GL	bil gal	percent
28	93	6.5	1.7	5
27	88	4.2	1.1	3
25	83	2.7	0.7	2

had been brought from conception to operation, their use had produced results, and these results had been implemented as a part
of the operating policy of the FCWA and the local governments.
Since then, the FCWA has implemented the streamflow-based technique on its own computer system and performs a position analysis
at the beginning of each month, or more often as necessary. An
excessive level of risk is taken as an indication that a drought may

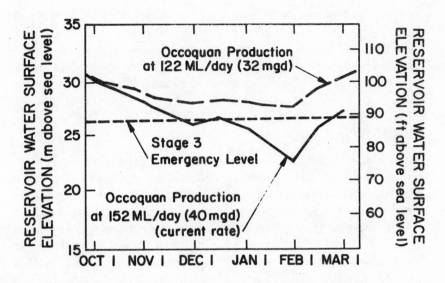

Figure 4. Reservoir levels at various production rates.

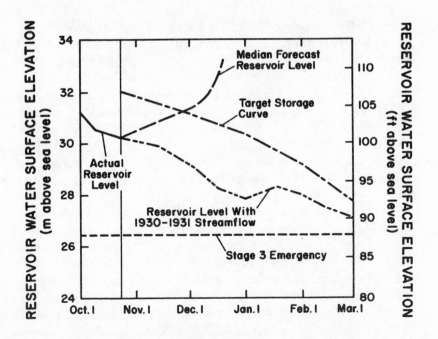

Figure 5. Occoquan target storage curve.

be imminent, although no formal mechanisms have been developed for integrating operational policy with the position analysis. The FCWA is continuing to upgrade its ability to use position analysis; for example, weekly instead of monthly flow records are being obtained from the USGS for use in the simulation.

COMPARISON OF THE USGS AND NWS TECHNIQUES

The USGS technique is by far the simpler of the two techniques developed. It requires considerably less data to implement than the NWS technique, and can be performed without the aid of a computer. For sources like the Occoquan Creek, where snowmelt is not a major factor and groundwater conditions are not influenced by events either outside the basin or totally independent of streamflow, the results produced are likely to correspond to those produced using the NWS techniques. In cases where snowmelt or groundwater is a factor, the analysis may be improved by the considered input of persons familiar with local conditions. Research is continuing on methods to incorporate serial correlation and other statistical properties of the historical record into streamflow-based position analysis (Hirsch, 1978). The author believes that streamflow-based position analysis should be the technique of first resort in estimating the risk of streamflow-induced drought.

The NWS technique has the potential to be the more precise of the two techniques, at the expense of considerably more data and effort. Still, neither the data requirements nor the computational burden are really excessive if it is desirable to implement the technique for future use. If, as is often the case in the United States, the data are available in digital form, the technique can be implemented with expediency.

In order to test the predictive ability of the NWSRFS model against the predictive ability of past streamflows, the author performed several multiple regressions using the Statistical Package for the Social Sciences program (SPSS) (Nie et al., 1975).

September flows were treated as the dependent variable, and two groups of variables—the values of three of the groundwater parameters in the NWSRFS model at the end of the preceding August and the preceding June, July, and August flows—were used as independent variables in two multiple regressions. The multiple regression coefficient for the model parameters was 0.39. For the flows in the three preceding months the multiple r was only 0.21, indicating that September flows were better related to modeled groundwater conditions than to previous flows.

Regressions were then run to test the ability of the two groups of parameters to predict droughts, again defining droughts as flows less than 190 ML/day (50 mgd) for the month of September. The dependent variable in these regressions was 190 ML/day (50 mgd), and 0 otherwise. Again the model parameters outperformed the previous flows, with multiple rs of 0.45 and 0.25, respectively. Transforming past flows by taking logarithms improved the correlation to 0.34, still lower than that achieved by the model parameters.

While these comparisons are far from conclusive, they indicate that even when snowmelt is not a factor, the model parameters may be a better predictor of future flows than past streamflows. When snowpack makes a significant contribution to future streamflows, the NWS technique has the tremendous advantage of accounting for the phenomenon explicitly.

The author also believes that techniques can be developed to effectively utilize long-range weather forecasts in the NWS technique. The attempts described here were crude, but nonetheless show great promise. As the precision of long-range forecasts improves, so will the precision of the risk estimates produced using the NWS technique.

The NWS technique attempts to account for soil moisture conditions without the need for field measurements of those conditions. (The information concerning soil type, soil depth, and other geologic data used in estimating the model parameters for calibration is generally available from the U.S. Geological Survey). Field measurements of groundwater conditions are often quite expensive. If the NWS technique is used, the value of such field data may be significantly reduced, obviating the need for such measurements.

CONCLUSIONS

The USGS risk analysis technique can be applied using only streamflow information, and should be the technique of first resort for assessing the likelihood of storage deficiencies. The NWS technique has the potential for being more precise, since it can explicitly incorporate snowmelt, long-range weather forecasts, and other factors. Furthermore, the groundwater parameters in the NWS model may be better predictors of future streamflows than are past streamflows. Both techniques provide information of great utility in public decisionmaking.

ACKNOWLEDGMENTS

This paper reports the work of many persons, in particular, the efforts of Robert Hirsch of the USGS and John Schaake, Eric Anderson, and George Smith of the NWS. Jay Corbalis, Floyd Eunpu, Fred Griffith, Warren Hunt, and Robert Etris of the Fairfax County Water Authority provided much assistance and encouragement while the risk analysis techniques were being developed and implemented. Paul Eastman of the Interstate Commission on the Potomac River Basin provided the encouragement necessary to produce this paper, which was prepared by Sandra Sayers. Finally, the author would like to thank Abel Wolman for his thoughtful and helpful review of the manuscript.

REFERENCES

Anderson, A.E., 1973: *National Weather Service River Forecast System, Snow Accumulation and Ablation Model.* NOAA Tech. Memo NWS HYDRO-17. Silver Springs, MD: National Oceanographic and Atmospheric Administration.

Burnash, R.J.C., R.L. Ferral, and R.A. McGuire, 1973: *A Generalized Streamflow Simulation System: Conceptual Modeling for Digital Computers.* Sacramento, CA: Joint Federal–State River Forecast Center, U.S. National Weather Service and the California Department of Water Resources.

Curtis, D.C., and G.F. Smith, 1976: *The National Weather Service River Forecast System—Updated 1976.* Hydrological Research Lab W23, July. Silver Spring, MD: National Oceanographic and Atmospheric Administration.

Greeley and Hansen, Consulting Engineers, 1976: *Safe Yield of Occoquan Reservoir.* Memorandum Rept. 900-3, 8 July. Arlington, VA: Fairfax County Water Authority.

Hirsch, R.M., 1978: *Risk Analysis for a Water Supply System—Occoquan Reservoir, Fairfax and Prince William Counties, VA.* USGS Open-File Rept. 78-452. Reston, VA: USGS.

Hirsch, R.M., J.C. Schaake, and D.P. Sheer, 1977: *Assessment of Current Occoquan Water Supply Situation.* October. Bethesda, MD: Interstate Commission on the Potomac River Basin.

Nie, N.H., C.H. Hull, J.G. Jenkins, K. Steinbrenner, and D.H. Bent, 1975: *SPSS, Statistical Package for the Social Sciences.* Second Edition. New York, NY: McGraw-Hill Book Co.

Seber, G.A.F., 1977: *Linear Regression Analysis.* New York, NY: John Wiley and Sons.

Twedt, T.M., J.C. Schaake, and E.L. Peck, 1977: *Proceedings of the 45th Annual Meeting, Western Snow Conference.* Fort Collins, CO: Colorado State University.

U.S. Department of Interior, 1977: *Design of Small Dams.* Appendix A, GPO 2403-0089. Washington, DC: Bureau of Reclamation.

Whorton, L., 1977: *Water Emergency Procedures.* Fairfax County Executive Memorandum to the Fairfax County Board of Supervisors, 21 September 1977. Fairfax, VA: Fairfax County Water Authority.

15

The Ogallala Aquifer and Carbon Dioxide: Are Policy Responses Applicable?

Donald A. Wilhite

INTRODUCTION

The implications of a global warming due to increasing atmospheric CO_2 and other trace gases such as methane and chlorofluorocarbons have stimulated discussion within the scientific and policymaking communities about actions that governments and others might take to prepare for such changes in climate. However, uncertainty about the nature and magnitude of climate change and its associated environmental and societal impacts makes it difficult to identify and select policy responses that may be appropriate under a climatic regime different from that of the past century. Ultimately, policymakers must choose policy responses that are not only technically and economically feasible but also socially and politically acceptable. Existing or proposed policy responses to contemporary environmental issues may offer analogues that can assist policymakers in selecting effective and acceptable options, while maintaining flexibility to cope with the impacts of unknown climate changes in the future.

General circulation models (GCMs) are currently being used to approximate regional as well as global climate changes that

Donald A. Wilhite is an associate professor of agricultural climatology at the Center for Agricultural Meteorology and Climatology at the University of Nebraska–Lincoln. He also has an adjunct appointment with the Environmental and Societal Impacts Group at the National Center for Atmospheric Research. He specializes in studies of the impact of climate on society, particularly climate's effects on agriculture. He is the author of numerous papers on drought management and planning.

might result from increasing atmospheric concentrations of CO_2 and trace gases. However, the outputs of these models disagree, particularly at the regional level, regarding the effects of increasing levels of CO_2 and trace gases on temperature and precipitation. From an agricultural perspective, it is clear that at least on a short- or medium-term basis, the potential biological productivity of some regions may improve while that of other regions may decline. For example, Manabe and Wetherald (1986) suggest that a portion of the Great Plains of the United States may experience a gradual decline in soil moisture during the summer months of June, July, and August in association with a doubling of pre-Industrial Revolution levels of atmospheric CO_2. This reduction in soil moisture was estimated to be on the order of one to three cm during the period June through August (statistically significant at the 10 percent level). Such a decline would further reduce the viability of rainfed agriculture in the already marginal agricultural area in the western part of the region. A reduction in the region's soil moisture regime would most likely place an even greater pressure on future development of the region's underground and surface water supplies.

The Ogallala Aquifer underlies approximately 580,000 square km (225,000 square miles) in the Great Plains, particularly in the High Plains of Texas, New Mexico, Oklahoma, Kansas, Colorado, and Nebraska (U.S. Fish and Wildlife Service, 1981) (Figure 1). This aquifer serves as the groundwater source for most of the region, providing water supplies for agricultural, domestic, and industrial purposes. Because of the ease of access to groundwater, Plains states have become dependent on it as a primary source of water, more so than in other parts of the nation. For example, groundwater provides only 20 percent of national water needs, but in Nebraska, Oklahoma, and Kansas, it provides 68, 61, and 86 percent, respectively, of their water needs (Wickersham, 1980).

The severe and prolonged drought of the 1930s and the availability of new and relatively inexpensive groundwater pumping and distribution technologies and low energy prices provided a major impetus to the regional development of this resource. Groundwater utilization has steadily increased since that time, rising from less than 8.6 billion m^3 (7 million acre-feet) of water in 1950 to 25.8 billion m^3 (21 million acre-feet) in 1980. Currently, more than 5.7 million ha (14 million acres) are irrigated from the Ogallala

Figure 1. The Ogallala Aquifer

Aquifer (High Plains Study Council, 1982). Irrigation development occurred first in the southern High Plains states, and as a result the aquifer has been depleted more seriously in that portion of the region than elsewhere (Firey, 1960; High Plains Associates, 1982). Today, there are more than 170,000 irrigation wells in the region (High Plains Study Council, 1982). The area underlain by the Ogallala Aquifer coincides closely with that portion of the Great Plains projected by GCMs (e.g., Manabe and Wetherald, 1986) to experience a decline in soil moisture because of climate change.

Glantz and Ausubel (in this volume) suggested that it may be useful to consider in concert the two environmental issues of the depletion of the Ogallala Aquifer and projected climate changes caused by increasing atmospheric CO_2 and other trace gases. Linking these two contemporary issues (one real and one projected) is a new approach. The issues differ with regard to the cause of the reduction in the existing regional water supplies (i.e., hydrologic versus atmospheric) and in other ways as well. They are similar, however, because they concentrate on societal responses to changes (e.g., reductions) in the water balance in the region.

The central concern of this chapter focuses on whether the policy responses that have been recommended and, in some cases, implemented because of the actual and projected continued depletion of the Ogallala Aquifer can be used as an analogue for the policy responses that may become necessary, appropriate, and acceptable in response to climate change. Ultimately, it will be the public's perception of these recommendations that will determine their acceptance or rejection. In addition, conclusions will be drawn about the comparability of these two environmental issues for the purpose of determining appropriate policy responses.

BACKGROUND

The Ogallala Aquifer is a geologic formation of semiconsolidated clay, sand, and gravel ranging in thickness from less than 30 meters to about 360 meters. The aquifer extends approximately 1,300 km from north to south and is highly variable in width from west to east, but is about 650 km at its maximum point. In 1977, the total water available in the aquifer was estimated to be about 3.75 billion m^3 (3.04 billion acre-feet). However, the amount of water available to each of those states underlain by the aquifer is highly variable. For example, in 1977 Nebraska was estimated to have 77 percent of the aquifer's available water supply, while Kansas and Texas each had about 8 percent; Colorado, 3 percent; Oklahoma, 2 percent; and New Mexico, 1 percent (High Plains Study Council, 1982).

For large portions of the aquifer, the pumping rate greatly exceeds the rate of recharge, although this rate is spatially quite variable. In some areas, particularly in the southern High Plains, recharge is negligible. Groundwater levels in some local areas have dropped so low that farmers have reverted to rainfed agriculture because of increased pumping costs resulting from greater lift, higher energy prices, and low crop prices. There is concern that the reversion of large amounts of irrigated land to dryland will have a substantial adverse impact on the national economy as well as on local and regional economies (High Plains Associates, 1982). Many have reported the concern of area residents and others that the loss of irrigated acreage, coupled with the decrease of oil and gas production and the depletion of oil and gas reserves in the area, poses a serious threat to the region's economic viability (e.g., Supalla et al., 1982; Kromm and White, 1985).

FEDERAL POLICY RESPONSES

Substantial efforts have been directed toward the study of the depletion of the Ogallala Aquifer and the identification and assessment of the various management alternatives and policy responses that exist for local, state, regional, and federal governing authorities to prevent or delay further decline of the aquifer. Most of these studies and their recommendations have been made in the past decade in response to increasing political pressure from those states that expect to be most affected by a continued depletion of currently available groundwater supplies.

The most notable response of the federal government thus far has been the provision of funding by the U.S. Congress for the High Plains Study, a $6 million project initiated in 1976 "to assure an adequate supply of food to the Nation and to promote the economic vitality of the High Plains Region" (High Plains Associates, 1982). Six states participated in the study: Texas, Oklahoma, New Mexico, Colorado, Kansas, and Nebraska. Responsibility for the administration of the study was given by Congress to the Economic Development Administration (EDA) of the U.S. Department of Commerce under Public Law 93-587. The U.S. Geological Survey completed a $5 million assessment of the region's water supplies, and the U.S. Army Corps of Engineers was given the responsibility under that law to plan and evaluate interbasin transfer of water from surplus to deficit areas.

The High Plains Study Council was formed by the six states and the EDA "to assure policy guidance for the Study, and to submit its conclusions and recommendations to the Secretary and Congress" (High Plains Associates, 1982). The objectives of the study were (1) to determine potential development alternatives for the High Plains Region; (2) to identify and describe the policies and actions required to carry out promising development strategies; and (3) to evaluate the local, state, and national implications of these alternative development strategies (or the absence of such strategies).

Five alternative management strategies were formulated by the council to carry out the congressional directives. Two of these strategies were alternatives designed to reduce water demands; the other three were designed to augment regional or subregional water supplies. These five alternative strategies were as follows: (1) voluntary water demand management, (2) voluntary plus mandatory water demand management, (3) local water supply management, (4) subregional intrastate importation supply management, and (5) regional interstate importation supply management. The strategies were believed to represent the range of options available to delay the depletion of the aquifer while maintaining the economy of the region. The results of investigations of these five strategies were then compared to a baseline (status quo) condition–that is, a continuance of current trends with no changes in public policy and no new programs that would affect water conservation or augmentation in the region (High Plains Study Council, 1982).

Supalla et al. (1982) have suggested that the initiation of the High Plains Study was based on four hypotheses and one broad policy inference. The *hydrologic failure hypothesis* suggests that the Ogallala Aquifer is being depleted at a rapid rate and recoverable supplies will be exhausted in the next ten to twenty years. The *economic failure hypothesis* is based on the premise that increased energy prices and increased pumping lifts will make it uneconomical in the near future to continue to withdraw water from the aquifer for irrigation purposes. The *regional economic and social impact hypothesis* suggests that there are substantial linkages between irrigated agriculture, energy production, and the regional economy of the area, and a reduction in any of these areas will have a major negative impact on the region. The *national agricultural commodity supply hypothesis* holds that the region's food and fiber crop production is large enough that a reduction in irrigated agriculture in the region would decrease national production sufficiently to affect exports, thereby driving commodity prices significantly upward.

The policy inference of these four hypotheses is that these concerns require public action through a variety of demand modification or supply augmentation alternatives at all levels of social organization in order to reduce the impacts on society of the depletion of the Ogallala Aquifer. This chapter focuses on the policy questions and conclusions of the High Plains Study. In addition, the recommendations emanating from the study that might lead to changes in public policy or to the development of new programs at various levels of government in response to the depletion of the Ogallala Aquifer will be assessed.

POLICY QUESTIONS AND CONCLUSIONS

Twelve policy questions arose as a result of the analyses carried out for the High Plains Study (High Plains Associates, 1982). These questions covered a wide range of political, social, and economic concerns regarding the need for and level of federal and state involvement in preserving the region's groundwater resources. For example, the study questioned whether depletions of the aquifer justified federal and state intervention in order to maintain necessary levels of agricultural production. Was it in the public's interest to restrict water use in the Ogallala region if those restrictions would result in significant near-term economic costs? Should investigation and planning of interstate and intrastate water transfers continue? Should water and energy conservation in irrigation enterprises be a major federal objective and program and, if so, how should it be implemented and enforced? What would be the role of state government in this activity? Should water supply augmentation measures be explored and who should be responsible for this activity? Is the federal government prepared to make a long-term participatory commitment to maintain the food and fiber production of the region? Questions about the preservation of wetland habitats and the equitable distribution of the costs between each of the states for water conservation and augmentation measures were also raised.

Supalla et al. (1982) analyzed the High Plains Study reports and concluded that, although serious local water supply and economic problems exist, the concern for maintaining the agricultural production and economic vitality of the region is unfounded. Their assessment produced five general conclusions: (1) the agricultural and economic problems associated with the depletion of the aquifer are localized, often within states; (2) diminishing water supplies could be dealt with effectively through education and research directed at improving agricultural productivity; (3) high economic costs are associated with regional application of mandatory pumpage restrictions, so only local problem areas should implement regulations; (4) large interstate and intrastate water supply augmentation projects are not economically feasible; and (5) the long-term solution to the problem of aquifer depletion lies mainly in education and research on dryland agriculture and through attracting non-water-intensive industries. Other authors (e.g., Beattie, 1981; Smith and Carlson, 1983) generally agree with the Supalla et al. assessment, particularly emphasizing the importance of generating an effort in public education and in developing water-conserving technologies.

Each state participating in the High Plains Study prepared separate reports in conjunction with those reports issued by the High Plains Associates (High Plains Study Council, 1982). The policy alternatives emanating from the High Plains Study were expected to be received differently by government officials and individual citizens within each participating state, reflecting individual views as well as the legal and institutional history of each state.

RECOMMENDATIONS OF THE HIGH PLAINS STUDY

The recommendations from the High Plains Study recognize that the regulation of groundwater withdrawal, and thus the management of the Ogallala Aquifer, is a state responsibility. Therefore, many of the suggestions relating to water conservation were considered to require individual state actions, including, in some cases, changes in state laws. The study also noted that the mechanisms for the management of water and natural resources at the local level varied from state to state. For example, four states (Nebraska, Colorado, Kansas, and Texas) have established local districts to oversee these management activities, while New Mexico and Oklahoma have authorized several state agencies to direct these functions.

Recommendations of the High Plains Study Council were grouped under seven major areas, proposing policy and program changes within both the public and private sectors: (1) water conservation technology, research, and demonstration; (2) public information, education, and technical assistance; (3) energy; (4) legal and institutional measures; (5) water supply; (6) environment; and (7) economic development (High Plains Study Council, 1982). They were made under each of these categories and are summarized in Table 1. The appropriate level(s) at which these actions were to be considered were classified as federal, state, local, regional institutions, and private. The recommendations are discussed briefly in the following paragraphs.

Table 1
High Plains Study Recommendations
(adapted from High Plains Study Council, 1982)

Recommendation	Federal	State	Local	Regional Inst.	Private
A. WATER CONSERVATION RESEARCH AND DEMONSTRATION					
(1) Increase funding for research on water-use efficiencies, erosion losses, and agricultural productivity.	X	X			X
(2) Expand demonstrations of irrigation techniques and soil/water conservation management systems.	X	X	X	X	X
(3) Increase funding for research, demonstration and market development of more water-efficient crops.	X	X	X	X	
B. PUBLIC INFORMATION, EDUCATION, EXTENSION AND TECHNICAL ASSISTANCE					
(1) Expand programs to promote water and soil conservation and to disseminate research results and management information to farmers.	X	X	X	X	X
(2) Conduct short courses and field tours to demonstrate cost-effective farm management methods.	X	X	X	X	
(3) Initiate a program to inform water users of ways to improve water-use efficiencies and conservation.	X	X	X	X	X

Table 1, continued

Recommendation	Federal	State	Local	Regional Inst.	Private
C. WATER SUPPLY					
(1) Expand research on and use of technology and programs to increase quantity and protect quality of water resources.	X	X	X	X	X
(2) Continue water transfer studies, considering needs of basins and states of origin of potential export waters.	X	X	X	X	X
(3) Provide funding to continue monitoring water quantity/ quality and projected effects of groundwater depletion.	X	X			
D. AGRICULTURAL ENERGY ALTERNATIVES					
(1) Demonstrate on-farm energy use efficiency and auditing methods and devices to increase energy efficiencies.		X	X		
(2) Increase research and demonstration programs to develop and use alternative energy sources.	X	X			X
E. LEGAL AND INSTITUTIONAL					
(1) Establish technical advisory committees in each High Plains state to oversee all aspects of water and energy use efficiency and conservation programs.		X			
(2) Provide financial incentives to encourage improved soil, water, and energy conservation.	X	X	X		

Table 1, continued

Recommendation	Federal	State	Local	Regional Inst.	Private
(3) States should evaluate water management, laws and institutions and suggest ways to improve management and water-use efficiency.		X			
F. ENVIRONMENTAL MAINTENANCE AND PROTECTION					
(1) Select and manage cropping systems, irrigation and farm management practices, and irrigated, dryland, and rangeland vegetation to conserve soil and water resources and wildlife habitats.		X	X		X
(2) Reestablish permanent vegetative cover on lands going out of cultivation to control erosion and restore habitat.	X	X	X	X	
(3) Include provisions to protect fish, wildlife, and related environmental resources in conservation plans.	X	X	X	X	X
G. ECONOMIC DEVELOPMENT OPPORTUNITIES					
(1) Help ongoing programs diversify the High Plains economy, develop less water-intensive enterprises, and improve the economic viability of dryland farming, ranching, and nonagricultural opportunities.	X	X	X	X	X

Water Conservation

Recommended actions focus principally on increased public and private sector support for research and demonstration projects to increase water-use efficiencies, decrease erosion losses, and improve agricultural productivity for both dryland and irrigated farms in the High Plains region. In general, it was suggested that these activities be carried out at each of the five levels referred to above, with the exception of funding for additional research. There was no expectation of funding at the local level or by regional institutions for additional research.

Public Information, Education, Extension, and Technical Assistance

Attention was given mainly to expanding publicly and privately funded programs to publicize the need for soil and water conservation improvements, and to better disseminate research results to farmers in the region through short courses and field tours. With the exception of those activities related to conducting short courses and field tours, the recommendations called for these activities to be conducted at all levels.

Agricultural Energy

Emphasis was placed on improving on-farm energy use efficiency and research and demonstration projects aimed at developing and using alternative energy sources for agricultural uses. These activities were considered mainly the responsibility of state government. However, demonstrations of on-farm energy use efficiencies were considered a local activity. Increased support for research in alternative agricultural energy sources was considered a responsibility of the federal and private sectors.

Legal and Institutional Measures

Suggestions included the establishment of technical advisory committees in each state in the region to provide guidance and coordination for research, demonstration, education and technical assistance programs for efficient use of water and of energy programs. There were also suggestions that would provide financial incentives to encourage the use of improved methods for conserving soil, water, and energy. States were requested to evaluate existing water laws and institutions and suggest necessary changes. Although the federal and local levels of government, along with states, were expected to provide financial incentives for improving methods for conserving soil, water, and energy, state government was expected to play the primary role in the establishment of the technical advisory committees and in the evaluation of existing state laws.

Water Supply

Suggestions emphasized research, planning, development and use of technology and programs for the augmentation of water supplies and the protection of water quality in the region. Additional feasibility and planning studies were suggested for regional-scale interstate water transfer projects. Continuation of groundwater quantity and quality monitoring programs was identified as important, with appropriate levels of federal and state funding. Recommendations included the need for estimating the implications of continued depletion of groundwater for the region and the nation. Funding for monitoring programs and studies of the effects of groundwater depletion were considered to be the responsibility of federal and state governments. The responsibility for the other proposed policy and program changes was to be shared by all levels.

Environment Maintenance and Protection

Recommendations included the selection of proper cropping systems and management practices for dryland, irrigated, and rangeland environments to ensure conservation of soil and water resources and wildlife habitats. This was considered the responsibility of state and local authorities. The need for programs to provide technical and financial assistance to reestablish a permanent vegetative cover on land being taken out of production was emphasized in order to control erosion. This was considered the responsibility of all levels of authority, except for the private sector.

Economic Development

Opportunities aimed at diversifying the economy of the region and promoting less intensive water activities were emphasized by the report as essential to the future of the region. All levels of government as well as the private sector were encouraged to participate in these types of activities.

Early implementation of these suggestions was considered a high priority and the High Plains states agreed to take the initiative in this activity. Although these actions were considered to be technologically and economically feasible by the High Plains Study Council, the real test of the suggested policy and program changes lies in their social and political acceptability.

HIGH PLAINS STUDY RECOMMENDATIONS: SOCIAL ACCEPTABILITY

The social acceptability of the proposed High Plains Study policy and program changes has been explored recently (Kromm and White, 1986). A mail survey of nearly one thousand respondents in fourteen counties within the region was completed in 1984. The rationale for selecting these counties was their dependence on groundwater from the Ogallala Aquifer for irrigation and other purposes.

The questionnaire was divided into three sections, focusing on social and economic information about the respondents; general awareness of the groundwater depletion problem; and the respondent's degree of preference for the major suggestions of the High Plains Study, possible adjustments to groundwater depletion, and the level of government to administer the suggested adjustments. The following discussion highlights the results of the last section of the questionnaire, which was concerned with the respondents' preferences.

The preferences of the respondents for twelve of the major recommendations of the High Plains Study were summarized in an earlier report (Kromm and White, 1985). Generally, respondents preferred those that focused on encouraging more research, increased planning, greater interaction between government agencies, and greater emphasis on water conservation. The respondents viewed two of the recommendations with less enthusiasm;

these concerned (1) actions to be undertaken by the federal government in order to preserve wildlife habitats, and (2) establishment of an institution to advise on water allocations. The respondents expressed a wide divergence of opinions on only one recommendation–the government's role in encouraging changes in crops as the transition from irrigated to dryland agriculture becomes more widespread.

Kromm and White classified the adjustments available to respond to the depletion of groundwater into five categories: (1) user conservation practices, (2) management policies, (3) financial incentives and disincentives, (4) technological fixes, and (5) other adjustments. User conservation practices most preferred were improving irrigation efficiency, conservation tillage practices, tailwater reuse, checks on well efficiency, the planting of shelterbelts, shifts to more water-efficient plant hybrids, and encouraging secondary recovery methods. Least preferred was a return to dryland farming methods.

Preferred management policies included the identification of intensive control areas, giving priority to municipal use of water, equitable apportionment of water, increased spacing between irrigation wells, restrictions on new irrigation wells, and the establishment of water quotas. Least preferred were limitations on the type of crops that can be irrigated and giving priority to industrial water use.

Financial incentives and disincentives were generally not strongly preferred by the respondents. The only response acceptable to respondents was charging for water permits. Least preferred were taxes on irrigated acreage and water use, government depletion insurance, a severance tax on groundwater, and a government subsidy for energy costs.

Technological fixes such as constructing reservoirs, building small recharge dams, improving weather modification techniques, and initiating intrastate water transfers were generally preferred. Least preferred was the importation of water from other states. Other adjustments acceptable to respondents included encouraging the development of new water conservation laws and educational programs in water law and use.

Institutional preferences for the administration of adjustments varied by adjustment category and by state. The choices provided to respondents were federal, state, county, local groundwater district, other, and no agency. The local groundwater district was preferred for the administration of user conservation practices and for some of the management policies. County government was preferred for only a few of the user conservation practices. State and federal governments were preferred mainly for adjustments in the technological fix category. No agency was preferred for all of the financial incentives and disincentives, for many of the management policies, and several of the user conservation practices. Variations in responses by state were primarily a reflection of the history of a state government's authority in water resources planning and management. Texas, for example, has typically discouraged governmental intervention in water resources management.

Several important conclusions can be drawn from this study regarding those policy responses or adjustments that are socially acceptable to residents of the High Plains region. First, residents prefer improving water conservation practices, particularly by improving irrigation efficiency and by employing conservation tillage practices. Second, schemes to build dams or recharge the aquifer are generally acceptable to residents. Third, residents seem to be strongly opposed to the use of financial incentives or disincentives to address the groundwater depletion problems of the region. Finally, local and state control of any practices or programs implemented in response to the depletion problem is preferred over federal involvement, except in instances such as the construction of storage dams.

CLIMATE CHANGE AND OGALLALA AQUIFER DEPLETION: COMPARABILITY OF ISSUES AND APPLICABILITY OF POLICY RESPONSES

Glantz and Ausubel (this volume) have compared the environmental issues of the depletion of the Ogallala Aquifer and the projected decline of water supplies in the region because of changes in climate due to increasing concentrations of CO_2 in the atmosphere. To what extent are these two issues comparable? To what extent are the policy responses that have been recommended to various levels of government and the private sector in response to the Ogallala depletion situation applicable to changes in regional water supplies caused by changes in regional climate?

Comparability

The Ogallala Aquifer and climate change issues are long-term, low-grade, cumulative environmental problems. Although the two issues result from different physical causes, the projected temporal and spatial dimensions are similar. The simultaneous occurrence of these environmental changes may magnify impacts and alter the future availability of water resources in the High Plains region.

The policy alternatives proposed to augment or conserve water in response to a hydrologic depletion appear, at first glance, to be useful for responding to reductions resulting from changes in the current climate regime due to atmospheric causes. If the depletion of the Ogallala Aquifer and changes in climate do occur simultaneously over the same geographical area, this may imply greater federal and state government involvement (i.e., more stringent policy responses)—an involvement that most residents of the region find unacceptable (Kromm and White, 1985). However, federal and state government involvement might be more palatable to residents if these issues occur simultaneously in their region. Both issues seem to provide enough lead time to formulate and evaluate a wide range of policy alternatives and to implement those that are deemed most acceptable.

These environmental issues, however, differ in several critical ways. First, the Ogallala depletion problem is perceived by residents to constitute a real and present danger to the economy of the region. The amount of water available in the aquifer and current withdrawal rates are generally known and aquifer life can be estimated for local areas. Likewise, alternative conservation and augmentation options can be evaluated for their effect on withdrawal rates and, thus, on the life of the aquifer. Although the estimates of the aquifer's life vary, the groundwater depletion problem in the High Plains is well documented and accepted by Plains residents. These are compelling reasons for society to respond to the Ogallala depletion problem.

Climate change is more speculative, and area residents are not well informed of the problem or the scope of related impacts. In fact, few are even aware of the problem or the scientific debate on the issue, not to mention the potential effects of a significant change in climate on the region. Under these circumstances, identifying policy responses that might be socially acceptable decades into the future is difficult if not impossible. It is essential that the scientific and policy communities establish a dialogue with area residents about the climate change issue and its likely implications. Only through public education can alternative policy responses, and the public's acceptance of those policies, be evaluated.

Second, when examined closely, the geographical scale of the two problems is different. Although the Ogallala Aquifer issue is considered to be a regional issue, from a management perspective it is principally a local one because of the different hydrologic characteristics of the aquifer (e.g., saturated thickness, variations in local withdrawal rates). Climate is more spatially homogeneous, changing gradually from east to west and north to south. Does this imply the need for greater homogeneity in the implementation of regional management measures in response to some future change in climate?

Third, the impacts of a change in climate will likely be more far-reaching than those associated with the gradual depletion of the Ogallala Aquifer. Impacts will be more ubiquitous, affecting all communities and economic sectors rather than only local regions or areas. The combined economic and environmental shock of projected changes in climate will be significant and will have major effects on local, regional, and national institutions. In addition, other regions will be affected simultaneously and will thus compound the difficulties associated with the federal government's attempt to respond in a timely and effective manner to the problems that arise in any one region.

Applicability

Are the policy responses that have been recommended in response to the Ogallala Aquifer issue applicable to the issue of climate change? Because the impact of both problems will result in diminished regional water supplies, the policy responses recommended by the High Plains Study Council and others might be appropriate for changes in regional water supplies resulting from an altered climate. Certainly, both cases will require governments and individual citizens to consider policy options that augment and/or conserve existing water supplies, as well as options to pursue alternate sources of income for the region.

Two factors need to be considered in a discussion of the applicability of policy responses—the spatial dimension of the issues and the possible simultaneous occurrence of the two environmental issues. With respect to the first point, although the Ogallala Aquifer problem is perceived to be a regional one, the results of the High Plains Study clearly demonstrate that alternative water management measures should be imposed only within local problem areas because of differences in aquifer thickness and water availability. However, climate change will likely result in regional-level impacts because of greater atmospheric than hydrologic homogeneity. Therefore, policy responses initiated because of climate change will need to be implemented over relatively larger geographical areas requiring a higher level of governmental involvement for administration and enforcement. Essentially, the Ogallala Aquifer depletion problem is local in cause and local in effect. The climate changes that might result from a CO_2-induced warming, on the other hand, will essentially be global in cause but local in effect.

If the two environmental issues were to occur simultaneously, the regional impacts would be exacerbated. Thus, stronger policy responses will be required to adequately manage diminishing water resources, which will in turn require a greater level of involvement by federal and state authorities than might be expected as a response to either of these environmental problems if they were to occur alone.

SUMMARY AND CONCLUSIONS

The availability of the Ogallala Aquifer as a resource has facilitated the expansion of irrigated acreage in the Great Plains. Because recharge rates for most of the aquifer are negligible (except for Nebraska), this resource is being depleted rapidly in some areas, especially Texas. Because of concern among area residents and others that depletion of this resource will force a substantial reduction in irrigated acreage, it has been suggested that the Ogallala Aquifer depletion and adjustments or policy responses to that depletion may be comparable to the drying of the Great Plains region as a result of the impacts on global and regional climate of increased atmospheric concentrations of CO_2 and other trace gases. Both problems are long-term and cumulative and may substantially alter the existing water balance in the High Plains.

Because the spatial and temporal dimensions of the two problems are similar, more drastic policy responses (i.e., greater state and federal government involvement) may be required.

Unlike the Ogallala Aquifer issue, the suggested climate warming and associated reduction in available water supplies in the region is somewhat speculative. Residents are generally unaware of the scientific debate over the climate change issue, the potential regional effect of that change on water supplies and economic activities, and the level of adjustments that might be necessary to respond effectively to the problem. Other critical differences between the issues include the geographical scale of the impacts from a management perspective and the nature, magnitude, and timing of the associated impacts.

Are the policy measures implemented in response to the Ogallala Aquifer depletion applicable to changes in regional water supplies that may result because of projected changes in climate due to increasing CO_2? By using the Ogallala Aquifer depletion problem as an analogue, scientists and policymakers can learn much about the process of evaluating and selecting socially acceptable policy responses to large-scale environmental issues. However, regional changes in climate of the magnitude suggested by GCM output may impose a new set of rules on society for choosing appropriate policy responses to climate-related environmental change.

REFERENCES

Beattie, B.R., 1981: Irrigated agriculture and the Great Plains: Problems and policy alternatives. *Journal of Agricultural Economy, 6,* 289–99.

Firey, W., 1960: *Man, Mind and Land: A Theory of Resource Use.* Glencoe, IL: Free Press.

High Plains Associates, 1982: *Six-State High Plains Ogallala Aquifer Regional Resources Study: Summary.* Austin, TX: Camp Dresser and McKee.

High Plains Study Council, 1982: *A Summary of Results of the Ogallala Aquifer Regional Study, With Recommendations to the Secretary of Commerce and Congress.* Washington, DC: Economic Development Administration.

Kromm, D.E., and S.E. White, 1985: *Conserving the Ogallala: What Next?* Manhattan, KS: Department of Geography, Kansas State University.

Kromm, D.E., and S.E. White, 1986: Variability in adjustment preference to groundwater depletion in the American High Plains. *Water Resources Bulletin, 22,* 791–801.

Manabe, S., and R.T. Wetherald, 1986: Reductions in summer soil wetness induced by an increase in atmospheric carbon dioxide. *Science, 232,* 626–7.

Smith, M., and D. Carlson, 1983: Study had shortcomings, also good ideas for saving the dwindling Ogallala Aquifer. *Denver Post,* 6 March.

Supalla, R.J., R.R. Lansford, and N.R. Gallehon, 1982: Is the Ogallala going dry? *Journal of Soil and Water Conservation, 37,* 310–4.

U.S. Fish and Wildlife Service, 1981: *High Plains–Ogallala Aquifer Study.* Albuquerque, NM: U.S. Department of the Interior.

Wickersham, G., 1980: Groundwater management in the High Plains. *Groundwater, 18,* 286–90.

16

Public and Private Sector Responses to Florida Citrus Freezes

Kathleen A. Miller

INTRODUCTION

Variability is an inherent feature of climate and includes the occurrence of "extreme" weather events. Decisionmakers expect the climate to be variable. In fact, adjusting for the risks associated with this variability is an important factor in decisionmaking for any climate-sensitive activity. Risk hedging strategies are incorporated into the organization of activities like citrus production where climatic variability exposes citrus growers and other industry participants to substantial risks. Therefore to fully assess the potential socioeconomic impacts of a climate change, it is necessary to understand its effects on the range of climatic variability and to understand how the physical impacts of climate anomalies, especially extreme meteorological events, might be filtered through ongoing and evolving risk-adjustment strategies.

We do not yet know what effect a CO_2/trace-gases-induced climate change will have on such factors as variability or persistence of temperature or precipitation and, therefore, on the occurrence of extreme events. In addition, global warming does not necessarily

Kathleen A. Miller is a scientist with the Environmental and Societal Impacts Group (ESIG) at the National Center for Atmospheric Research. She completed her Ph.D. in Economics at the University of Washington in 1985. Her dissertation research examined the economics of water rights institutions in the Western United States. In ESIG, her research has included work on the economics of natural resource use in the presence of climatic variability and climate change.

imply regional warming or a decreased frequency of freeze events in important agricultural areas. Under our current climate regime, an anomalous "cold-outbreak" can send freezing temperatures into Florida, despite an otherwise warmer than normal season. In fact, damage to citrus trees in 1983 was intensified by the fact that unusually warm weather immediately preceding the freeze event had prevented the normal build-up of cold hardiness.

We will not necessarily recognize climate change as it occurs. Despite recent suggestions to the contrary (e.g., Broecker, 1986), climate change may not occur suddenly. It might manifest itself as a gradual increase in the frequency of unusual weather events. It is therefore useful to examine the societal impacts of and responses to past anomalous climatic periods. An examination of the immediate responses and long-term adjustments to such anomalous periods may allow us to foresee the types of responses and adjustments that are likely to receive consideration as the societal impacts of climate change are increasingly felt. Future policies for dealing with the impacts of climate change may be better informed if strengths or weaknesses in the responses and adjustments to past periods of climatic anomaly can be identified and evaluated.

Florida's recent set of back-to-back severe freezes represents such an anomalous period. The impacts, immediate responses and longer-term adjustments resulting from this period can be seen as analogous to the potential effects of and responses to future anomalous periods that might result from a CO_2/trace-gases-induced climate change. This chapter first describes the setting within which the impacts of this anomalous period occurred. Pre-existing patterns of adjustment to the climatic risks are detailed. The impacts of these freezes as well as both governmental and private-sector responses to these impacts are outlined. Longer-term grower adjustments are shown to be affected by several factors, including increasing competition from the Brazilian citrus industry and increased opportunities for non-agricultural land development, as well as altered perceptions of climatic risks as a result of the recent freeze damage. Finally, conclusions are drawn regarding the potential relevance of this case analogy to future periods that may be affected by CO_2/trace-gases-induced climate change.

THE SETTING

Historical Overview

Citrus fruit has been produced commercially in Florida (Figure 1) since the 1820s. During the past century and a half, the industry has repeatedly been affected by freeze episodes. The first serious freeze, following the development of a commercial citrus industry, occurred in 1835. It killed the majority of Florida's citrus trees which at that time were located primarily near St. Augustine and along the St. Johns River, well to the north of today's commercial citrus growing areas (Brey, 1985). The industry gradually recovered, with new plantings taking place further to the south, particularly along the Indian River. It has been estimated that by the 1893–94 season, just prior to another devastating freeze, Florida's citrus production exceeded five million boxes* (Florida Crop and Livestock Reporting Service, 1985). Severe freezes in December 1894 and February 1895 destroyed most of Florida's citrus trees but spared those along the Indian River. During the following ten years, a pattern of recurring freezes swept through Florida's traditional citrus growing areas. There were at least eleven freezes of varying intensity between 1894 and 1906 and, in most areas, temperatures during the freeze of 1899 were as cold as or colder than those during the 1894 and 1895 freezes (Florida Crop and Livestock Reporting Service, undated). The severity and back-to-back nature of these freezes resulted in a southward migration of the industry and the permanent abandonment of the former planting center near St. Augustine (Brey, 1985).

By 1909, Florida's citrus output had finally recovered to the level achieved prior to the freeze of 1894. In the ensuing decades a relative absence of severe freezing weather allowed a steady increase in Florida's citrus output. Over time, Florida's Central Ridge, which crosses Lake and Polk Counties, became a favored location for citrus groves, despite the presence of freeze risks in' that area. The well-drained sandy soils of the ridge proved ideal for citrus trees, allowing high per-acre yields and a relative absence of diseases such as foot-rot. Furthermore, the numerous lakes along the ridge provided some protection from the cold.

---

* A standard Florida citrus box contains approximately 90 lbs. of fruit.

FLORIDA

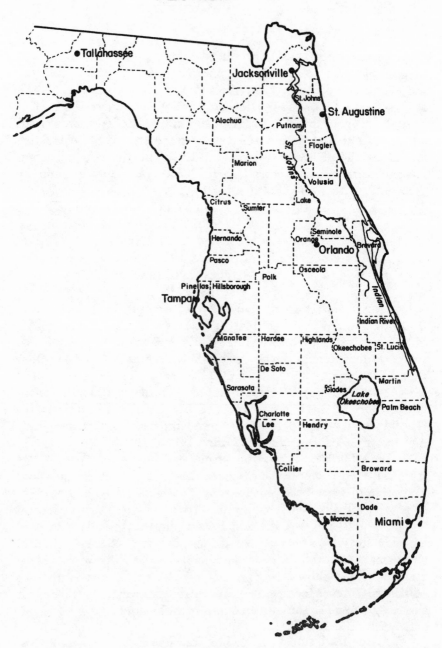

Figure 1.

The expansion of orange production accelerated sharply following the Second World War as a result of technological breakthroughs that launched the frozen concentrated orange juice (FCOJ) industry. The processing sector soon became the dominant market outlet for Florida's oranges while production for the fresh orange market became increasingly dominated by California and Arizona growers. As of the 1980–81 season,* 95 percent of Florida's orange crop was processed, with the remaining 5 percent sold on the fresh fruit market. By the late 1970s Florida produced 79 percent of the U.S. orange crop and the state accounted for 90 percent of the U.S. output of orange juice and other processed orange products.

The output of Florida oranges exceeded 100 million boxes for the first time in the 1961–62 season, but during the following season, in December 1962, Florida was hit with its most severe freeze since the turn of the century. This freeze killed a large number of orange trees and resulted in a dramatic reduction in output. This reduction led to sharp increases in prices for oranges and for FCOJ. Growers in southern Florida, whose groves had not been affected by the freeze were suddenly enriched and entrepreneurs in Brazil seized this opportunity to enter the industry. The previous high level of Florida's output was not regained until four seasons later.

In the 17 seasons following the 1962 freeze, Florida's orange output increased rapidly, with short-term setbacks resulting from a late frost in 1967 and freezes in January 1971 and January 1977. The price increases following each of these events yielded increased profits for Brazil's FCOJ producers and for those Florida growers whose groves had escaped major freeze damage. Throughout this period, the Brazilian FCOJ industry was growing very rapidly, and by the end of the 1970s it was Florida's major competitor for the dominant position in the world market.

January 1981 ushered in the first in a set of four severe freezes in a five-year period (January 1981, January 1982, December 1983, and January 1985). The January 1981 freeze and the freeze of January 1982 were considered to be more damaging than the 1962 freeze and were therefore considered to be the worst events that had been experienced since the beginning of the century. Their

---

* The citrus season runs from September to the following June.

impacts, however, were relatively minor compared to the destruction associated with the subsequent freezes of December 1983 and January 1985.

Together, the 1983 and 1985 freezes are estimated to have killed approximately one third of the trees in commercial citrus groves statewide, with much heavier losses near the northern end of the citrus belt. Lake County, which had the second largest citrus acreage in the state at the beginning of 1983, is estimated to have lost 89 percent of its commercial citrus trees as a result of these two freezes. The freezes of the 1980s resemble those of the 1890s in terms of the magnitude of the damage and in their back-to-back nature. The period of adjustment following these freezes is described in detail later in the chapter.

In addition to these major freezes, since the turn of the century there have been at least 24 other freeze events of sufficient intensity to be categorized as severe to moderately severe and another 24 moderate to slight freezes (Florida Crop and Livestock Reporting Service, undated; 1986). With approximately fifty freezes of varying intensities in Florida's north-central citrus growing areas since 1900, growers have had to make adjustments to the ever-present possibility of freeze damage.

Growers' Adjustments to Freeze Risks

The organization of Florida's citrus industry has been affected by the fact that the industry developed in the presence of freeze risks in the northern portions of the citrus belt. Among the elements affected have been grove locations, patterns of grove ownership and management, variety selection and the form of contractual arrangements for fruit sales.

Where to locate groves is a seemingly obvious decision for adjusting to freeze risks, but the issue is not as trivial as one might suppose. While it is generally true that freeze risks diminish as one moves southeastward along the Florida peninsula, other environmental conditions become considerably less favorable for citrus in those relatively freeze-free southerly locations. Although freeze risks are present along Florida's Central Ridge, particularly toward its northern end, the sandy soils there have traditionally resulted in higher per-acre yields and lower land preparation costs. Furthermore, orange prices have typically been high enough and expected freeze damage slight enough to make north-central Florida citrus groves an acceptable risk.

It should be noted that actual freeze risks on the Central Ridge can vary greatly between locations separated by only a few miles. For example, in Lake County, located at the northern end of the Central Ridge, the terrain is rolling and studded with numerous lakes. Land in close proximity to one of the larger lakes remains warmer during a freeze event than land further from that body of water. The lake effect is particularly strong just to the southeast of a sizeable lake because the prevailing wind direction is from the northwest during advective freeze events.* Such locations are therefore especially favored for citrus production. In the more common radiative freezes* low-lying land becomes considerably colder than land on a slope or on a ridge top, but north-facing slopes and ridge tops are particularly at risk during windy advective freezes such as the first nights of the 1983 and 1985 events.

Low-lying land that is not protected by the lake effect is commonly referred to as a "cold pocket." The location of these relatively risky "cold pockets" has long been well known to local growers. Such pockets are much less likely to be planted with citrus than more favorable locations nearby, but some growers have occasionally gambled on extending their plantings into these locations, particularly when the prospects for profits appeared favorable. Likewise, some growers have been willing to undertake more risk than others by pushing their citrus plantings further north into increasingly cold-prone areas. In general, these individuals have been seen as high-risk gamblers and recurring freeze damage in these marginal locations has come to be accepted as a routine fact of life in the citrus industry.

Diversification is a common strategy for citrus growers throughout Florida. At the latitude of Lake County and northward, citrus acreages tend to be smaller than they are further to the south and grove owners appear to derive a larger proportion

---

* The primary source of cooling during a radiative freeze is the vertical flow of infrared radiation from the surface to the atmosphere. Advective freezes, however, are the result of a sudden movement of the Continental Arctic air mass southward into Florida. (For more information about the differences between these types of freezes and their impacts see IFAS, 1985a.)

of their income from other activities. A survey of citrus growers found that 53 percent of the respondents in the northern counties derived at least half of their income from other activities, while 44 percent of the respondents elsewhere in Florida received half or more of their income from other sources (*Florida Grower and Rancher*, 1986). This diversification strategy may involve combining citrus operations with other agricultural or non-agricultural activities.

In Lake County, for example, it is common for grove owners to live away from their groves, sometimes in nearby towns where they may hold other jobs. Some groves are owned by out-of-state parties and many owners contract with local grove management companies for all harvesting, pruning, and other grove maintenance services. Despite the fact that the tax-shelter advantages of owning a citrus grove have been eroded over the years due to tax reforms,* it is not uncommon for groves to be owned by wealthy professionals and investors. The relatively risky but potentially lucrative nature of citrus groves in north-central Florida makes them an attractive investment for such individuals, because the smaller the weight of an activity in the individual's income portfolio, the greater the risk that the investor will generally accept. Furthermore, the fact that routine operation and management activities can readily be carried out by professional custom management companies makes citrus groves more attractive to absentee grove owners than other types of agricultural activities which might require more constant managerial attention.

---

* The Tax Reform Act of 1969 generally required agricultural capital expenditures, including grove development, costs to be depreciated rather than expensed (Powe and Langham, 1980; Brey, 1985). The depreciation period for citrus trees was set at the expected useful life of the grove (about 20 to 30 years), thus greatly reducing the usefulness of citrus holdings as a tax shelter for other income. Although a shorter depreciation period was introduced with the 1981 tax reforms, the most recent (1986) tax reforms are quite restrictive to the use of citrus investments as tax shelters (personal communication, Ron Muraro, Economist, Citrus Research and Education Center, Lake Alfred, Florida, 9 September 1987).

Another strategy for managing freeze risks involves the method by which the fruit is sold for processing. Growers in freeze-prone areas generally insure themselves against the total loss of freeze-damaged fruit by selling their fruit through a cooperative or participation plan. Entering into such an arrangement, rather than selling the fruit on the spot market, assures the grower that his fruit will be salvaged quickly if it should be freeze damaged. On the other hand, growers who sell their fruit on the spot market will receive higher prices during freeze years for their undamaged fruit.

Growers in the more freeze-prone areas also deal with that risk by planting a larger proportion of early bearing varieties such as Hamlins. These varieties are more likely to escape cold damage than the later bearing Valencias, but they also generally sell for somewhat lower prices.

Governmental Involvement in the Florida Citrus Industry

The ongoing pattern of governmental involvement in Florida's citrus industry provided the framework within which governmental responses to the recent freezes were carried out. In general, it can be argued that responses were focused on pre-existing agency-clientele relationships, and that the strength of any specific response appears to have been a function of the strength of the pre-existing relationship.

It can also be argued that the patterns of pre-existing governmental involvement as well as governmental responses to the freezes of the 1980s were strongly affected by traditionally accepted notions of liability for freeze damages. It appears, for example, that growers are generally viewed as having willingly accepted the freeze risks inherent in the location of their groves. However, it appears that unemployed citrus industry laborers have been treated by the public sector as innocent victims of freeze events rather than as willing risk-takers.

At the federal level, there has been relatively little ongoing agency involvement with Florida's citrus industry. Unlike many agricultural commodities, Florida's citrus has not been the beneficiary of direct federal price support programs. Profits in the industry have traditionally been high enough that there has been little pressure for such a program. The lack of an active price support program is significant in that it means that there is no federal agency directly concerned with managing output, prices, or profitability for the Florida citrus industry.

Florida's FCOJ industry has, however, benefited from substantial tariff protection. The tariff rate, which stands at 35 cents (U.S.) per single-strength gallon equivalent, has insulated Florida producers from expanding Brazilian output. Without this tariff to buffer them, Florida producers would have felt the pinch of Brazilian competition much sooner, prices for oranges and for FCOJ would be lower, and the profitability prospects for Florida grove expansion or replanting would be considerably less favorable. In the past, citrus growers have also benefited from the generally favorable tax treatment of agricultural investments. The current tax code, however, is considerably less favorable to investments in agriculture, including citrus.

The Federal Crop Insurance Corporation allows growers to insure their crops against a variety of weather risks including hail, hurricanes, tornadoes, and freezes. The rates vary by county, type of citrus, and the specific terms of the policy. It appears that agents may also exercise discretion in issuing policies to reflect site-specific risks (personal communication, Ray Stallings, District Director, Federal Crop Insurance Corporation, Bartow, Florida, 26 June 1987). It therefore cannot be assumed that an individual who plants a grove in an obvious cold pocket can obtain freeze insurance on the same terms as a neighbor who selected a less risky grove location.

Insurance can be purchased for seven different types of citrus, with the rate for each set to reflect the susceptibility of that type to the weather risks covered in the policy.* The tailoring of insurance rates to approximate variations in the riskiness of alternative citrus operations represents an attempt to overcome the adverse selection problem that would otherwise arise in this type of industry. In other words, if the rates did not mirror variations

---

* Different insurance rates are available for the following: (1) early and mid-season oranges, (2) late season oranges, (3) grapefruit for processing into juice, (4) tangelos and tangerines, (5) murcotts, (6) lemons, and (7) grapefruit to be sold as fresh fruit (personal communication, Ray Stallings, 26 June 1987).

in relative riskiness, growers in the most risky locations would perceive the rates as being subsidized and that would result in an artificial inducement to plant groves in relatively risky areas.

No figures are readily available on the proportion of total citrus acreage that was covered by freeze insurance prior to the onset of the recent set of freeze events. According to a Lake County extension agent, relatively few growers in that part of the state had been insured against freeze damage, primarily because of the high cost of the insurance (interview with John Jackson, Lake County Extension Service, Tavares, Florida, 26 March 1987).

Other ongoing federal programs include the National Weather Service's agricultural frost warning service at Ruskin, Florida, the U.S. Department of Agriculture's fruit inspection and canker eradication programs, and environmental regulation. The latter includes provisions of the Clean Air Act which prohibits the use of many of the less expensive forms of grove heating for freeze protection.

At the state level, the Florida Department of Citrus carries out regulatory, research, and advertising functions. It is a state agency that is completely financed by per-box taxes on citrus output.

One of the functions of the Florida Department of Agriculture is to maintain detailed statistics on the extent of citrus acreage, yields, and prices, and to make this information readily available to the interested public. These data are collected and published by the Crop and Livestock Reporting Service of the Florida Department of Agriculture.

An important intergovernmental program is the Florida Cooperative Extension Service–Institute of Food and Agricultural Sciences (IFAS), University of Florida. This program is jointly funded by the state through the University, by the federal government through the U.S. Department of Agriculture, and by individual counties through their support of county extension offices. IFAS performs three functions: teaching, research, and extension. Teaching and research are conducted primarily at the University and at a number of research and education stations located throughout the state. Some research is also carried out by county extension personnel, often in cooperation with local growers who have requested technical assistance in designing experiments to test new cultivation practices or new varieties (personal communication, Pat Cockrell, Florida Farm Bureau, Gainesville, Florida, 25 June 1987).

The extension function involves the dissemination of information and advice directly to growers and other members of the interested public. This function is carried out primarily by the county extension agents as well as by the faculty at the decentralized research and education centers. The information provided by extension agents to citrus growers includes the results of research on citrus cold protection, cultivation practices, the characteristics of alternative rootstock-scion combinations, and grove-level economics. Routine day-to-day information on efficient grove maintenance operations is also provided. Some extension offices, such as the Lake County office in Tavares, provide special information services to growers throughout the night during freeze events. Recently, this information has included real-time satellite temperature data made available through a direct down-link from the GOES satellite to the IFAS Climatology Lab and from there to color graphics terminals in the county extension offices.

The cooperative extension service is in a rather unique position in that its agents are continually in direct contact with growers and are therefore intimately acquainted with their interests, concerns and future plans. Furthermore, the traditional functions of the service have included assisting growers in adapting to freeze risks and helping them to maintain their operations.

Workers in the citrus industry, including fruit pickers, truck drivers, processing plant and packing shed employees, and grove maintenance workers, make up an important class of other parties potentially affected by freeze events.

A variety of ongoing programs are available for workers who become involuntarily unemployed. Unemployment insurance and food stamps are among the most important of these programs. The unemployed, especially those who have become unemployed through no direct fault of their own, are seen as having a legitimate claim to assistance from government at the federal, state, and local levels.

IMPACTS OF THE 1980s FREEZES

Grove Owners

Florida's citrus growers lost fruit and some trees during the 1980 and 1981 freezes, but the most serious impacts occurred as a result of the 1983 and 1985 freezes. The direct impacts of these freeze events on growers include lost output during the freeze year, reduced output from damaged trees during subsequent years, the capital losses resulting from freeze-killed trees, and increased operation and maintenance costs during and following the freeze event.*

The financial impacts of the crop losses resulting from these freezes were exacerbated by the fact that orange prices had become much less responsive to reductions in Florida's output than had been the case only a few years earlier. This can be traced to the expansion of Brazilian FCOJ output and to the large carryover stocks of FCOJ that were on hand in both the United States and Brazil following the record 1979–80 crop. The increased availability of Brazilian FCOJ now provides a ready substitute for shortfalls in Florida's production. This increases the elasticity of demand facing Florida's producers, thus reducing the price response to any given reduction in output.

Figure 2 displays the record of Florida's orange output and the real (i.e., inflation adjusted) on-tree prices received by Florida orange growers. It can be seen that the output reductions in the two crop seasons following the January 1977 freeze resulted in a substantial increase in orange prices, whereas reduced output during the 1980–81 and 1981–82 crop seasons had little or no impact on prices. The continuation of depressed Florida output through the subsequent two freeze events finally resulted in higher orange prices so that by the 1984–85 crop season, orange prices were at their highest level in nearly 20 years. In terms of the total value of Florida's orange crop, Figure 2 shows that the output reductions

---

* If a tree has been killed, the change in the capital value of the land incorporates the lost future output from that tree. Drummond (1984, 4) suggests that one should count both the reduced land value and the value of lost future output from dead trees to estimate the total cost of tree losses. That suggestion would result in double counting.

388

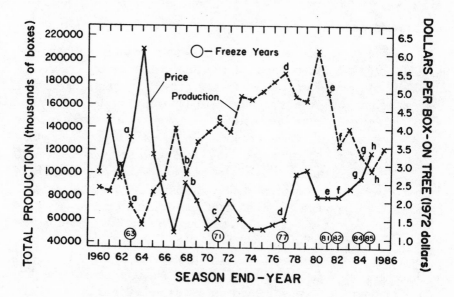

Figure 2.

following the 1977 freeze actually resulted in a dramatic increase in the real total value of the crop, while the subsequent freeze shocks have not had the same result. Nevertheless, those Florida growers whose groves have escaped freeze damage have profited from the recent set of freeze events because real orange prices have been consistently higher than they were during the early 1970s.

The capital losses incurred by north-central Florida citrus growers were especially severe following the December 1983 and January 1985 freezes. As previously noted, approximately one-third of Florida's commercial citrus trees were destroyed by these two freezes. Groves in which the trees were killed by the freezes suffered a decline in market value on the order of $5,000 to $7,000 (U.S.) per acre. This loss in value represents the combined effects of the lost capital value of live trees and the impact of recent events on the perceived future productivity of that land for citrus. Overall, it appears that lost output, tree losses, and expenses incurred in an effort to protect or salvage trees imposed costs on the order of two to two-and-a-half billion dollars on Florida grove owners.

Citrus Industry Employment

The number of citrus industry employees who lost work time as a result of these freeze events can be estimated from monthly citrus industry employment data available from the Division of Labor Statistics of the Florida Department of Labor or from Disaster Unemployment Assistance (DUA) applications, when they are available. The two sources of this information yield somewhat different results, making it difficult to ascertain the true impact of the freezes on citrus industry employment. Industry employment is strongly seasonal, and workers who would ordinarily have been seasonally unemployed may qualify for DUA payments. This may explain why the number of applications for DUA payments following the 1983 and 1985 freezes exceeded the apparent decline in peak season citrus industry employment following those freeze events. Neither the 1981 nor 1982 freezes were declared to be federal disasters, so the employment impacts of those events can only be surmised from the total citrus industry employment statistics.

Table 1 provides a summary of Florida citrus industry employment by peak and off-peak seasons.* The peak season is defined here as December through May and the off-peak season as June through November. Within this seasonal pattern, January is consistently the peak employment month and August is the lowest employment month. The August workforce in the citrus industry is usually about one-third the size of the January workforce. In 1980, for example, prior to the onset of the recent set of freezes, January citrus industry employment was 79,235 while it was 25,943 in August.

It appears from Table 1 that the freeze of January 1981 reduced peak season monthly industry employment by approximately 3,000 workers and off-peak season employment by 2,000. Employment continued to decline in subsequent years with the largest drop following the January 1985 freeze. Peak season monthly employment dropped by 4,000 during the 1983–84 season following the December 1983 freeze, and by another 9,000 the

---

* Computed from unpublished raw data series, monthly Florida citrus employment, Bureau of Labor Market Information, Florida Department of Labor and Employment Security, Tallahassee, Florida.

Table 1

Florida State Average Monthly
Citrus Industry Employment
(in thousands)

Period	Peak Season Dec.–May	Off-Peak Season June–Nov.
Dec. 1979–Nov. 1980 (no freeze)	74	38
Dec. 1980–Nov. 1981 (freeze Jan. 1981)	71	36
Dec. 1981–Nov. 1982 (freeze Jan. 1982)	68	35
Dec. 1982–Nov. 1983 (no freeze)	64	34
Dec. 1983–Nov. 1984 (freeze Dec. 1983)	60	31
Dec. 1984–Nov. 1985 (freeze Jan. 1985)	51	27

following year. Mean monthly peak season employment in that year was approximately 23,000 below the 1980 level, a decline of approximately 31 percent.

Federal DUA figures are available for the 1983 and 1985 freezes, which were both officially declared disasters by the President. As a result of the 1983 freeze, Florida's Division of Unemployment Compensation received a total of 11,052 applications for DUA from workers in all affected industries. The following year, separate statistics were kept on citrus industry DUA applications and a total of 12,006 of these workers applied for assistance following the January 1985 freeze (personal communication, David

Bagley, Division of Unemployment Compensation, Florida Department of Labor, 29 October 1987). This number is slightly larger than the decline of 9,000 workers in monthly peak season citrus employment between 1983–84 and 1984–85, as reported by the Division of Labor Statistics.

Related Business Enterprises

Many other business enterprises were affected by these freezes. While the impacts were detrimental in many cases, some enterprises saw an increase in demand for their products or services as a result of the freeze damage. Nursery owners, for example, saw both negative and positive effects. While many lost their stock or were faced with a sharp increase in costs as a result of their efforts to protect it, the freezes did bring about an increase in demand for young trees for replanting (called resets). Efforts to expand production in response to that demand were, however, stymied by an outbreak of citrus canker in several nurseries in the summer of 1984. The ongoing canker eradication program led to the destruction of millions of young trees in those nurseries. Some of the nurseries producing resets rooted on cold-tolerant Swingle Citrumello rootstock were especially hard hit. The canker outbreak affected that rootstock, which temporarily dampened grower interest in it and thereby postponed replanting efforts.

Custom grove management companies were also affected. While the first three of the 1980s freezes may have generated some additional pruning work for these companies, by the end of the 1985 freeze, so many trees were dead in the north-central part of the state that there were few surviving groves in need of their services. Some of these companies have since gone out of business, while others have reduced their workforce or have redirected their activities toward clearing land of dead trees and preparing it for replanting.

Processors and packers were also affected by reduced access to local citrus. Some processing plants and packinghouses were forced to close because of the unavailability of local fruit as well as the high costs of transporting fruit from areas not affected by the freezes. Researchers at the University of Florida (IFAS) conducted surveys of packinghouses and processing plants in the eleven-county area most severely affected by the 1983 freeze

(Polopolus et al., 1985). These surveys were carried out just prior to the January 1985 freeze.

The survey of processing plants found that the 1983 freeze contributed to the closing of three of the 14 responding plants. The survey also found that many of the processors who responded to the survey intended to mitigate the impact of the reduction of local fruit supplies by increasing their utilization of Brazilian FCOJ. The Brazilian product is imported in bulk, in a more concentrated form than is standard for retail FCOJ sales. Florida processors increased their use of Brazilian FCOJ during this period, blending it with Florida concentrate and with fresh Florida juice to achieve the desired flavor characteristics.

The survey of packinghouses found that 16 of the 34 responding plants planned to be closed during the 1984–85 season. Six of those closures were to be permanent, although half of the permanent closures cited reasons other than the freezes for their closure. No comparable surveys appear to be available following the 1985 freeze.

PUBLIC-SECTOR RESPONSES

There appears to have been little public-sector response to the 1981 and 1982 freezes, other than increased research on cold protection at the IFAS agricultural experiment stations and at the USDA Horticultural Research Laboratory in Orlando. The severity of the 1983 and 1985 freezes, however, prompted the governor to request presidential disaster declarations thereby making federal disaster relief programs available to businesses and individuals affected by these freezes.

Public-Sector Responses to Losses of Growers and Other Businesses

Although grove owners suffered the largest monetary losses, they received very little direct assistance from the state. County assessors did, however, lower property tax assessments on groves that had sustained serious damage from the freezes (Polopolus et al., 1985). These lower assessments reflected the reduced capital value of these properties.

The benefits provided by the standard federal disaster relief programs were quite small in relation to the losses incurred.

For example, the Small Business Administration's (SBA) Physical Disaster Loan Program provided subsidized credit on $19,969,900 in loans to 252 Florida citrus growers and other parties suffering physical losses as a result of the 1983 freeze. An additional $9,011,800 in low-interest loans were made to 39 such individuals following the January 1985 freeze. Economic Injury Disaster Loans were made available to other businesses such as custom grove management companies, packinghouses and processing plants that lost business as a result of the freezes. Ninety-six such loans were approved following the 1983 freeze for a total of $7,041,100; another 15 loans were approved following the 1985 freeze for $1,820,300 (personal communication, Ron Ammerman, SBA, Jacksonville, Florida, 25 June 1987).

The Farmer's Home Administration (FmHA) also has a disaster loan program, but the rules regarding repayment schedules and loan amounts have made this program relatively unsuitable for financing the replanting of citrus groves destroyed by the freezes. At the time of the December 1983 freeze, the regulations of the FmHA Emergency Loan program stipulated that all loan funds must be dispersed within the first year and that repayment must begin within three years. Since it takes approximately five years before a newly planted citrus grove will bear any fruit, these restrictions would have required any citrus grower accepting such a loan to also have had an alternate source of financing. These restrictions resulted in very few loan applications, and the rules were changed following the 1985 freeze. The new rules allow loans to be dispersed over a three-year period and extend the deferral period to five years. An additional provision states that "loans may be made ... in areas which are determined not to be freeze-prone areas...." The practical effect of that provision is unclear.

The maximum loan amount is, however, limited to the difference between the pre- and post-disaster values of the property. Since each freeze is counted as a separate disaster, growers whose groves were severely damaged by both the 1983 and 1985 freezes (as was commonly the case) could only borrow the difference between the post-1983 freeze value of their groves and their post-1985 value (personal communication, Mr. Neyaert, FmHA, Gainesville, Florida 24 June 1987). According to Neyaert, grove values in much of north-central Florida fell from approximately $10,000 per acre to the neighborhood of $4,000–$5,000 per acre as a result of the

1983 freeze. The additional damage generated by the 1985 freeze caused average grove values to fall even further to approximately $3,000 per acre. Thus, the maximum amount that a typical frozen-out citrus grower could borrow under this program after the 1985 freeze was often inadequate to finance the replanting of the grove. As a result, very few loans have been closed under this program despite the 1985 rule changes.

The total loan amounts under both the SBA and FmHA programs represent only a fraction of the economic losses incurred by Florida growers and related business enterprises, and the actual subsidy to these parties is only a fraction of the total loans. The "gift" or transfer payment embodied in these loans is equal to the difference between the subsidized interest payments charged on the loans and the interest that would have to be paid to a private-sector lender.

The loan programs are intended to help businesses get back into operation quickly and to foster rapid recovery to the pre-disaster economic conditions in the affected counties. Application deadlines and repayment schedules are designed to promote rapid reconstruction. The use of the funds is also restricted to serve that purpose. For example, SBA Physical Disaster Loans are to be used "only ... to restore property affected by the disaster..." (SBA, 1985). Farmer's Home Administration Emergency Physical Loss Loans to citrus growers were to be used exclusively to "... rehabilitate and/or reestablish their damaged or destroyed citrus groves for production of similar type(s) citrus crops grown during the disaster year" (FmHA, 1985). The disaster recovery loan programs thus restricted grower-participants to replanting citrus. This may not have been viewed as an appropriate goal by many frozen-out citrus growers. Many of these individuals apparently adopted a "wait-and-see" strategy, leaving former grove land idle while sorting out their options. Such a strategy does not mesh well with the deadlines, repayment schedules, and other restrictions inherent in the disaster loan programs, and many of these individuals apparently opted not to apply for this type of assistance.

The IFAS and the Cooperative Extension Service responded to the 1983 freeze by establishing the Central Florida Freeze Recovery Task Force. The Task Force was composed of about 50 research and extension faculty. It was established to study the extent of citrus damage in the eleven-county area most severely affected by that freeze and to assemble and disseminate information that could be useful in assisting individuals, businesses, and government entities in dealing with freeze related problems. The Task Force, established in May 1984, included two analysis teams and three working groups. The analysis teams focused on economic impact assessment and natural resources assessment, and the working groups focuses on citrus recovery, unemployment, and alternative commodities.

As part of its work, the Task Force conducted a survey of growers and grove managers in the eleven-county area to assess damage, determine grower intentions, and establish the type and form of information desired by these individuals (Central Florida Freeze Recovery Task Force, 1984). The respondents, representing nearly 90 percent of the pre-freeze citrus acreage in the area, indicated a considerable degree of enthusiasm for retaining their land in citrus production. Although 97 percent of this acreage had been damaged to some extent by the freeze, respondents indicated their intention to return 69 percent of the damaged acreage to citrus production. About 40 percent of the respondents indicated an interest in alternative commodities.

In response to this indication of grower intentions, the Citrus Recovery Working Group prepared a "Cold Protection Guide" summarizing the results of research on cultivation practices, freeze protection techniques, cold hardiness, meteorology, the climatology of north-central Florida, cold pockets, and other site selection considerations (IFAS, 1985a). Several publications were also made available on the economics of grove replanting or rejuvenation. A number of workshops were conducted to convey this information to growers.

The other working groups also compiled existing reports and prepared new publications to be made available to interested parties through the County Extension programs. While the immediate work of the Task Force focused on collecting and disseminating the results of previous research, there was also a renewed interest in cold protection research within IFAS and the extension community, and the freeze hazard has been a central target of citrus research since the 1983 freeze.

The final report of the Central Florida Freeze Recovery Task Force, submitted in January 1985, had apparently been completed before the January 1985 freeze. The fact that the Task Force had planned to conclude its work prior to the end of the next cold-weather season and a statement in the report calling the 1983 freeze "a once-in-a-century type freeze" (Central Florida Freeze Recovery Task Force, 1985, 2) indicate that the Task Force members felt no reason to suspect that the destruction of the preceding season would be immediately repeated.

The 1985 freeze did not lead to a resurrection of the full Task Force. The results of existing research had already been made available in convenient form through existing extension channels, and new research initiatives were already under way. IFAS did, however, conduct a new survey of growers and grove managers. The new survey indicated that the more serious destruction associated with the 1985 freeze further diminished (but did not extinguish) grower interest in re-establishing citrus groves in north-central Florida. The response to this second survey suggests that growers in the eleven-county area intended to return about one-half of the original (pre-1983) acreage to citrus, but that considerable uncertainty remained about the fate of the other freeze-killed acreage. Fear of damage from future freezes was cited as the major factor inhibiting replanting in the area. Grower intentions were also strongly influenced by their proportional losses. For example, grove managers from Hernando County indicated that they had lost more than 90 percent of their trees to the 1983 and 1985 freezes, and they expected that 20 percent of the grove land under their management would remain or be restored to citrus production. Grove managers responding from Brevard County, on the other hand, indicated that less than 15 percent of the acreage under their control had been lost to these two severe freezes, and they expected 88 percent of the land under their management to remain in citrus (IFAS, 1985b).

Public Sector Responses to Unemployment

Five counties were included in the federal disaster area following the December 1983 freeze, and 20 counties were counted as part of the disaster area after the January 1985 freeze. DUA payments were made available to workers in those counties covered by the disaster declaration who had become unemployed as a result of the freezes. The DUA program provided payments to 11,052 individuals who were unemployed as a result of the December 1983 freeze. Following the January 1985 freeze, payments were made to 12,006 unemployed individuals in the citrus industry alone. Total DUA payments amounted to approximately $6.4 million for the 1983 freeze and $7 million for the 1985 freeze. Many workers were also eligible for regular unemployment insurance payments. This provided relief for workers outside the declared disaster counties as well as for those who could not qualify for DUA but who had nevertheless lost work as a result of the freezes.

In addition to direct monetary payments, many families received food stamps and other forms of assistance from community service programs. The County Extension programs, for example, provided additional nutrition counseling services and financial management advice to the newly unemployed.

Other Public-Sector Actions Affecting
Citrus Industry Recovery

In addition to direct governmental responses to freeze-related problems, other governmental actions and policy debates could have an impact on Florida's citrus industry during and following future freeze episodes. For example, the Reagan Administration has repeatedly attempted to eliminate funding for the National Weather Service's agricultural frost warning service at Ruskin, Florida, which produces minimum temperature forecasts for the local Florida forecast zones. The Reagan Administration wants to privatize this particular service (personal communication, Fred Crosby, National Weather Service, Ruskin, Florida, 25 June 1987).

Funding constraints have also led to an attempt to charge user fees for on-line access to the satellite temperature data now being made available to citrus growers through the Lake County Extension Office. According to an IFAS researcher who is closely connected with the satellite downlink system, that effort at privatization has had somewhat disappointing results. Relatively few subscribers have been willing to pay for the service (personal communication, David Martsolf, 27 March 1987).

Recent changes in the Federal Income Tax code represent a rather different type of governmental policy change that may affect Florida citrus growers in the future. The subchapter S Family Corporation Code was recently changed to reduce the financial burden of recovery from federally declared disasters. The change, which received the support of the Florida Farm Bureau, allows growers who suffer freeze losses in a federal disaster area to spread the loss to an outside partner who can finance up to 49 percent of the cost of replanting. Unlike other agricultural investments, the cost of restoring trees lost in a federally declared disaster can be expensed rather than depreciated, making this arrangement a tax shelter for the outside party. Furthermore, the replanted acreage can be located anywhere, allowing growers to choose a less risky location for their new groves (personal communication, Ron Muraro, Lake Alfred Citrus Research and Education Center, 9 September 1987, and Pat Cockrell, Florida Farm Bureau, 25 June 1987). Other recent tax law changes, however, have reduced investment incentives.

PRIVATE-SECTOR RESPONSES

Although Florida citrus growers suffered tremendous financial losses as a result of the recent freezes and received relatively little financial assistance from government programs, Florida's citrus industry is beginning to show strong signs of recovery. Much of this recovery results from private-sector initiatives. Replanting efforts have been financed mainly by growers and private lending institutions. Private financing has allowed greater flexibility in terms of timing and grove relocation decisions than would have been possible under the federal disaster loan programs. A spokesman for the Florida Farm Bureau (a private organization of farmers and ranchers) indicated that a major public relations campaign by the Lake County Farm Bureau, touting the expected future profitability of Lake County citrus groves, may have enhanced the willingness of these lending institutions to extend credit to the owners of freeze-damaged groves (personal communication, Pat Cockrell, Florida Farm Bureau, 25 June 1987).

Private businesses have also undertaken much of the work of translating research results into new products, making cold protection devices affordable, and disseminating information about new cold protection techniques to growers. For example, research by members of the IFAS Extension faculty has shown that microsprinklers and tree wraps can provide substantial protection to tender young trees during freeze events. These findings have been quickly translated into a wide variety of products, many of which are manufactured locally and marketed through a number of industry publications. Although the research is very recent and is still ongoing, it is already rare to see a newly replanted grove in north-central Florida in which the young trees are not protected by one of a wide variety of newly available commercial tree wraps. Although they are not used as universally as the wraps, microsprinklers are also increasingly in evidence in recently reset groves.

Citrus growers have also been actively experimenting with other methods of coping with freeze risks. In cold-prone areas, new cold-tolerant rootstock and scion combinations are replacing the more tender varieties that had formerly been widely used. For example, Rough Lemon rootstock was formerly very popular because it allowed large yields and rapid tree growth. It is now being abandoned because it renders trees more susceptible to both cold damage and blight than alternative rootstocks such as Sour Orange and Swingle Citrumello.

Another trend in cultivation practices that may provide greater protection against freeze damage has been a movement toward more densely planted groves. When trees are planted closer together, the denser canopy will tend to hold more heat in the grove. If used in conjunction with microsprinklers, this technique might significantly increase the survival chances of groves in cold-prone locations. Even if the physical chances of survival are not much improved, such groves are more likely to be financially successful in a freeze-prone environment than old-fashioned groves with fewer trees per acre. This is because denser planting allows higher output per acre during the early years of the grove's life but lower output per acre at full maturity. This results in a shorter payback period, meaning that a densely planted grove can be expected to be financially successful (i.e., have a positive net present value) with a shorter expected life than would be required for a grove with widely spaced trees. The movement toward denser plantings has been noted throughout Florida. This may be explained by the fact that the practice also economizes on land costs. In southern Florida, where much of the new planting activity is occurring, dense planting is used because a great deal of expensive drainage work must be done to prepare the land for planting citrus.

The southward movement in the location of new citrus acreage is another way in which Florida growers are adjusting to freeze risks. While a gradual southward shift has been under way for a number of years (Brey, 1985), the trend now appears to be accelerating. As of December 1986, the largest acreages of new citrus trees (set within the previous four years) were in Polk, Highlands, and Hendry Counties, in that order. Highlands County experienced a 32 percent increase in orange acreage between 1980 and 1986, and Hendry County experienced a 43 percent increase. While Polk County continues to be Florida's largest citrus producer, the rapid expansion of acreage in such southern Florida counties as Highlands and Hendry is a relatively new phenomenon. Figure 3 is a map of Florida showing the percentage increase or decrease in orange acreage by county between 1980 and 1986.

There are indications that this relative southward shift will become permanent as land owners at the northern end of Florida's citrus growing area readjust their hedging strategies against freeze risks in light of both their recent experiences with freeze damage and the price stabilizing effects of increased Brazilian competition. An extension agent in Lake County estimates that perhaps one third of the freeze-killed acreage will be permanently converted to non-agricultural use while perhaps another one-third will be restored to citrus (personal communication, John Jackson, Lake County Extension Office, 26 March 1987). It also appears that many growers in heavily freeze-damaged areas intend to limit the proportion of their assets reinvested in citrus in those freeze-prone locations. Some plan to sell a portion of their land to developers in order to finance replanting of the remaining portion. Others plan to replant only a portion of their land to citrus and to devote the remainder to other agricultural uses. For example, a small portion of the former grove land in Lake County has already been planted with other tree crops. Some growers, whose holdings had formerly been located entirely in heavily freeze-damaged counties, have also recently acquired existing groves in the southwestern portion of the state. They have done this in part to facilitate the fulfillment of pre-existing fruit-delivery contracts and in part to diversify their holdings in order to provide financial protection against the likelihood of losses from future damage to their north-central Florida groves.

One can only speculate about the true motivations behind the southward shift in Florida citrus plantings and the apparent reduction in the willingness to gamble on citrus groves in north-central Florida. These changes could be explained either by an

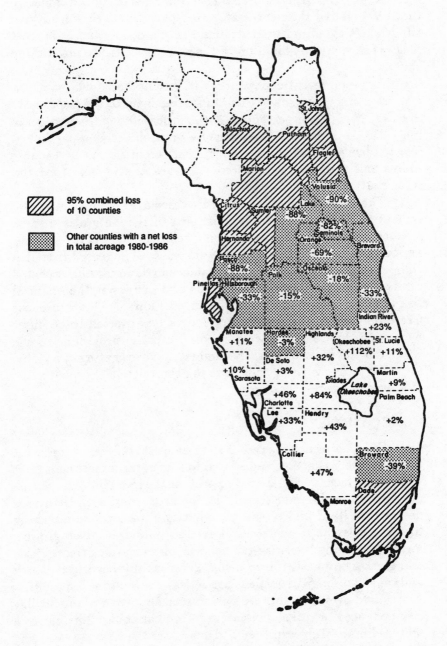

Figure 3.

increase in the perceived probability of freezes or by other factors that have altered the relative expected payoffs and risks from new citrus plantings in north-central and southern Florida. It is most likely that some combination of these factors has produced these shifts.

Growers' recent experience with serious freeze-related financial losses may have made them more cognizant of the freeze risks in their areas. Their perception of the probability of freeze damage may also be increased either because of an underlying psychological tendency to place more weight on recent events or because individuals explicitly see the recent freezes as evidence of a trend or as part of a cycle.

The relative payoffs and risks associated with planting or replanting citrus groves in various parts of the state may also be affected by such factors as changing relative land values for non-agricultural uses. The price-stabilizing effect of increased Brazilian competition could also affect the balance by lowering the expected returns on groves throughout Florida and increasing the financial risk of citrus operations in north-central Florida. This is because freeze damage to the crop can no longer be expected to be offset by price increases. Higher fuel costs and the provisions of the Clean Air Act have also largely eliminated grove heating as a viable response to freeze risks, thus lowering the relative expected value of groves in freeze-prone areas.

CONCLUSIONS

Whether or not the recent anomalous set of Florida freezes in the early 1980s is in any way related to long-term climate change, it will clearly have a long-lasting impact on the industry. The resulting changes in the industry will be the product of both the freeze destruction itself and of the way that land owners are adjusting to these extreme meteorological events. Florida's citrus growers are restructuring their risk-taking strategies regarding freeze risks, and greater emphasis is now being placed on diversifying that risk and on expanding citrus plantings in less freeze-prone locations.

Land owners and other decisionmakers hedge risks and this risk hedging is a rational response to the variability inherent in a stable climate. How can they hedge against the risk of a changing climate as well? When an anomalous weather event occurs, it is not possible to tell at that moment whether the event is simply a random occurrence from a stable underlying process or a manifestation of a climate change.

There is a widely noted tendency for decisionmakers to discount extreme events in the distant past while placing apparently disproportionate weight on recent extreme events. This may be a manifestation of some inherent tendency to see trends in nature, or decisionmakers may hold the belief that climatic events tend to be autocorrelated or that long-term climate change is under way.

While such tendencies would result in biased perceptions and therefore in biased decisionmaking if the true climate is stable, the tendency to react as if recent events were part of a trend would be valuable if a real change were indeed under way. This suggests that perception and behavior that may appear biased, if one assumes that the underlying climate is stable, may actually be a useful way of hedging against the risk that climate is changing.

In the case of Florida's citrus freezes, it appears that private decisionmakers have acted to adapt their investment strategies and cultural practices to protect themselves more fully against freeze risks in the northern portion of the traditional citrus belt. This type of flexibility would be valuable in coping with climate change as well.

In opposition to the flexibility shown by the private sector in responding to the freezes, some of the governmental responses to the freezes can be characterized as rigidly tied to the goal of re-establishing the previous status quo. Such rigidity would be a serious handicap in coping with a changing climate. In particular, the federal disaster recovery loan programs were rigidly committed to the re-establishment of the freeze-killed citrus groves in their original locations. These programs were not widely used, in part because they required an immediate decision on the part of the land owner, and in part because their goal may have been seen as inappropriate.

Despite the rigidity of some government programs, this case makes it clear that public-sector programs can promote flexible responses. The Cooperative Extension Service's focus on the examination of alternatives, experimentation, and the dissemination of information provided a valuable service to growers who were suddenly faced with the necessity of making decisions about the future use of their land.

404

This case analogy suggests that in planning public policy responses to CO_2-induced climate change, we would do well to recognize the strengths of private methods of adapting to climatic risks. Policies will be more likely to promote effective adaptation to climate change to the extent that they are flexible and avoid an obsession with preserving the status quo.

REFERENCES

Brey, J.A., 1985: *Changing Spatial Patterns in Florida Citriculture, 1965–1980*. Ph.D. Dissertation. Madison, WI: University of Wisconsin–Madison.

Broecker, W.S., 1987: Unpleasant surprises in the greenhouse? *Nature, 328*, 123–6.

Central Florida Freeze Recovery Task Force, 1984: *Citrus Freeze Survey, Summary Report: A Producer Assessment of Freeze Damage, Future Intentions, and Information Needs*. Gainesville, FL: Institute of Food and Agricultural Sciences, University of Florida.

Central Florida Freeze Recovery Task Force, 1985: *Final Report, Institute of Food and Agricultural Services*. Gainesville, FL: University of Florida.

Drummond, H.E., 1984: *Estimated Citrus Tree Loss Caused by the December 1983 Freeze*. Staff Paper 266. Gainesville, FL: Institute of Food and Agricultural Sciences, University of Florida.

Florida Crop and Livestock Reporting Service, undated: Freezes 1886 thru 1967 and Brief Statement of Citrus Damage. Orlando, FL: Florida Crop and Livestock Reporting Service.

Florida Crop and Livestock Reporting Service, 1985: *Florida Agricultural Statistics: Citrus Summary*. Orlando, FL: Florida Crop and Livestock Reporting Service.

Florida Crop and Livestock Reporting Service, 1986: *Florida Agricultural Statistics: Commercial Citrus Inventory*. Orlando, FL: Florida Crop and Livestock Reporting Service.

Florida Grower and Rancher, 1986: FG&R Survey Shows Favorable Outlook for Citrus, 29–34 (April).

FmHA (Farmer's Home Administration), 1985: *Instruction 1945-D, Emergency Loans for Citrus Grove Rehabilitation and/or Reestablishment, Exhibit D, Revision 1* (November), 3. Gainesville, FL: FmHA.

IFAS (Institute of Food and Agricultural Sciences), 1985a: *Cold Protection Guide, 1985 Revision.* Gainesville, FL: University of Florida.

IFAS (Institute of Food and Agricultural Sciences), 1985b: *Citrus Freeze Update, Summary Report: A Producer Assessment of Freeze Damage, Future Intentions, and Information Needs Resulting from January 1985 Freeze.* Gainesville, FL: University of Florida, in cooperation with *The Orlando Sentinel.*

Polopolus, L.C., R.P. Muraro, W.F. Wardowski and S. Moon, 1985: *The Effects of the December–January 1983–84 Freezes upon the Citrus Industry in Eleven Florida Counties: Responses from Packinghouses, Processing Plants, Property Tax Assessors, and Lending Institutions,* Staff Paper 267. Gainesville, FL: Institute of Food and Agricultural Sciences, University of Florida.

Powe C., and M. Langham, 1980: *Toward a Policy Testing Model for the Florida Orange Industry,* Technical Bulletin 815. Gainesville, FL: Institute of Food and Agricultural Sciences, University of Florida.

SBA News, No. 85-14 (March 1985), SBA Declares Disaster Area, 1.

17

Summary
Michael H. Glantz

Looking at the preceding case studies as well as the chapters of Section I, many common themes and significant findings arise. These are highlighted in this chapter. Following the general findings, key findings of each chapter are presented, underscoring the insights as well as lessons that these case studies provide regarding potential societal responses to a global climate change.

GENERAL FINDINGS

- Because we are concerned about the local and regional effects of and responses to a global climate change, there is a need to identify how well societies have in the past dealt with local climate-related environmental changes, regardless of cause. In order to add to the body of knowledge about societal responses to possible global climate change, it is important to know what that body looks like with respect to societal responses to climate variability and to extreme meteorological events.

- The involvement of the local and state levels of government in climate-related environmental problems is extremely important. These are the levels at which such changes will ultimately occur. These are the levels at which societies will most likely respond, at least initially.

- The importance of the local nature of societal impacts and responses to those impacts also underscores the need to generate awareness among decisionmakers at that level about what a global warming might mean to the expected continuance of the present-day activities of a specific region.

- There is a problem in some instances with conflicting signals. While the levels of the Great Salt Lake and the Great Lakes are now at or near record-setting highs, the scientific community suggests that with a global warming the levels of those lakes should decline. How credible can alarms about a global warming be to decisionmakers at the local level, when environmental changes at that level are seemingly contradictory to scientific projections about what should be happening?

- In each case study a catalyst prompted action by people and their governments. In the case of Louisiana's sea-level rise/coastal subsidence, action was not prompted by the gradual rise in sea level or the gradual pace of coastal erosion. The catalyst was the realization that two Louisiana parishes (counties) would most likely be under water in about a century. In the Great Salt Lake situation that catalyst was not a gradual rise in lake level (which had been occurring since the early 1960s) but the sharp annual increases in lake level since 1982.

- Each case study raised the issue of intergenerational equity. For example, several of the GCM models have suggested that there will be a "drying out" of the Great Plains in the event of a global warming. Users of the water from the Ogallala Aquifer must decide at what rate to drawdown the aquifer today (to produce crops that are in surplus and that bring low prices) or whether to save that finite water supply for use by future generations, especially in those parts of the region overlying the aquifer where recharge rates are low.

- In all of the cases ad hoc responses were favored over longer term planned responses. As a result, there has been a tendency to "muddle through." This has not necessarily been an inappropriate response, but it is probably more costly in the long run than putting a long-term strategy together in order to cope with climate-related environmental change.

- Several of these cases show that ad hoc decisions made in response to an environmental change have often built into the existing social structures an additional degree of rigidity that would in the long term decrease society's ability to respond to future changes.

- The Florida case study points out that climate variability can adversely affect the economic competitiveness of climate-dependent industries. Such severe freezes as Florida witnessed in 1962 and in the early 1980s served as catalysts to accelerate economic and social changes at the local level, changes that may already have been underway but at much slower rates.

- Coalition-building is an important part of creating an awareness of as well as coping with a climate-related problem. The Louisiana case study provides an excellent example of how a few concerned individuals in a parish were able to develop broader statewide interest in a set of problems that seemingly threatened only local populations and local political units. Coalition-building in Louisiana provides an interesting example to other states (such as Florida) that also face a variety of climate-related problems.

- Societies are constantly changing and they will continue to do so regardless of whether the global climate changes. It is important to take societal changes into account when considering societal responses to the impacts of climate variability, climate change, and extreme meteorological events.

- It has been argued in the climate change literature either (1) that everyone will lose if the climate changes and therefore we should act now, or (2) that we must wait to identify what the regional impacts of such a change might be *before* we proceed to make policy to deal with climate change. The Colorado River study suggests, however, that when the winners and losers have been identified there will be little interest on the part of the winners to alter their status in order to compensate the losers.

SPECIFIC CASE STUDY FINDINGS

EVOLUTION OF AN AWARENESS (Kellogg). The impacts of increasing levels of carbon dioxide in the atmosphere is a contemporary environmental problem that has generated considerable scientific and, now, political concern internationally, nationally, and even regionally. Interest in the topic has varied over time with a steady increase in interest and concern beginning in the mid-1970s. Several major scientific reports issued in this period support the belief that increased amounts of carbon dioxide and other radiatively active trace gases in the atmosphere will most likely result in a global warming of between 1.5 and 4.5°C by the middle of the next century.

- The first signs of an understanding of the role played by infrared-absorbing gases in maintaining the warmth of the surface of our planet were voiced early in the nineteenth century, and by the turn of this century quantitative calculations had been carried out that demonstrated how a change in atmospheric carbon dioxide would alter the earth's mean temperature.

- There is now a strong consensus that the observed increase in the atmospheric concentrations of carbon dioxide and other infrared-absorbing trace gases is indeed warming the atmosphere, and that this change is caused by human activities. The determination of the regional patterns of the changes to be anticipated, especially in terms of rainfall and soil moisture, is now being seriously addressed in national and international forums.

- There is a serious debate between activists, who would take action worldwide to avoid the climate change (or at least slow its advance), and those who would simply wait and see what happens and perhaps take what local measures are necessary to mitigate the effects.

POLITICS AND THE AIR AROUND US
(Glantz). Compared to the degree of consensus in the scientific community on the causes of the global warming, there has been relatively little political interest in this issue, with some notable exceptions. There are many reasons for this lack of political action; there are also a number of factors that promote action by society to prevent or cope with a global warming. A comparison between societal responses to the ozone depletion issue and the CO_2-induced global warming issue may be instructive, although there are many differences between them.

- Despite large volumes of scientific information on the global warming issue, one can still find disagreement over the projections of future climate change that are expected to occur globally as well as regionally.

- Warnings by scientists regarding a global warming as a result of fossil fuel burning have been issued since at least the end of the 1800s and early 1900s. The question arises: when will decisionmakers have enough information to take deliberate action on this issue (inaction is also a form of action)?

- The issue of a global warming is much clearer today than at any time in the past few decades, but important scientific uncertainties remain. Those uncertainties must be put into proper perspective (are they major or minor?), lest they be used to paralyze any action on the issue.

- A global warming of several degrees Celsius will prove to be a boon to some and a bane to others, at least in the short term. Some areas will become drier while others become wetter. Some areas will seemingly benefit from, and therefore desire, those changes, while others will not. The perception about the potential benefits as well as losses of the yet-to-be-identified impacts of a global warming will constrain international cooperation on the CO_2 issue.

- The list of reasons for not doing anything about global warming is long and specific, while the list of reasons pushing toward some degree of international action or at the least cooperation is ostensibly much shorter and more diffuse.

- We must avoid a "rush to judgment" on creating new international institutional arrangements to deal with the CO_2/trace-gases warming issue until we have investigated the potential contributions and roles of existing institutions.

GLIMPSING THE FUTURE (Jamieson). Scenarios are one way of attempting to get a look at the future; numerical models are another way. Scenarios also include analogies. These can be referred to as case scenarios. They provide a way of "telling the story" about the future that model-generated scenarios cannot provide. Certain dangers are associated with the use of analogy-based scenarios, including stretching an analogy too far or selecting an analogy that is inappropriate. Appropriate use of scenarios can provide useful "stories" about the future and about how human populations respond to variations in environmental conditions.

- If global warming is now occurring, it is not just something that is happening *to* people: People are implicated in bringing it about. In response to global warming, we can expect various modulations of human behavior, which will in turn affect atmospheric conditions, which in turn will affect human behavior, and so on. One consequence of this feedback between human behavior and atmospheric conditions is that in order to answer our question—how are humans likely to respond to environmental changes at the regional level brought about by a carbon dioxide/trace-gases-induced global warming—we must gain insight into the interactions between climate and behavior.

- The notion of scenario is widely used in the climatology literature. Unfortunately, it is often used in vague and misleading ways. The concept of a scenario is a rich one, and of great utility in a number of different areas of investigation. Scenarios are sketches or outlines of stories rather than abstract sets of statements or propositions. They are constructed in order to serve some purpose, and they are told from a point of view. They bring together diverse information, and engage our imagination about a natural or expected course of events.

- Good analogical reasoning does not concern the number of similarities two objects share, but rather the significance of the important similarities. Identifying important similarities involves pragmatic considerations regarding contexts, interests, and purposes. These considerations cannot be taken up in any purely structural account.

- There are four advantages of the case scenario approach; wealth of detail, integration of a broad range of knowledge, multiplicity of perspectives, and communicability and usability. There are, however, dangers to be avoided when using the case scenario approach: lack of definition, straining an analogy, and failure of analogy.

STATISTICS OF CLIMATE CHANGE: IMPLICATIONS FOR SCENARIO DEVELOPMENT (Katz). Keeping in mind the nature of the information about future climate that would be needed for societal impact studies, the adequacy of the output from climate experiments based on general circulation models (GCMs) requires evaluation. The prospects for resolving the so-called "first-detection" problem, concerned with identifying and attributing any climate change that might have taken place by subjecting the recent observational record of climate to statistical analysis, is also an important issue. It is inherently more difficult to make statistical inferences about changes in climate variability than about changes in climate means. The problem of how best to estimate the probabilities of various climate events can be examined from a decision-analytic viewpoint.

- GCM climate experiments do not currently produce information in a form that is useful for societal impact studies.

This should not be surprising because GCMs were developed with the intention of aiding in basic research about the atmosphere, rather than with the needs of climate impact research in mind. It is not at all clear that GCMs will be able to better meet these requirements in the near future.

- The projection of future climate on the basis of currently observed trends is not necessarily justified. This suggests a fundamental quandary concerning the best way to estimate the likelihood of climate events. Should decisionmakers retain the "stationarity" hypothesis that the climate is not changing in any permanent fashion and estimate the probabilities of occurrence of future climate events using the observed frequencies of occurrence of these events in the historical record? Or should these probability estimates be based only on the relatively recent historical record or on the extrapolation of an apparent trend or cycle?

- Several fundamental issues need to be resolved before impact studies should rely on scenarios of future climate based on GCM climate experiments or based on the extrapolation of observed climate change. However, the utility of the case scenario approach is not solely dependent on whether the climate event considered is analogous in some respects to an anticipated future climate change. Instead, the value of this approach rests more in providing information about how society has dealt with climate events in the past, regardless of whether the specific event being examined is at all analogous to events of interest in the future. Given the lack of a reliable way of projecting future climate change, perhaps the reliance on such case studies is a viable approach to climate impact assessment.

CO_2/OGALLALA COMPARISON: FORECASTING BY ANALOGY (Glantz and Ausubel). Two contemporary long-term, low-grade but cumulative environmental problems are considered together. One problem reflects a real change (i.e, the Ogallala Aquifer drawdown), the other a potential change (i.e., the potential change in regional rainfall and soil moisture patterns) in the water balance in the U.S. Great Plains region. If both of these changes were to take place, as many authorities are currently predicting, their combined effects on regional water supplies (and therefore on agricultural output) could prove to be severe.

- There is value in assessing the similarities and differences between responses to seemingly similar environmental changes in order to develop at least a first approximation about how societies might respond to environmental changes that are expected to occur in the near- to mid-term future.

- Contemporary environmental problems are often considered in isolation, even though there is evidence that the impacts of many of these problems might converge, making adverse situations even worse. It is important to identify those potentially convergent problems so that society can better assess as well as prepare for future environmental risks.

- It is important that new research efforts benefit from appropriate past experience. Just as there is sometimes insufficient attention paid to the lessons of the past within a particular field, there are often unused opportunities to learn from those with whom one may be working in parallel. For example, several environmental issues such as ozone depletion, soil erosion, acid rain, CO_2/trace-gases increases, and groundwater depletion may have strong "family" similarities.

CHANGING LEVELS IN THE GREAT LAKES (Cohen). The Great Lakes Basin's water resources are important to the economic vitality of the American states and Canadian provinces in the region. Recent climatic fluctuations have caused lake levels to deviate significantly from average, thereby affecting commercial shipping companies, hydroelectric utilities, shoreline property owners, and others. Responses of policymakers to past fluctuations may provide some insights into possible future responses to projected climatic warming.

416

- Some of the water resource management policies for the Great Lakes have traditionally developed in an ad hoc manner, often in response to perceived crises. The lake levels oscillate due to climate variability. Variations in climate occur over different time and space scales and, therefore, need to be defined according to local conditions. For example, even during a wet regime (1965–1987), local droughts occurred.

- Control measures need to show a greater understanding of how that system oscillates and what the impacts of those oscillations might be and how they may change. This will be especially true because of demographic changes in the Great Lakes Basin. If control measures do not reflect the lakes' variability, those measures could exacerbate the adverse effects of the variability.

- It appears that the "struggle" between proponents of ad hoc or long-term responses, operational or structural changes, belief or disbelief in a global warming will continue in the Great Lakes Basin until there is more convincing evidence that there is a dire need for action. Thus, maintaining flexibility is a key element in any policy to cope with future climate-related environmental changes in the region.

- With regard to the global warming issue, some scientists suggest that lake levels should decline rather than rise with a global warming. The rising lake levels and the fact that the 1959–85 annual mean flow (measured at Cornwall, Ontario) is higher than the longer term 1861–1985 mean create confusion among the general public about whether a warming is indeed occurring and whether there is a need for long-term strategy development.

RISE IN THE LEVEL OF THE GREAT SALT LAKE (Morrisette). Utah's Great Salt Lake has risen 12 feet since 1982, flooding valuable lakeshore property and forcing resource managers to address the environmental and societal impacts and the problem of variability in lake levels and climate. Policymakers have been reluctant to address the long-term implications of the problem; instead they have opted to respond incrementally to mitigate impending crises, while hoping that the lake would soon recede and "normal" conditions would prevail.

- Decisionmakers tend to rely on traditional approaches to environmental problems, even when faced with new or unusual conditions. In this particular instance society tended to rely on ad hoc technological fixes and rigid, structural responses that could be completed in the short term. For example, railroad beds were repeatedly raised several feet at great cost each time, as the water level continued to rise. The most recent and most costly response ($60 million) was the construction of the west desert pumping station. Even this technological fix exposed the belief of decisionmakers that the water levels of the lake would not rise much beyond the present high stage, because the pumping station would not have been able to cope with inflow on the order of the 1983–86 period.

- This study underscores society's tendency to define "normal" in unscientific, often misleading, ways. The considered responses spanned the gamut from adaptive to mitigative to preventive measures. Adaptive measures were favored at first, followed by mitigative measures. This study shows that structural responses to climate-related environmental crises can create a false sense of security and can build into society a rigidity that precludes optional responses in the face of future climate-related variability.

- Because the global warming is likely to be gradual, it is probable that local and regional resource managers will have difficulty in recognizing the initial environmental and societal impacts of a carbon dioxide/trace-gases-induced global warming. Even if decisionmakers recognize that a long-term problem exists, they are still often reluctant to deal with it.

- Hindcasting is considerably easier than forecasting. When assessing responses to climate-related crises, we must try to recapture "the spirit of the times" to avoid misinterpretation of societal responses to past events.

FUTURE SEA-LEVEL RISE AND ITS IMPLI-
CATIONS FOR CHARLESTON, SOUTH CAROLINA
(Davidson and Kana). A large part of Charleston, South
Carolina, is located on a peninsula less than five feet above
mean sea level. The average elevation of the metropolitan
area of Charleston is approximately ten feet. The region's
topography is characterized by a number of tidal rivers
and creeks, expansive marsh resources, and large areas of
freshwater wetlands. The region is prone to flooding during periods of heavy rainfall, particularly when rainfall
events coincide with high tides or strong easterly winds.
It is reasonable to conclude that a 3–5 ft rise in sea level
could seriously impact Charleston. Global sea-level rise
produces direct impacts along coastal areas as well as
some less obvious ones. These include flooding, increased
beach erosion, marsh destruction by saltwater intrusion,
increased risks from hurricanes, and destabilization of waterfront property and activities.

- The sophistication of information provided by the scientific
  community will play an important role in determining how
  and when local communities will prepare for and respond to
  the impacts associated with sea-level rise. More detailed and
  less variable projections are needed (including very localized
  impact assessments) in order to give policymakers the information they need to make responsible decisions about sea-
  level rise. To the extent that certainty cannot be provided,
  and translated to inexpert and untrained decisionmakers, societal responses to the impacts of sea-level rise are likely to
  be postponed.

- As the magnitude of the effects of sea-level rise increase, the
  likelihood of national and state regulatory agency involvement also increases. These levels of government possess both
  greater fiscal resources and the necessary political jurisdiction to deal with complex environmental issues. Local governments, however, have the tools available to set their own
  policy directions. These include zoning ordinances, building
  codes and local property tax structures.

- It is generally accepted that technology is available to provide structural solutions to sea-level rise. The question is, however, what property value is sufficient to justify the very large cost of structural protection.

- To a large extent, fiscal constraints dictate the substance of the responses to sea-level rise. Local governments will have to take a pragmatic approach in deciding their range of commitments in the allocation of scarce resources. Private-sector investment will play a large part in the decisionmaking process as it will dictate to a large extent the development patterns that occur once the impacts of sea-level rise begin to become manifest.

- As with many environment-related issues, a catalyst may lead to government response where efforts to initiate progressive planning have failed. In the case of Charleston, the state and local communities only began to appreciate the potential impacts of sea-level rise after an unusual meteorological occurrence which precipitated elevated water levels, marked erosion of waterfront properties, and substantial public concern with the future of South Carolina's valuable beaches. Even so, the close parallel between that occurrence and the impacts of sea-level rise has to be continually drawn, for the issue has been only indirectly addressed.

SEA-LEVEL RISE AND COASTAL SUBSIDENCE IN LOUISIANA (Meo). Coastal Louisiana is affected by two concurrent problems: sea-level rise and coastal subsidence. An important aspect of sea-level rise is that multiple impacts are likely to affect various regions differently, be typically cumulative and generally irreversible. Because the impacts are varied, coalition building becomes difficult and at the same time very important.

- It appears that those concerned about the implications (political, economic, social, environmental) of wetlands losses, regardless of the causes, agree in general on what needs to be done. What they seem unable to agree on is when and how fast to take action, with some actors wanting to take immediate action and others, feeling a less pressing need to act, wanting to "wait and see" whether environmental changes continue to occur in the future. Nevertheless, at the local and state levels coalitions have formed to deal with the loss of wetlands and marshes. This case study is a good example of coalition building at the local level in response to local environmental changes.

• The coalition that developed in response to wetland losses served to catalyze public demand for action to stop these losses and to strengthen local and state institutional capacity for the management of the wetlands. For example, the state has created a Coastal Protection Trust Fund, with a budget in the tens of millions of dollars. In addition, there is a Coastal Protection Task Force.

• Cumulative sea-level rise and coastal subsidence impacts will occur over a long period of time and will not occur in a uniform fashion along coastal areas. This means that there will also be a low probability for the development of a comprehensive strategy to cope with these impacts on a larger than local (or state) basis, and coalition building on a regional level might be a more difficult process.

CHANGES IN THE MISSISSIPPI RIVER SYSTEM (Koellner). The Mississippi River is linked to the Gulf of Mexico, Great Lakes, as well as several other waterways. This navigation system is extremely important for the movement of commodities, such as grain, agricultural chemicals, cement and stone, coal, and petroleum products. It is also widely used for recreation. The system's capacity is directly related to the climate of the entire region. Large changes in water level (too little or too much) can disrupt the flow of trade or at least greatly reduce the efficiency of the system's operation, affecting all parties dependent on commercial navigation.

- Most societal actions are, in fact, responses to a given regional stress. Thus, ad hoc reactions occur. Generally speaking, extended periods of extremely low water in the Mississippi River system had not been experienced until the summer of 1987, nor have extended periods of excessively high water. Interestingly, we have recently witnessed the period of greatest historical stress, 1959 to 1986, the wettest 28-year period in the climate history of the Mississippi River Basin.

- There is really no good long-term analogue in the historic hydrologic record of the Mississippi River system for societal response either to a wetter regime or a drier one. Most responses to hydrologic changes (usually climate induced) in the Mississippi River system have been made in an ad hoc manner in the form of crisis management.

- It appears that there is little value placed on drought planning strategies because of the belief that future situations (hydrologic or demographic) will be so different from past situations as to render them useless.

- To date, the U.S. Corps of Engineers has worked from existing records and depended for planning on the identification of historical precedents. If the climate is changing, to what extent will the past historical record serve as a reliable guide to the future with regard to large navigational systems? Until we know what a climate change will look like at the regional level, we may not be able to answer that question with a high degree of confidence.

CLIMATE VARIABILITY AND THE COLORADO RIVER COMPACT (Brown). The Colorado River is a highly regulated river that flows primarily through a semiarid region in which the water demands by society are great, and in which the societal impacts of climate variability can be major. However, the initial agreement drawn up in 1922, after a few decades of abundant precipitation, to divide the waters of the Colorado River between the Upper and Lower Basin states, essentially ignored the

potential effects of climate on the magnitude and timing of river flow.

- Important lessons for the future can be found by reviewing the effects of the Colorado River Compact of 1922. First of all it is unwise to ignore the variability that is inherent in natural systems. Average conditions do not provide a complete story. A mechanism for the periodic rechecking of the rainfall and streamflow data should be built into systems responsible for the long-term management of water resources.

- It is important not to ignore changes that will occur in social systems. This is especially true for multipurpose resources, such as water, where there are constant changes in competing demands. Even with the same amount of water flowing through the Colorado River system, there will be changing needs for and values of water resources.

- The decline in streamflow in the Colorado River Basin can be regarded in retrospect as a climate change. We know that the original estimates of annual streamflow were erroneous, but to date nothing has been done to redress the prospects of shortages which the Upper Basin states must absorb.

- In retrospect, decisions that bring rigidity to a management system can ultimately generate more problems than they resolve. For example, dividing up the water supply in absolute terms as opposed to proportionally has locked the states into certain patterns of interaction. Even though we now know that there is less water in the system today than was originally estimated, and even though there are now clearly identified winners (the Lower Basin states) and losers (the Upper Basin states), there is no compelling reason for the "winners" to renegotiate their position.

CLIMATE VARIABILITY AND WATER RESOURCES IN A CALIFORNIA RIVER BASIN (Gleick). Because of the growth of agricultural production and commercial and industrial development in the Sacramento

River Basin, it has become increasingly vulnerable to fluctuations in the magnitude and timing of precipitation. The societal responses to the 1976–77 drought and to flooding in the early 1980s include changes in the physical structure of resource management systems (such as reservoirs and water transfers), changes in the operation of these systems (such as reservoir rule curves), and a range of socioeconomic alternatives (such as pricing and market mechanisms, institutional initiatives, and regulatory responses).

- This study raises concern about the value of existing information, about the frequency and intensity of extreme events, about the magnitude and timing of snow melt, changes in seasonal evaporation demands, and so on. Water resource planning standards, as we know them today, could become less reliable in the face of a changing climate. Thus, we need a better understanding of climate variability and climate change, including changes in the frequency and range of extreme meteorological events.

- Some of the longer-term responses to the mid-1970s drought led to an unintended increase in the vulnerability of the system to high water levels. In particular, changes were made in reservoir operating rules in the late 1970s to increase the ability of the system to provide water under low-flow conditions. These changes worsened the effects of flooding in the early 1980s when unusually high flows occurred.

- Like climate, social systems (especially demographic trends) are also changing. As a result, future climate anomalies on the same order of magnitude even in the same locations as past events may have increasingly larger impacts on society and the economy. As demographic patterns change, options that are available today may no longer be useful or even available in the future. For example, in 1976–77, southern California temporarily relinquished its use of a portion of the water normally transferred from northern California. This permitted northern California to use that water to meet its demands during the drought. Southern California was willing to do this only because they could make up the difference by temporarily increasing withdrawals of Colorado River water. Yet, in the near future the waters of the Colorado River system will be fully appropriated and there will be little chance for surplus.

WATER SHORTAGES IN METROPOLITAN NORTHERN VIRGINIA (Sheer). The Fairfax County Water Authority is the sole water purveyor supplying some 650,000 people in Fairfax County, Virginia, a suburb of Washington, D.C. The primary source of water is a small dam and reservoir on the Occoquan Creek, a tributary of the Potomac River. In 1977 the reservoir had been dropping steadily throughout the summer, and by the end of August it had receded to a record low level. A technical effort was organized to analyze the risk of extreme water shortages and to evaluate the policies necessary to reduce those risks to acceptable levels. The techniques and procedures analyzed could be applied to future drought problems.

- Actions to accommodate water resource problems must be taken at the local level; there is usually debate over appropriate actions; and it takes a united front of political and technical leaders to implement a conservation plan.

- There is rarely sufficient information available to quantify the risk of dire consequences in a credible manner. In the Occoquan case such information was available and proved to be invaluable in formulating an effective response. The earlier the risk is recognized, the milder are the measures required to control it.

- It is critical that society continually monitor its vulnerability to changes in climate and reassess which measures are most appropriate to reduce that vulnerability. Because the rate of climate change is expected to be small in relation to normal climate variability, society should be able to monitor at least partially its vulnerability to climate change by monitoring its vulnerability to the "normal" variations in climate. An assessment of what constitutes "normal" climate must also be continually updated, based on the most recent meteorological experiences.

- In the Occoquan case it took both a drought and an increase in water use in society to threaten the reliability of the supply. Societal vulnerabilities change as society changes. The measures appropriate to reduce that vulnerability must also change as both social conditions change and as short-term changes in meteorological conditions occur. Appropriate emergency measures must be thought out and accepted as far in advance as possible to maximize their effectiveness.

DEPLETION OF THE OGALLALA AQUIFER (Wilhite). The depletion of the aquifer represents an important change in the water balance of the Great Plains region. The thickness of the aquifer that underlies the region is not uniform. Thus, people in the Texas panhandle, where the aquifer is relatively thin and has in the past been heavily used, are much more directly affected by a drawdown of groundwater than those in Nebraska, a state which possesses a major share of the aquifer's resources and high aquifer recharge rates. In 1982 a \$6 million federally funded project was completed with the objective of determining potential development alternatives for the Great Plains and identifying policies and actions required to carry them out.

- The drawdown of the Ogallala Aquifer raises an important issue that permeates discussions about the social and political responses to a global warming: discounting the future. It provides a good example of a choice that society must make—consume the groundwater resource today or save it in the ground for future generations, for a time when climate in the region might not be as favorable to agricultural production as it is today. At which time would that groundwater be of most value? And to whom? Today, other factors have slowed down the rate of drawdown, such as higher energy prices, low crop prices, large stockpiles of grain and so forth. Nevertheless, the issues of intergenerational equity should be addressed now when there is less pressure to decide one way or another.

• How concerned should national leaders be about the depletion of this regional aquifer? Is the Ogallala Aquifer drawdown in fact a national problem? While there is regional interest in generating national concern about the "problem," there is also a strong desire by those in the region to keep the control of the aquifer and its management at the local and state levels. Due in part to the variable thicknesses and longevity of different parts of the aquifer, people in the region prefer local responses to local changes in the aquifer. They do, however, accept the responsibility of the state in the over-all management of the aquifer. In fact, a pecking order of preferences emerges from inhabitants of the region: state involvement is preferred over federal and local over state; water conservation measures are preferred over other more drastic changes such as reverting to dryland farming practices.

• Societal responses to relatively short-term climatic extremes often spark a search for technological changes on which societies then become dependent. A good example of this would be recurrent droughts in the Great Plains which ultimately led to a regional dependence on groundwater exploitation from the Ogallala Aquifer for irrigated agricultural production.

INSTITUTIONAL AND PRIVATE SECTOR RESPONSES TO FLORIDA FREEZES (Miller). The Florida citrus processing industry, a major producer of frozen concentrate orange juice (FCOJ), was adversely affected by four freezes in five years in the early 1980s. This set of freezes killed nearly a third of the trees in commercial groves statewide, with much heavier losses in the northern counties.

- This study shows that grove owners are willing to take risks with respect to climate variability. It appears that they adjust their risk perceptions in response to climate-related events, perhaps weighting recent events most heavily. Thus, the recent set of freezes altered their perceptions of the risks that their groves face, as witnessed by their tendency to replace only a portion of their freeze-damaged trees and their desire to develop new groves at the southern edge of the current citrus-producing area. However, the most recent years were without freezes and could begin to cause grove owners to again change their assessment of risk to freezes and to view the most recent good years as a return to pre-1980 "normal" conditions. This situation presents an excellent setting for an assessment of perceptions about climate variability, climate change and the occurrence of extreme meteorological events.

- This case study suggests that the most effective and immediate response to a climate-related problem often occurs at the local level and in this particular case through the private sector and not at the state or federal levels. The University of Florida's Institute for Food and Agricultural Sciences (IFAS) was one of the main public-sector actors to identify the freeze impacts, suggesting responses to them. Federal disaster programs were not widely used because their goals were deemed rigid as well as inappropriate, as they tended to foster replanting citrus groves in the same locations as those decimated in the freezes.

- While the citrus industry is a large one, it is only a relatively small percentage of Florida's economy and perturbations in it have even smaller repercussions for the federal government. Thus, it seems to have essentially remained as a local and a state problem, in that order.

Each chapter in this book provides ideas about dealing with societal and environmental responses to climate variability and climate change. They can only hint at how society might respond to future changes in the climate. However, they do provide the reader with examples from different geographical locations and different economic sectors of how well societies (from local to regional to national) are prepared to deal with the variability of climate today. If we are to add to the body of knowledge regarding climate-related impacts on society and the environment, then we must have a better understanding of the existing knowledge on this topic. It is hoped that this collection of essays proves to be a step in that direction. It is also hoped that this approach will stimulate other attempts to get a glimpse of the future and to identify ways in which societies around the world might better cope with their climates.